WORLD HEALTH ORGANIZATION

INTERNATIONAL AGENCY FOR RESEARCH ON CANCER

AND

THE CANCER REGISTRIES OF SCOTLAND

ATLAS OF CANCER IN SCOTLAND

1975 — 1980

INCIDENCE AND EPIDEMIOLOGICAL PERSPECTIVE

Edited in Scotland by I. Kemp & P. Boyle

Edited in Lyon by M. Smans & C. Muir

In collaboration with

L. Cameron, M.H. Elia, C.R. Gillis,
M.A. Heasman, G. Innes, J.S. Scott,
G. Venters

IARC SCIENTIFIC PUBLICATIONS No. 72

Lyon, 1985

The International Agency for Research on Cancer (IARC) was established in 1965 by the World Health Assembly, as an independently financed organization within the framework of the World Health Organization. The headquarters of the Agency are at Lyon, France

The Agency conducts a programme of research concentrating particularly on the epidemiology of cancer and the study of potential carcinogens in the human environment. Its field studies are supplemented by biological and chemical research carried out in the Agency's laboratories in Lyon, and, through collaborative research agreements, in national research institutions in many countries. The Agency also conducts a programme for the education and training of personnel for cancer research.

The publications of the Agency are intended to contribute to the dissemination of authoritative information on different aspects of cancer research.

Distributed for the International Agency for Research on Cancer by
Oxford University Press, Walton Street, Oxford OX2 6DP, UK

London   New York   Toronto
Delhi   Bombay   Calcutta   Madras   Karachi
Kuala Lumpur   Singapore   Hong Kong   Tokyo
Nairobi   Dar es Salaam   Cape Town
Melbourne   Auckland

Oxford is a trade mark of Oxford University Press

Distributed in the United States
by Oxford University Press, New York

ISBN 92 832 1172 3

Frontispiece: Physical geography of Scotland indicating relief, railways, roads, rivers, cities and larger towns.

(By kind permission of John Bartholomew and Son, Cartographers, Edinburgh)

CONTENTS

LIST OF TABLES

LIST OF TEXT FIGURES

CONTRIBUTORS

Production of this monograph on cancer incidence in Scotland would not have been possible without the data, supporting descriptive text and comment so willingly provided by the directors and staffs of the five regional cancer registries of Scotland:

Northern Cancer Registry
Radiotherapy Department
Raigmore Hospital
Inverness IV2 3UJ

Dr M.H. Elia (Director)
Mrs M. Palmer

North-eastern Cancer Registry
Department of Clinical Oncology
Aberdeen Royal Infirmary
Foresterhill
Aberdeen AB9 2ZB

Dr G. Innes (Director)
Miss M. Clark
Dr A.E. Whitter

Eastern Cancer Registry
Radiotherapy Department
Ninewells Hospital
Dundee DD2 1UB

Dr L. Cameron (Director)
Dr J.S. Scott (Director)
Mrs M. Cannon

South-eastern Cancer Registry
Liberton Hospital
Lasswade Road
Edinburgh EH16 6UB

Dr G.A. Venters (Director)
Mrs C.A. Aikenhead
Mr D.J. MacLean
Mr R.C. Sayers

Western Cancer Registry
Ruchill Hospital
Glasgow G20 9NB

Dr C.R. Gillis (Director)
Mr P. Boyle
Mr D.J. Hole
Mrs A. Graham

The Scottish Cancer Registry
Trinity Park House
South Trinity Road
Edinburgh EH5 3SQ

Dr I.W. Kemp (Coordinator)
Mr P. Trotter
Miss A. Munro

The Scottish Cancer Registration System is coordinated by the Committee of Directors of Scottish Cancer Registries, which is chaired by Professor K.C. Calman, Cancer Research Campaign Professor of Clinical Oncology, University of Glasgow[1]. The members include the Directors of the Registries and Dr M.A. Heasman, Director of Scottish Health Service Common Services Agency, Information Services Division, Dr W. Forbes, Senior Medical Officer, Scottish Home and Health Department (who provided helpful information for the atlas), Dr F. Lee, Consultant Pathologist, Western Infirmary, Glasgow, and Dr I.W. Kemp, Coordinator, Scottish Cancer Registry.

The editors are deeply grateful for the support, interest and constructive advice obtained from the committee.

---

[1]Currently Dean of Post-Graduate Medicine, University of Glasgow

FOREWORD

Epidemiology has been defined as use of knowledge on the occurrence and distribution of disease in the search for determinants.

One of the ways in which the distribution of a disease can be portrayed is on a map. This atlas of the incidence of cancer in Scotland, the first of a series of cancer atlases to be produced jointly by the International Agency for Research on Cancer and national bodies, illustrates once again the validity of the saying that a single picture is worth a thousand words.

As the Editors point out, cancer occurs in people and not in geographical areas. The aim of this atlas is to indicate those areas in Scotland where a given cancer is unduly common or unexpectedly infrequent, thus providing a point of departure for further analytical studies to uncover possible determinants of risk.

There is evidence of considerable variation in risk throughout Scotland for many cancers. While the Editors have provided an interpretation of the observed patterns of distribution for several cancers they have not been able, in the light of present knowledge, to explain anything like all of the variation. Further, several of their suggestions about cause, notably in relation to diet, remain to be tested.

To set their findings and suggestions in perspective, the Editors have provided descriptions of many facets of the Scottish scene, emphasizing those related to the collection and accuracy of incidence data and those features of the environment that may influence cancer risk.

It is to be hoped that the differences demonstrated will be followed up by further studies of etiology as, without knowledge of cause, there can be no effective or rational prevention.

L. Tomatis   Sir John H.F. Brotherston
Director   Emeritus Professor of Community Medicine
IARC, Lyon   University of Edinburgh
France   Scotland

PREFACE

Statistical and epidemiological studies are often viewed as dry, sterile and uninteresting. The fact that this is clearly untrue can be seen from a glance through this atlas. The visual effects, the technical skill used in its presentation and the questions raised by the information it contains make this book exciting reading. Scotland's national poet, Robert Burns, once wrote: 'Facts are chiels that winna ding an downa be disputed'. The sentiments expressed in these words assert that facts provide the basis for discussion. The possession of information allows the generation of hypotheses and, as they are tested, the development of new knowledge.

We are proud in Scotland that this will be the first of a series, which will be bench marks for future epidemiological studies. The techniques developed and the method of presentation are innovative and forward thinking. In addition to the basic data presented in this book, its production has generated a considerable amount of unique methodological research, which will be available for future publications.

But why should we need such a book? The planning of health care priorities, the appropriate use of resources and the recognition of health services research as a legitimate area of study all point to the need for better information on which to base judgements. Incidence, mortality, geographical variation, the social setting, occupation and age are some of the many factors that are required to provide the basis for decision making. Just as important, however, are the questions that this book raises. These questions require careful consideration and, where appropriate, the development of experimental protocols to study them further.

This project could not have been completed successfully without the collaboration and active participation of a variety of individuals and institutions. First, there are the clinicians and pathologists from all over Scotland who have provided the information over the years; then there are the cancer registry staff from the five regional registries who, with hard work and devotion, have collated the basic data. Finally, the staff of the Information Services Division, in their turn, have worked with their colleagues in Łyon to produce this volume. Our thanks must go to all who have been part of this important venture.

K.C. Calman
Chairman
Committee of Directors
Scottish Cancer Registries

Chapter 1

CANCER AND THE SCOTS

Cancer has long been of interest to the Scots. Despite their legendary stoicism, the Scots have always shown a strong humanitarian interest in the consequences of the disease. It is difficult to imagine a more moving account of the surgical treatment of breast cancer before anaesthesia than that found in *Rab and His Friends* in which Brown (1858) described Ailie's operation for breast cancer. The Society for Investigating the Nature and Cause of Cancer (1806) published the findings of a questionaire about the disease in the *Edinburgh Medical and Surgical Journal* in 1806, commenting: 'With regard to cancer, it is not only necessary to observe the effects of climate and local situation, but to extend our views to different employments, as those in various metals and manufactures; in mines and collieries; in the army and navy; in those who lead sedentary or active lives; in the married or single; in the different sexes, and many other circumstances. Should it be proved that women are more subject to cancer than men, then we may enquire whether married women are more liable to have the uterus or breasts affected; those who have suckled or those who did not; and the same observations may be made of the single.'

This atlas, in essence, reports on the current geographical distribution within the country; we still have to 'extend our views' to explain the pattern observed. Once the causes are known, rational prevention becomes more likely.

THE DISTRIBUTION OF CANCER

The study of the distribution of disease, in an attempt to uncover factors which influence risk, is the science of epidemiology.

Cancers occur in people, not in geographical regions. While much cancer risk is influenced by personal habits such as diet, use of alcohol and tobacco, age at marriage, etc., these are, in turn, often governed by occupation, which frequently determines place of residence, socio-economic status, food eaten, personal habits, work exposures, atmospheric pollution, etc. Thus, the Editors have not only provided a description of the regional cancer patterns of Scotland, but have also attempted to place them in the framework of a nation, emphasizing those aspects of geography, economic activity and diet which may touch the causes of cancer (see Chapters 2-4).

Differences in the distribution of disease have frequently led to the discovery of causes and, from time to time, prevention. A classical and much-quoted example is that of London (Snow, 1936), where John Snow in 1854, by recording and mapping the addresses of the victims of a cholera epidemic was able to show that the disease was much commoner in those drinking water supplied by the Southwark and Vauxhall Water Company and, more specifically, from the pump in Broad Street, than in those obtaining their supply from other sources. Removing the handle from the pump effectively stopped the epidemic.

Nonetheless, epidemiologists do not like to use the word 'cause', preferring to talk about risk factors, etiological agents or determinants — in other words, the exposures that increase (or decrease) the risk of cancer. Such risk factors can range from well-defined agents such as cigarette smoking (see Lung cancer) to the age at which a woman has her first full-term pregnancy (see Breast cancer).

CANCER MAPS AND ATLASES

The earliest surviving multicoloured map of the distribution of disease comes from the City of Glasgow (Perry, 1844), and portrays the distribution of an infectious disease, probably influenza. The condition was most frequent where housing was poor and crowded (the map is technically interesting in that the published copies were hand-coloured by patients at the Gartnavel Royal [Mental] Hospital).

The first cancer atlas ever published described female mortality in England and Wales and drew attention to increased risk of cancer in the north of England (Haviland, 1855). Stocks (1928), presenting his results for the counties and county boroughs of England and Wales in 1928, pointed out two regions of high and low death rates and concluded that geographical influences were concerned in the causation of cancer. Stocks (1936, 1937, 1939) subsequently published a series of 33 such maps in the Annual Reports of the British Empire Cancer Campaign. The regional distribution of cancer mortality for the UK was first presented by Howe (1970) (now at the University of Strathclyde, Glasgow). Black and white shading and hatching were used to illustrate areas of low and high cancer mortality.

The British atlases described so far, including that published for England and Wales (Gardner *et al.*, 1984), all presented information about deaths (mortality) from cancer. However, a much clearer idea of the burden of cancer in a population can be obtained from information on incidence, i.e., the rate at which newly diagnosed cases of cancer occur in the population at a particular time. Incidence, unlike mortality, is not influenced by variations in survival from cancer. A large number of persons with cancer do not die from the disease.

Information about newly diagnosed cases of cancer is collected by cancer registries. Scotland is particularly fortunate in that the entire nation is covered by five regional cancer registries, located in Aberdeen, Dundee, Edinburgh, Glasgow and Inverness, whose work is coordinated by the Scottish Health Service, Information Services Division, in Edinburgh. These registries, and the work of central coordination, are described in greater detail in Chapter 4.

Although the cancer registries of Scotland have published their findings over the years (Doll *et al.*, 1970; Waterhouse *et al.*, 1976, 1982), the patterns and the significance of differences in the geographical distribution of the various types of cancer are not, as a rule, easily seen from printed tables. Their presentation in the form of maps helps to bring out the differences and similarities, which may suggest possible causes and, once such causes are established, soundly based means of prevention.

Before the maps are presented, their setting is outlined by descriptive chapters on geography, climate, soil and vegetation, the people, economic activity, communications, health care, cancer registration, etc.

It will become apparent that little work has been done in Scotland on the causes of human cancer. This weakness was pointed out in a report of the Scottish Health Service Planning Council (1979) entitled *Cancer Services in Scotland*, which recommended a high priority for epidemiological research. It is hoped that this atlas will emphasize that recommendation, pointing once again to the need for expansion of research in this critical area.

Chapter 2

THE SCOTTISH HABITAT

This chapter describes some of the features of topography, climate, soil and vegetation which influence population distribution, agriculture and diet in Scotland. Communications are also discussed.

PHYSICAL PROFILE

The geography of Scotland appears on the first coloured map (Frontispiece). The country forms the northern part of the island of Great Britain and is situated between latitudes 54°38′ and 60°51′N and longitudes 1°45′ and 7°40′W. It is bounded on the west and north by the Atlantic Ocean and on the east by the North Sea, while in the south the land border with England runs for about 100 km along the line of the Cheviot Hills. Around the coast are some 790 islands, of which 130 are inhabited. The country covers some 78 772 sq km, including many fresh-water lochs (lakes). From north to south, mainland Scotland is about 440 km, while the maximum width is 248 km. Few parts of the country are more than 64 km from salt water. The three main geographic regions are, from north to south, the Highlands, the Central Lowlands and the Southern Uplands.

The Highlands, whose mountains and lochs are renowned for their beauty, account for more than half the total area of Scotland. The Central Lowlands fall roughly between the Highland Boundary Fault, which runs from Dumbarton in the south-west to Montrose in the north-east, and another parallel fault further south, which runs from Girvan in the south-west to Dunbar in the north-east. Dury (1973) noted that the Central Lowlands contain all the mineral wealth of Scotland, except slate and granite, virtually all the coal and ironstone, all the large industries and four-fifths of Scotland's population. Few major towns lie outside this belt, and most lie in the western half of the Central Lowlands, with Glasgow as a centre, this city alone accounting for nearly one-sixth of the Scottish population.

The Southern Uplands are characterized by fertile farmland and magnificent scenery. The terrain is gentler than that found in the Highlands, the highest peak being only 815 m.

A more detailed description of the 56 local authority and islands areas that appear in the maps of cancer distribution is given in Chapter 6.

CLIMATE

While Scotland can be divided from north to south into three main topographical zones, the climatic separation is between east and west. Despite its northern situation (on the same latitude as Labrador), the climate of Scotland is temperate, the west coast being bathed by the warm Gulf Stream. There is no month in which average temperatures fall below freezing point at sea level. There is, however, great climatic variation within the country.

The climate of the west of Scotland may be described as mild, wet and windy. There is little frost, and snow seldom lies long. The mean winter temperature of the island of Tiree (off the west coast) is 5°C, and the mean summer temperature 13.9°C. The east, however, tends to be colder in winter and warmer in summer. Dundee on the east coast has mean temperatures of 15°C in the warmest month and 2.8°C in the coldest month. Inland areas towards the north and north-east have a more extreme climate. In northern lowlands, snow falls on an average of 30-35 days each year. The average Scottish rainfall ranges from about 56 to 100 cm a year, the east coast being markedly drier than the west. On the east coast around the Moray Firth, the average annual rainfall is about 60 cm, while in the mountains of the north and the west it exceeds 250 cm.

Particularly in the west, the climate tends to be dominated by the passage of Atlantic depressions, which have many effects on living conditions. Even before air pollution was drastically reduced by the Smoke Control Regulations introduced in 1959 (Glasgow is now 80% smokeless), long-lasting fogs in windswept Glasgow were relatively infrequent; since the late 1960s, smogs have been very rare indeed. However, the combination of temperature and high humidity in the west poses problems for the heating of homes, and many are still inadequately heated for the prevailing conditions.

SOME CHARACTERISTICS OF SOIL AND VEGETATION

There is great diversity in the character of the soil (Dury, 1973). Good soil for arable farming is found in the Lowlands, the north-east coastal plain, Caithness and the Orkney Islands. By contrast, in the north-west,

Hebrides and the Shetland Islands, the soil is poor and cultivation is largely confined to river estuaries, some glens and along the coastal strip.

Approximately two-thirds of Scotland's total area is rough grazing, virtually treeless, unimproved heather-grass-sedge moorland that is described in other parts of the world as 'open range' land. Rough grazing comprises all those areas that are not woodland and where the environment has made cultivation impossible physically and economically. Several major types of plant flourish in the rough grazing; one of these is heather — like the thistle, a 'national emblem' of Scotland.

Heather

Heather (*Calluna vulgaris*) moorland covers approximately 1.2 million hectares, some 25% of the total area of unimproved land. Heather has been used over the centuries for a wide variety of purposes: as a source of tannin and vegetable dye, of nectar for honey, of bedding and roofing material and as the main diet of the grouse. Heather is also important in the diet of sheep, hill cattle and red deer.

Bracken

About 180 000 hectares of rough grazing are covered by bracken (*Pteridium aquilinum* var. typica), which is widespread in the south-west Highlands and the Galloway hills but less common in the north of the country. The geographical distribution is governed by a preference for areas with moderate to high rainfall and freely drained soil.

Bracken has been used for a variety of purposes: thatching, bedding and even fuel, glass and soap manufacture. There have been trials of its potential as a livestock feed, but this is limited by its unpalatability and by the 'bracken poisoning' that may ensue.

Peat

On moors and hills there is extensive peat, thought to cover approximately 820 000 hectares, or almost 11% of the total land area of Scotland. (Peat is here defined as a surface organic layer not less than 30 cm thick containing more than 80% organic matter on a dry-weight basis; Tivy *et al.*, 1973). It occurs where soils are water-logged. The current geographical distribution of peat is to the west and north, the Outer Hebrides having very extensive areas and 60% of Caithness being peat-covered. Peat is used as domestic fuel and in horticulture.

COMMUNICATIONS

There are two main land crossing points between Scotland and England — one on the eastern seaboard and the other on the western. Scotland's relative inaccessibility by road from the large English and continental population centres and markets not only causes some economic disadvantages, but may well have contributed to differences in lifestyle between Scotland and her southerly neighbours. For example, eating patterns may have been influenced, including differences in the consumption of fresh fruit and vegetables (see Chapter 3).

The internal road system extends to more than 46 000 km. Despite gradual improvement, much of it, especially large stretches of single-track road in the Highlands, is inadequate for the increasing traffic. Communication between centres of population has become much easier in recent years. Since 1964, the Forth and Tay road bridges and the Kingston and Erskine bridges over the Clyde have been opened, and a network of motorways and dual carriageways has been developed in the central belt of the country. Although the road links northward have been improved, there is no continuous dual carriageway to Aberdeen or Inverness.

Rail travel between central Scotland and London is swift. The west coast route has been electrified recently, the journey between London and Glasgow (576 km) currently taking some five-and-a-half hours, and the east coast route from London to Edinburgh taking only a little longer. The journey between Aberdeen and London takes some eight hours, and that between Inverness and London eight hours-and-a-half. The internal rail system has been considerably reduced in recent years because of poor economic viability. Good rail links still cross central Scotland.

There are major airports at Glasgow, Prestwick, Edinburgh and Aberdeen. Transatlantic flights leave from Prestwick, while Glasgow, Edinburgh and Aberdeen are used for internal UK flights and direct flights to Europe. One of the effects of the North Sea oil industry has been development of air services to and from Aberdeen, and the city currently has the busiest helicopter airport in the world. The Highlands and the Islands have eight small airfields, with services connecting with Edinburgh and Glasgow. Inverness has direct flights to Aberdeen, Edinburgh, Glasgow and London as well as to the Western Isles.

Chapter 3

THE SCOTTISH PEOPLE AND THEIR WORK

The name Scotland originated in the eleventh century, when the name Scotia was given to the south-western tract of the country settled by the tribe of Scots who had migrated from Ireland. From 1603, Scotland and England had a common sovereign, and Scotland became an integral part of the UK on the Union of the Parliaments of Scotland and England in 1707. Although it has no separate Parliament, Scotland retains vestiges of ancient sovereignty in its own legal and educational systems, a national church and a separate administration. While its economy is integrated with that of Britain, it has managed to preserve its identity in many ways. At different periods in history, Scotland has been in tne forefront of the fields of philosophy, economics, physics, literature, medicine and engineering. Some of its products are internationally known, and generations of emigrant Scots have spread the awareness of a distinctively Scottish culture.

Following a brief statement on the historical origins of the Scottish people, this chapter gives details of those characteristics of the population of Scotland that may influence cancer risk.

ORIGINS

The distribution of races within Scotland in early times was the direct result of invasion or occupation. Early history is conjectural but there are numerous traces of Mesolithic and Neolithic man, including a village at Skara Brae on Orkney of 10 stone-built houses buried around 1400 BC by a sand storm. Smout (1972) noted that the longest residents were the Picts, an ancient Celtic people described by Tacitus as having red hair - still regarded as a characteristic Scottish feature. The Picts settled along the eastern side of the country from Fife to the Moray Firth. In the south-west of the country were the Britons, driven there by the warring Teutonic Angles when the Romans withdrew from England. The Angles, invading from Northumberland (Northern England), were settled in south-east Scotland by the seventh century. Another Celtic people, the Scotti, from Ireland, had invaded the west of Scotland two centuries earlier, occupying roughly the area of present Argyll, while a second branch of the Scotti settled in Galloway in the south-west. The last of these early invasions occurred at the end of the eighth century

when the Norse settled along the northern and western coasts, the shores of the Pentland Firth and parts of the western seaboard from Wigtown to Sutherland. In contrast with England, where Teutonic stock predominated, the people of Scotland at this time were of two major derivations: the Picts, the Scotti and the Britons of Celtic origin and the Teutonic Angles.

Since those early times there have been many incursions. In the twelfth century, Normans and Britons followed David I to Scotland and settled in Lothian, while Normans and Flemings colonized the Strathclyde area in the time of Malcolm IV.

In the nineteenth century, a considerable number of people from Ireland settled in the west of Scotland. In the early part of the twentieth century, peoples from central and southern Europe immigrated mainly into central Scotland. Since the Second World War, migrants from India, Pakistan and the West Indies have settled in Scotland, principally in the west; but this more recent immigration has been comparatively small.

One important trend over the last two centuries has been the migration of Scots to other English-speaking countries. There have been various estimates, varying from 25 to 40 million, of the number of persons of Scots descent living abroad. The emigration occurred as a result of industrialization, changing agricultural practices, expulsion of crofters (small farmers with special tenancy), a growing population and the opening of new lands. These movements may have had some effect on the morbidity and mortality rates of the Scottish population, since the migrants are likely to have been from the active and healthy sector of the population. The effects of emigration are most noticeable in the Highlands, where considerable changes took place in the population levels due to clan subjugation following the ending of the Jacobite uprising in 1745, and where the population was cleared to make way for sheep farming. In 1851, the Highland region had around 250 000 inhabitants, but by 1951 depopulation had reduced them to 150 000. Many went overseas, but sizeable numbers moved to Glasgow. In recent years there has been a movement out of Glasgow to new towns (see Chapter 4).

In common with many other countries, Scotland has lost segments of the male population as a result of wars, particularly in the present century, the consequences of which are difficult to quantify.

A high proportion of the present Scottish population was born in the UK. This is remarkably uniform throughout the country. In no region does the percentage fall below 97.

PRESENT POPULATION

According to the 1981 census (Registrar General Scotland, 1982; General Register Office, 1983a), 5 035 315 people are usually resident in Scotland (Table 3.1). The country thus has some 64 people per square kilometre. But since three quarters of the population live in the Central Lowlands, in extensive areas of the country there is no more than one person to two square kilometres (Fig. 3.1).

In 1981, one-third of the population was aged 45 or more, and 14% was aged 65 or over. Men aged 65 and over form just over 5% of the total population, while women constitute almost 9%.

CURRENT SOCIAL CLASS CHARACTERISTICS

There is strong evidence from both England and Wales (Registrar General for England and Wales, 1978) and Scotland (Registrar General Scotland, 1982) that social class has some effect on mortality for many types of cancer. Death from lung cancer, for example, occurs proportionately much more commonly among members of social classes consisting of unskilled or partly skilled workers, while cancer of the testis is commoner in professional groups.

Almost one-quarter of Scotsmen and one-fifth of Scotswomen belong to the professional and intermediate classes (Registrar General Scotland, 1982; General Register Office, 1983a,b) — Social Classes I and II, respectively (Table 3.2). However, within these groups, most women belong to Class II, which includes school teachers and nurses.

Table 3.1. Home population of Scotland by sex and age, 1981 census

Age (years)	Male	Female	All persons
0-4	158 288	150 075	308 363
5-9	176 331	168 077	344 408
10-15	263 717	250 465	514 182
16-19	182 210	175 371	357 581
20-24	199 755	194 557	394 312
25-29	172 216	170 013	342 229
30-34	180 453	178 274	358 727
35-39	149 502	150 796	300 298
40-44	141 274	146 854	288 128
45-49	139 458	146 194	285 652
50-54	139 552	150 059	289 611
55- 9	138 444	151 312	289 756
60-64	114 442	136 832	251 274
65-69	105 683	135 120	240 803
70 74	83 201	120 329	203 530
75-79	50 910	91 578	142 488
80-84	22 547	55 797	78 344
85 and over	10 489	35 140	45 629
Total	2 428 472	2 606 843	5 035 315

[a]From General Register Office (1983a)

Figure 3.1. Population density (per 100 hectares) by Local Government District, Scotland, 1981 census (Registrar General Scotland, 1982)

1

2000
1000
400
100
66
30
15
5
1

Table 3.2 Social class^a **distribution of economically active population by sex, Scotland, 1981**^{b}

	Males		Females	
	Number^c	Percentage	Number^c	Percentage
Social class I (professional)	7 312	5	967	1
Social class II (intermediate)	26 229	18	19 019	20
Social class III (N) (skilled, non-manual)	14 347	10	34 813	37
Social class III (M) (skilled, manual)	52 731	36	7 946	8
Social class IV (partly skilled)	25 067	17	20 187	21
Social class V (unskilled)	0 787	7	7 748	8
Armed forces and inadequately described	8 322	6	4 660	5
TOTAL	144 795	100	95 340	100

^{a}Classification of the Registrar General

^{b}Usually resident, economically active population, 1981 Census (General Register Office, 1983b)

^{c}10% sample

By far, the largest proportion of men and women are in Social Class III (skilled workers). The majority of the men in this class are manual workers while the women follow non-manual occupations; many of the men work in manufacturing industries, the women in skilled clerical occupations.

Over one-quarter of the economically active and retired population, both men and women, are in the partly skilled or unskilled categories (Social Classes IV and V). Figure 3.2 shows the proportion of men and women in Social Classes IV and V for the administrative areas of the country which are used to depict cancer incidence patterns (areas described in detail in Chapter 6). It is clear that these social classes are concentrated in and around Glasgow and the southern-most districts, and hence a higher incidence of cancer related to social class in these areas would not be unexpected. A detailed analysis of the incidence of several sites of cancer in Scotland by social class has been published by the Scottish Health Services (Information Services Division, 1981). Some of the published and unpublished results which show effects of social class on the incidence of different cancers are given in Figure 3.3 for males aged 15 years and over. No social class gradation is seen for the incidence of colon cancer, although there is such a gradation for cancer of the rectum.

Such differences in cancer incidence between the social classes offer possibilities for uncovering risk factors; most of the social class differences in lung cancer are likely to be associated with variations in tobacco use.

THE SCOTTISH DIET

Blaxter (1977) pointed out that there are many misconceptions about the diet of the Scots. They do not habitually consume haggis[1], brose[1], stovies[1], porridge[1] or malt whisky. The Scottish diet is, in fact, very similar to that of the rest of the UK. However, differences do exist which may have some effect on the health and cancer pattern of the two regions. The *Household Food Consumption and Expenditure Surveys* (National Food Survey Committee, 1976-1980), show that the Scots eat considerably less fruit and less than half the quantity of fresh green vegetables consumed by the English (Table 3.3; Fig. 3.4). They also tend to eat more cereal products, potatoes, eggs, beef and beef sausages (veal is rarely eaten). Surprisingly, the Scots eat less mutton and lamb than the English, in spite of the volume of sheep farming.

---

[1]Haggis: the heart, lungs and liver of a sheep, calf, etc., finely minced with suet, oatmeal, onion, seasoning and boiled in the animal's everted stomach; brose, stovies, porridge: traditional dishes the main constituents of which are, respectively, peasemeal, oatmeal, or potato, and oatmeal.

Figure 3.2. The proportion of all economically active households in Social Classes IV and V (General Register Office, 1983b)

34.0
31.5
29.5
28.5

24.0 ————— 24.7

21.5

15.5
14.0

2

Figure 3.3. Incidence of selected cancers, by social class, age-standardized rates for males aged 15 years and over, Scotland, 1970-1972

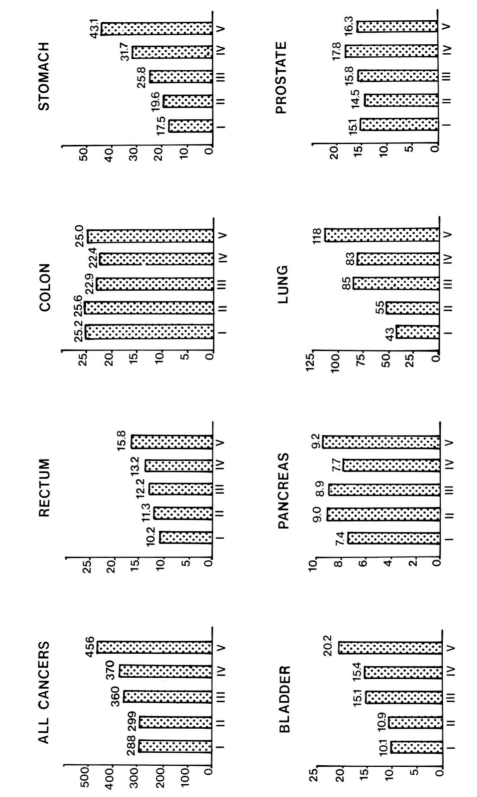

Table 3.3 Quantities of major categories of food consumed by households in Scotland and in England, 1976-1980 [a]

Food[b]	1976		1977		1978		1979		1980	
	Scotland	England	Scotland	England	Scotland	England	Scotland	England	Scotland	England
Cereal products	58.7	57.5	60.1	57.2	61.5	55.6	61.6	55.1	60.1	54.9
Fresh fruit	15.0	18.7	15.3	17.6	13.6	18.7	14.7	20.3	17.4	21.1
Fresh green vegetables	5.5	11.9	6.2	12.7	6.9	14.1	5.2	11.5	6.2	13.0
Fresh potatoes	35.7	35.1	41.6	40.9	53.9	42.6	47.8	42.2	43.5	40.1
Fish	4.1	4.6	4.1	4.2	4.0	4.3	3.9	4.6	4.9	4.8
Eggs (number)	4.3	4.1	4.2	4.0	4.5	3.9	4.4	3.8	3.8	3.7
Butter	4.7	5.2	4.7	4.7	4.6	4.5	4.7	4.4	3.6	4.1
Margarine	2.7	3.1	3.4	3.5	3.1	3.6	3.3	3.7	3.7	3.8
Total fats	9.2	11.1	10.2	11.0	10.0	11.2	10.4	11.0	9.5	11.3
Beef (and veal)	11.9	7.3	10.7	8.1	10.7	8.2	11.6	8.2	13.0	7.8
Mutton and lamb	1.8	4.4	2.6	4.1	1.9	4.1	1.9	4.4	2.2	4.8
Pork	1.1	3.1	1.9	3.5	1.6	3.5	1.8	3.8	2.3	4.3
Bacon and ham (cooked and uncooked)	4.1	5.1	5.0	5.4	5.0	5.4	4.8	5.5	5.1	5.2
Sausages, pork (uncooked)	0.6	2.0	0.9	2.0	0.7	1.9	0.9	2.0	1.1	1.8
Sausages, beef (uncooked)	3.7	1.3	3.3	1.4	4.8	1.4	3.7	1.5	3.5	1.3
Meat pies and sausage rolls (ready to eat)	0.4	0.8	0.5	0.8	0.5	0.8	0.4	0.8	0.5	0.7
Cheese	3.4	3.8	3.6	3.8	3.5	3.8	3.8	3.9	3.3	3.9
Liquid milk (pints)	4.8	4.7	4.6	4.5	4.5	4.5	4.2	4.3	4.4	4.1
Coffee (beans, ground, instant)	0.5	0.6	0.4	0.4	0.4	0.5	0.5	0.6	0.5	0.7
Tea	1.7	2.3	1.9	2.1	1.9	2.0	1.8	2.1	2.0	2.1
Soups (tinned)	6.2	3.0	5.3	2.6	6.3	2.4	5.9	2.7	5.0	2.5
Pickles and sauces	1.5	1.7	1.7	1.7	2.2	1.7	2.1	1.8	1.8	1.8
Salt	0.9	0.7	1.0	0.8	0.9	0.8	1.0	0.9	1.1	0.9

[a]National Food Survey Committee (1976-1980)

[b]ounces per person per week, unless otherwise stated; 1 ounce = 28.35 g

The English eat more pork, bacon and ham. Despite their apparent popularity in Scotland, meat pies and sausage rolls are consumed at a slightly higher rate in England. The amount of milk and cheese taken is approximately the same for both regions. English consumption of tea and coffee is a little higher than that of the Scots, while twice as much tinned soup and slightly more salt are consumed in Scotland.

Fruit and vegetables

The reasons for the lower intake of fruit and vegetables are not difficult to deduce. The climate cannot easily sustain the cultivation of fruit. While soft fruit grows in abundance in the Tay and Clyde Valleys, the impact of soft fruit on the diet is rather small because of its short growing season. Green vegetables

Figure 3.4. Consumption of fresh green vegetables in Scotland and England, 1976-1980 (ounces per person per week)

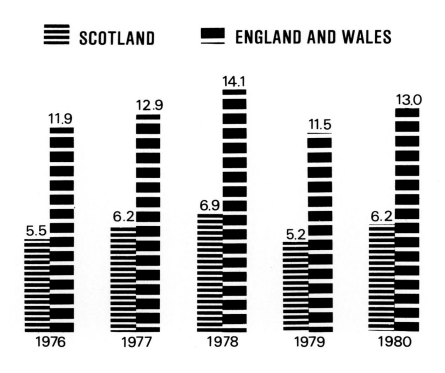

Figure 3.5. Average annual consumption of all tobacco products per adult in Scotland and England and Wales, 1956-1978

are grown in quantity in certain parts of Scotland, notably the south-east, but the supply is clearly insufficient. Blaxter (1977) estimated that Scotland is only about 25% self-sufficient in vegetables (for fruit only 7% self-sufficient). This would not matter greatly if transport were cheap and quick from the point of production or from the point of importation in the UK. Although some fruit and vegetables are imported directly, Scotland depends mainly on the long overland haul from the major fruit and vegetable markets of the south of England. The disparity in the volume of vegetables grown in different parts of Scotland is likely to lead to inter-area differences in their consumption. The Highland region, in particular, has a relatively low production, and this, coupled with its isolation due to an indifferent road and rail system, affects price, quality and availability. The transport system may also influence the availability of other fresh dietary items in the more isolated rural areas. Since fresh fish is such a perishable commodity, it is possible that regular consumption is greater in the cities and areas close to the fishing ports (Table 3.3), although statistics on this point are not available.

Alcohol

There is some evidence that the consumption of alcohol is higher in Scotland than in England. The 1973 Family Expenditure Survey (Department of Employment, 1974) showed that the average weekly expenditure on alcoholic drink in Scotland was £2.4 compared with £2.0 in the UK as a whole. In the 1980 survey (Department of Employment, 1981), the expenditure for Scotland was £6.1, and in the UK, £5.7. Much of this increase is due, of course, to inflation and increases in excise taxes. Further, the numbers in the sample are relatively small, and expenditure on alcohol is known to be understated.

Another measure of excess alcohol consumption is the admission rate to psychiatric institutions due to alcoholism or alcoholic psychosis. In 1980 in Scotland, the age-standardized rate per 100 000 for male patients admitted for the first time with alcoholic dependence syndrome was 58.4, while in England the rate was 11.0 (I.W. Kemp, 1984, unpublished data).

Some indications of differences in alcohol consumption in various areas of Scotland (I.W. Kemp, 1984, unpublished data) are also shown in the first admission rates to psychiatric hospitals by area of residence. The north-west has particularly high rates.

While it may be more convenient to admit people to hospital than to give domiciliary care in isolated areas, in view of the large differences in admission rates, this is unlikely to be the complete explanation. The evidence suggests higher alcohol consumption in the north-west of Scotland, and also in Scotland as a whole, compared with England. This has been questioned (Crawford *et al.*, 1984; Katcham *et al.*, 1984).

Tobacco

Tobacco consumption is also believed to be high in Scotland. The 1973 Family Expenditure Survey (Department of Employment, 1974) showed that the average weekly expenditure on tobacco per household in 1973 was £2.0 in Scotland and £1.6 in the UK as a whole; while in 1980 (Department of Employment, 1981), Scottish expenditure was £4.5 compared with £3.5 for the UK

These findings are supported by unpublished data from the General Household Survey (1982), which shows that the proportion of current cigarette smokers in Scotland is higher than in Great Britain as a whole (Table 3.4). Data are not available for consumption in different areas of Scotland. Further, the average annual consumption of all tobacco products per adult of either sex has been consistently higher in Scotland than that in England and Wales for at least 20 years (Fig. 3.5).

Table 3.4. Percentage of current smokers in Scotland and in Great Britain

Current smokers	Scotland (%)	Great Britain (%)
Male	45	38
Female	39	33

ECONOMIC ACTIVITY

The level and nature of the economic activities of a nation determine its prosperity, the distribution of the social classes and exposures due to lifestyle, occupation and other factors which may influence cancer risk.

In the 1970s, a population of just under 1 989 000 was employed in Scotland, of whom 1 194 000 were males (Scottish Information Office, 1974). The present economic recession has diminished those numbers considerably. At that time, some 32% of the employed population worked in the manufacturing industries — one of the highest proportions in Europe — while 2% worked in mining and quarrying and 1% in gas, electricity and water supply. Some 3% were employed in agriculture, forestry and fishing, while 8% were in the construction industry. The remaining 54% of the work force were in the service industries, the largest single group (15%) being those in professional and scientific services, followed by the distributive trades (12%). Since the Second World War, tourism has become increasingly important in Scotland and in the Highlands in particular, directly or indirectly involving many workers in the service, food and drink industries.

Industry

Some 75% of industries are concentrated in the Central Lowlands. Many are long established, some for hundreds of years.

Coalmining — Coal is Scotland's chief mineral resource and has been mined for centuries in the Central Lowlands. However, the coalfields in the west central area are almost exhausted, and the main mining areas are now in the Lothian and Fife regions (see Frontispiece and Chapter 6) and the south-west coastal area of Strathclyde. With the closure of many collieries, production has been concentrated on a much smaller number of large mechanized pits.

Steel production — Another long-established industry is steel production, which is concentrated in the west, in the Motherwell and Monklands Districts, but is now considerably diminished despite investment in a huge strip steel rolling mill and an ore terminal.

Ship-building — In former days, much of Scotland's international industrial reputation rested on ship-building and heavy engineering. In spite of re-organization and a switch from traditional ship-building to other marine engineering, such as oil-rig construction, there has been a severe decline.

Whisky distilling — Whisky distilling has had economic vicissitudes but remains one of the nation's major industries. Scotland contains the greatest concentration of malt whisky distilleries in the world; of the 100 or so, about 60 are in the Grampian region.

Although whisky forms a major part of Scotland's exports, the number of people employed in the industry is relatively small.

Newer industries - New industries include electronics and light engineering, chemicals and pharmaceuticals. The aluminium smelting industry has contracted somewhat, while in the last thirty years, hydro-electricity and, more recently, commercial nuclear energy, have become established. Further, the discovery of oil in the North Sea has led to the creation of a number of companies involved in supply and maintenance, notably in Grampian region.

Agriculture, forestry and fisheries

The distribution of persons employed in agriculture, forestry and fisheries is given in Figure 3.6. As in most of the rest of Europe, the proportion of the economically active population following these pursuits has fallen over the past 70 years (Alderson, 1981).

Agriculture: Three-fifths of Scotland's 7.7 million hectares are given over to rough grazing and one fifth to crops and grass.

Dairy farming is concentrated in the south-west and central areas — those with the highest rainfall — while arable farms are found on the fertile lands of the east coast where the rainfall is comparatively low. The poorer quality land of the hills and uplands is used for cattle and sheep rearing. Most crofters combine farming with other occupations, including fishing, home weaving and knitting or catering for tourists. The Southern Uplands and the Highlands sustain 5000 hill sheep farms while the north-east supplies most of the country's beef, pig meat and barley.

Table 3.5 shows the areas used to grow vegetables (excluding potatoes), in different regions of Scotland, in Scotland as a whole and in England. For ease of comparison, the areas under cultivation are shown as rates per 100 000 population. In Scotland, most vegetables are grown in the south-east, where over 6800 hectares were cultivated in 1981. Only in this region does the rate exceed that for England as a whole. With knowledge that road and rail communications are comparatively poor in certain parts of Scotland, particularly in the north-west, it can be concluded from Table 3.5 that the consumption of foods such as green vegetables, which are not produced locally in these areas, is likely to be relatively low.

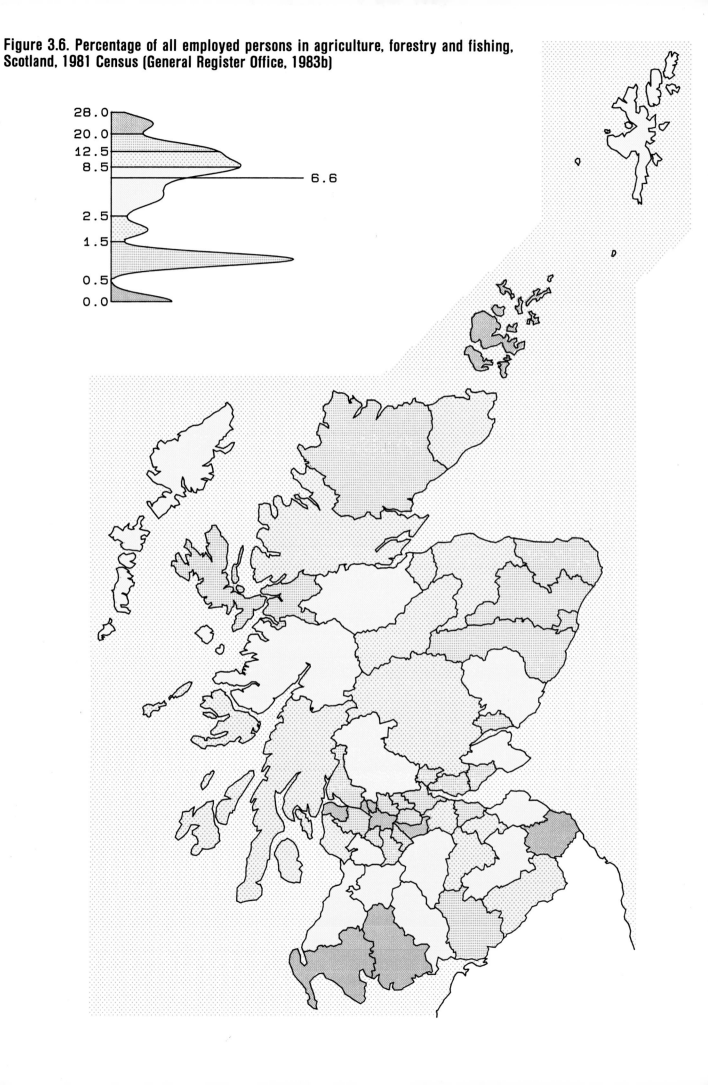

Figure 3.6. Percentage of all employed persons in agriculture, forestry and fishing, Scotland, 1981 Census (General Register Office, 1983b)

Table 3.5. Areas devoted to growth of vegetables for human consumption (excluding potatoes) within Scotland and in England, 1981

Area	Hectares	Hectares per 100 000 population
North-west Scotland (Shetland Islands, Orkney Islands, Western Isles, Highland region)	134	51.8
North-east Scotland (Grampian region)	398	86.0
South-east Scotland (Tayside, Fife, Lothian and Border regions)	6 882	450.4
South-west Scotland (Central, Strathclyde, Dumfries and Galloway regions)	311	11.2
Scotland	7 726	153.1
England	168 162	367.4

[a]Department of Agriculture and Fisheries for Scotland (1982)

Forestry: Forestry is making an increasingly significant contribution to the Scottish economy. By 1982, there were over 800 000 hectares of forest in Scotland, well over half of which were planted by the Forestry Commission, a State body which covers Great Britain (Forestry Commission, 1983).

Fishing: Fishing in Scotland accounts for about 50% of the total UK catch, the industry operating from 65 listed ports all round Scotland, many of them small (Mackay, 1983). It is largely concentrated in the Grampian (e.g., Peterhead, Aberdeen and Fraserburgh) and Highland regions (e.g., Ullapool), and in Argyll and the Shetland Islands. However, the industry is also important for communities in Fife (e.g., Pittenweem), Tayside (e.g., Arbroath), parts of Strathclyde (e.g., Ayr), the Orkney Islands and the Western Isles. Although fishing has always been a fundamental way of life in north-east Scotland, the industry is in a state of decline and change. Fish farming, notably salmon, is of increasing importance.

The effects of industrial development and decline in the west of Scotland

The effects of industrial development in Scotland and of its subsequent decline are seen most clearly in the west, in the Glasgow area. The great growth of population during the nineteenth century industrial boom drew migrant labour from rural Scotland and from the north of Ireland. During the twentieth century, the industrial boom began to slow, to be replaced in the 1930s by economic run-down and unemployment, which continues to this day. At the same time, the city's housing stock had aged to the point where massive clearance and rebuilding had become a necessity. Both economic and social re-vitalization were required.

In the early 1950s, major public policies initiated the transfer of large numbers of people out of the Glasgow slums to new towns around the city and beyond. The first phase took place between 1950 and 1970, and incorporated several kinds of resettlement. New towns (Chapter 6) were proposed not only to rehouse Glaswegians in the best possible environment, but also to provide attractive sites for new kinds of industry. All have gradually absorbed large numbers of Glaswegians. The other types of resettlement sites for overspill were newly built housing estates at the edge of the city and 'expanded towns' elsewhere in Scotland. This transfer of population allowed major redevelopment of the inner city, and a proportion of the original residents eventually returned there.

In addition, a substantial number of people (sometimes estimated as an equivalent number) migrated independently out of the city in pursuit of better homes and jobs. The solution of the housing problem came within sight by 1970, but thereafter, other economic and social reasons forced a change of policy. These factors included a rise in building costs, the continuing migration from the country as a whole and the repercussions from sudden depopulation of the city's younger and more skilled groups. The policy now focussed on improving the social and economic health of the city core, in the form of massive rehabilitation of houses, .reclamation of derelict industrial land, building of new factories and training workshops and overall environmental improvement.

These population movements may have had an effect on geographical cancer patterns, as the young moved out and the elderly tended to stay behind.

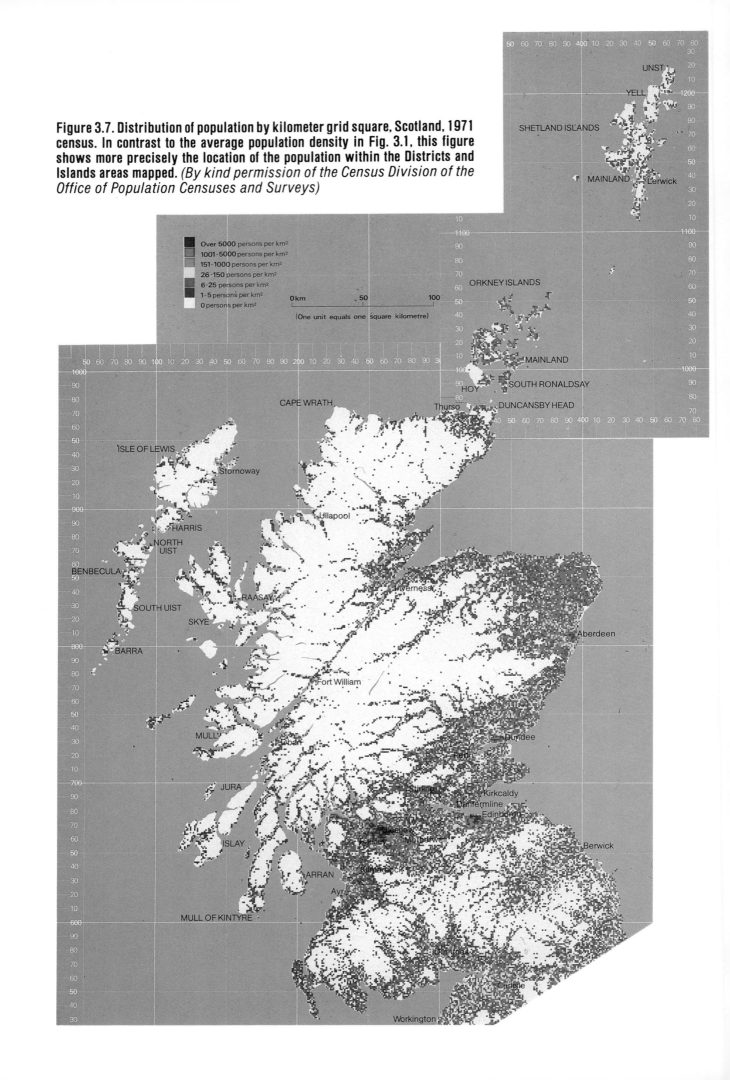

Figure 3.7. Distribution of population by kilometer grid square, Scotland, 1971 census. In contrast to the average population density in Fig. 3.1, this figure shows more precisely the location of the population within the Districts and Islands areas mapped. *(By kind permission of the Census Division of the Office of Population Censuses and Surveys)*

Over 5000 persons per km²
1001-5000 persons per km²
151-1000 persons per km²
26-150 persons per km²
6-25 persons per km²
1-5 persons per km²
0 persons per km²

0km   50   100

(One unit equals one square kilometre)

Chapter 4

HEALTH CARE IN SCOTLAND

This chapter outlines the present administration and provision of health care throughout Scotland, preceded by a brief resumé of the historical development of medicine in Scotland.

EARLY LANDMARKS IN MEDICINE IN SCOTLAND

Hamilton (1981) has described the history of medicine in Scotland. The Scottish universities developed early and are among the oldest in the world. Three were founded in the fifteenth century - St Andrews in 1411, Glasgow in 1451 and Aberdeen in 1494; Edinburgh was founded in 1582. With Cambridge and Oxford, these were the only universities in Great Britain until 1832, when Durham was opened.

Medical education began in Aberdeen in 1497. By the sixteenth century, the long tradition whereby Scottish physicians travelled to the Continent for training, notably in Paris and Leiden, was well established. Landmarks in the sixteenth century were the establishment of the Incorporation of Surgeons and Barbers in Edinburgh in 1505, from which grew the Royal College of Surgeons of Edinburgh, while in 1599 the Faculty of Physicians and Surgeons of Glasgow received its charter, followed in 1681 by The College of Physicians of Edinburgh. In the early eighteenth century, Aberdeen, Edinburgh, Glasgow and St Andrews began almost simultaneously to set up formal medical teaching. The institutional care of the sick was taken up afresh in the eighteenth century and began with the poor, the parish authorities allocating some of the new poorhouses built in that century for the purpose. In addition, institutions set up by voluntary subscription were established for the non-pauper sick, e.g., the Royal Infirmary of Edinburgh (1729) and Aberdeen Royal Infirmary (1742).

The great output of the Scottish medical schools had enormous influence throughout the world on medical practice and on the design and operation of new medical schools. A stream of medical books emanated from Scotland, e.g., Cullen's (1777) book on medical practice which was used until the nineteenth century. Other developments included the rise of scientific societies; the first purely clinical society was the Medical Society in Edinburgh, founded in 1731.

The nineteenth century saw further enormous developments. During the first half of the century, about 95% of doctors in Britain with a university medical degree had been educated in Scotland. In that century, Scotland made crucial contributions to

surgery. The use of chloroform, which was to rival ether as an anaesthesic was introduced by Simpson in Edinburgh in 1847, while Joseph Lister in Glasgow solved the problem of surgical wound infection by means of antiseptic techniques. Some 50 years later, McEwan instituted rigorous asepsis in the operating theatre, while in the field of cancer Beatson in Glasgow introduced oophorectomy for the treatment of inoperable breast cancer.

CURRENT ADMINISTRATION AND PROVISION OF HEALTH CARE

Central administration

Responsibility for domestic administrative affairs within Scotland rests with the Scottish Office, which is headed by the Secretary of State for Scotland, a Cabinet minister. Some functions, however, are carried out by departments that cover the whole of the UK. These include social security, employment, taxation and certain aspects of trade and industry. Scottish control extends over agriculture and fisheries, education, health, social work, planning and development. Health is one of the major responsibilities of the Scottish Home and Health Department. This Department controls the overall policy of the National Health Service in Scotland, which includes hospitals, public health services, general medical practitioners (family doctors), dentists, pharmacists and opticians.

National Health Service administration

The country is split into 15 health administrative units called health board areas, each of which is autonomously responsible for the medical services within its boundaries. Each area covers the same territory as its equivalent local government region or Islands area, except Strathclyde, which comprises Greater Glasgow, Argyll and Clyde, Lanarkshire and Ayrshire, and Arran. Forth Valley Health Board area is equivalent to the Central region. Districts vary widely in environmental and social characteristics (Chapter 6) and, consequently, in their health problems. One of the main aims of the Scottish Home and Health Department is to ensure that, as far as possible, services throughout the country are provided to meet needs. This is achieved by a resource allocation formula that takes into account such factors as mortality levels, the use made of medical services and movement of patients across the borders of health board areas for hospital care elsewhere.

In addition to the health board area organization, a further health administrative body, the Common Services Agency, is responsible for a number of nationally based services, such as blood transfusion, hospital building and supplies, ambulances, communicable disease, health education and information. The Information Services Division of the Common Services Agency coordinates the collection, processing and dissemination of data on health services, including cancer registration.

Health care in relation to cancer

Almost the entire Scottish population is registered with family doctors or general medical practitioners (GPs) who undertake to provide complete primary health care — preventive measures as well as treatment. There are presently about 3250 GPs covering the population of some five million. GPs frequently initiate laboratory and X-ray investigations, but a number of the more specialized investigations required to establish many cancer diagnoses, such as contrast media X-rays and endoscopy, are carried out in hospital under consultant supervision. The hospital consultant service provides the GP with fully trained specialist opinion and help, accepting total responsibility for patient care; within the National Health Service, the patient normally has access to this service only through his GP. Hospitals vary somewhat in the services provided, depending on size. The mainstay is the district general hospital, which provides general hospital services and treats most of the cancers encountered, including those of lung, breast, gastrointestinal tract and female reproductive tract. There are also 'teaching' hospitals, which provide not only general services for their local populations, but also certain specialist services for much larger populations, e.g., radiotherapy, neurosurgery, plastic surgery and the investigation and treatment of blood and lymphatic cancer.

Although the treatment of cancer is becoming more complex, often involving consultants from more than one discipline, the primary responsibility for the patient remains with the GP. When the initial, intensive investigation and treatment are over, continuing care often reverts to the primary care services, although follow-up consultations and treatment are usually given by the hospitals.

Cancer care at Health Board level

It is possible to give only a broad description of the health services provided by the health boards for the care of cancer (Fig. 4.1). Table 4.1 presents a summary of the distribution, by health board, of the numbers and rates per 100 000 local residents of hospital beds, hospital medical staff by specialty and family doctors (Information Services Division, 1982). Although more recent information is available, data for 1980 are given, since they are more relevant to the period covered by this atlas. In interpreting the rates, it should be noted that a number of health boards provide services for patients resident in other health board areas. This inflow of patients is particularly marked in health board areas with teaching hospitals, e.g., Grampian, Tayside, Lothian and Greater Glasgow.

In the Grampian Health Board Area, the teaching hospitals are all in the City of Aberdeen and include Aberdeen Royal Infirmary (752 beds), Woodend General Hospital (392 beds) and the Royal Aberdeen Children's Hospital (167 beds). The Department of Clinical Oncology at Aberdeen Royal Infirmary is responsible for all radiotherapy and much of the cancer chemotherapy given in the health board area. A Malignant Disease Advisory Committee promotes collaboration in care and in planning of services, and helps coordinate research. The entire Grampian Health Board area is served by the University Department of Pathology in Aberdeen. Cancer patients from the Orkney and Shetland Islands receive specialist services in Aberdeen.

The Highland Health Board and the Western Isles are served by several district hospitals with a total complement of 2634 beds; Raigmore Hospital, Inverness, with 410 beds, serves as the regional hospital. The regional histopathology and the radiation and oncology services are located there, and most specialties are represented. There is no medical school, but medical students from Aberdeen University take part of their tuition at Raigmore Hospital.

In the Tayside Health Board area, the teaching hospitals, Ninewells Hospital (784 beds) and Dundee Royal Infirmary (287 beds), are both associated with the University of Dundee. The Department of Radiology and Oncology and the regional laboratory services are located at Ninewells Hospital. A continuing care unit is provided in Dundee for the management of terminal malignancy. Dundee hospitals also supply specialist services to the population of North Fife.

The major teaching hospitals in Lothian Health Board area are the Royal Infirmary of Edinburgh (978 beds) and the Western General Hospital, Edinburgh

Figure 4.1 Health Board Area Boundaries, National Health Service, Scotland[a]

The Health Board Area co-incides to a considerable degree with the Local Government Regions. Forth Valley is synonymous with Central Region, Strathclyde Region covers the Health Board Areas of Greater Glasgow, Lanark, Argyll and Clyde and Ayrshire and Arran (see text in Chapter 6).

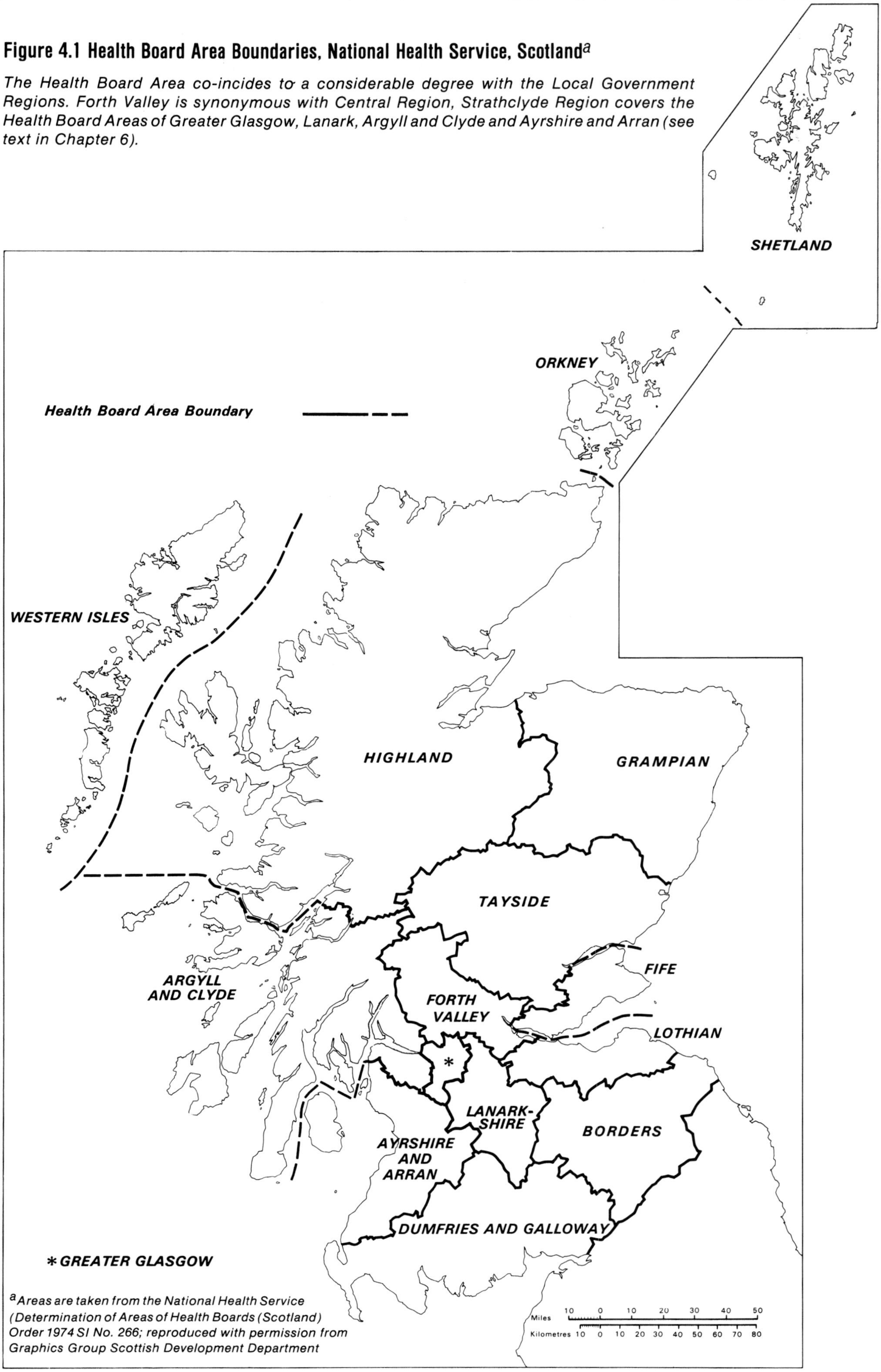

SHETLAND

ORKNEY

Health Board Area Boundary

WESTERN ISLES

HIGHLAND

GRAMPIAN

TAYSIDE

FIFE

ARGYLL AND CLYDE

FORTH VALLEY

LOTHIAN

*

LANARK-SHIRE

BORDERS

AYRSHIRE AND ARRAN

DUMFRIES AND GALLOWAY

*GREATER GLASGOW

Miles 10 0 10 20 30 40 50
Kilometres 10 0 10 20 30 40 50 60 70 80

[a]*Areas are taken from the National Health Service (Determination of Areas of Health Boards (Scotland) Order 1974 SI No. 266; reproduced with permission from Graphics Group Scottish Development Department*

Table 4.1. Average available staffed beds for all and selected specialities, and hospital clinicians and family physicians, for Scotland and each health board area.[a, b] Numbers and rates per 100 000 resident population, 1980[c]

Registry		Scotland	North		North-East			East
Health Board			Highland	Western Isles	Grampian	Orkney	Shetland	Tayside
1. Average available staffed beds, all specialities	No.	58 208	2 422	212	5 391	126	159	5 522
	Rate	1 130	1 267	714	1 142	699	713	1 383
Selected specialities								
2. General surgery	No.	3 901	221	36	309	40	44	350
	Rate	76	116	121	66	222	197	88
3. Urology	No.	456	-	-	33	-	-	36
	Rate	9	-	-	7	-	-	9
4. General medicine	No.	4 499	195	26	372	-	18	412
	Rate	87	102	88	79	-	81	103
5. Gynaecology	No.	1 318	39	8	98	-	-	99
	Rate	26	20	27	21	-	-	25
6. Radiotherapy (consultative)	No.	324	_[d]	-	27	-	-	37
	Rate	6	<1[d]	-	6	-	-	9
7. Paediatric surgery	No.	325	-	-	24	-	-	-
	Rate	6	-	-	5	-	-	-
8. Paediatric medicine	No.	627	32	-	32	-	-	74
	Rate	12	17	-	7	-	-	19
Staff								
9. Hospital clinicians	No.[e]	4 944	147	10	441	4	5	491
	Rate	96	77	33	93	21	23	123
10. Family physicians	No.	3256	165	25	300	21	19	255
	Rate	63	86	84	64	117	85	64

[a]Each Health Board covers the same territory as the equivalent local government region, except the Greater Glasgow Health Board, Lanarkshire, Argyll and Clyde and Ayrshire and Arran Health Boards which, combined, cover the Strathclyde region. Forth Valley Health Board is equivalent to the Central region.

[b]The current cancer registry coverage of these health board areas is also given in Table 5.2

[c]Information Services Division (1982), Tables 6.4a, 6.4b, 9.3a

[d]No beds were allocated for radiotherapy or oncology at this time, patients being housed by the referring physician or surgeon.

[e]Numbers of hospital staff are whole-time equivalent

Table 4.1 (contd)

Registry		South-East				Western				
Health Board		Lothian	Fife	Borders	Forth Valley	Dumfries and Galloway	Greater Glasgow	Lanark-shire	Argyll and Clyde	Ayrshire and Arran
1.	No.	8 389	2 913	897	3 616	1 595	14 084	5 536	4 523	2 822
	Rate	1 121	856	901	1 333	1 120	1 392	968	985	751
Selected specialities										
2.	No.	519	166	60	156	120	911	372	366	233
	Rate	69	49	60	56	84	90	65	80	62
3.	No.	60	22	-	32	-	179	55	-	40
	Rate	8	7	-	12	-	18	10	-	11
4.	No.	851	183	56	179	88	1 136	409	363	213
	Rate	114	54	56	66	62	112	72	79	57
5.	No.	213	58	-	69	30	403	139	108	55
	Rate	28	17	-	25	21	40	24	24	15
6.	No.	106	-	-	-	-	153	-	-	-
	Rate	14	-	-	-	-	15	-	-	-
7.	No.	98	13	-	33	12	123	22	-	-
	Rate	13	4	-	12	8	12	4	-	-
8.	No.	129	19	7	9	18	183	61	33	30
	Rate	17	6	7	3	13	18	11	7	8
Staff										
9.	No.[e]	976	168	42	159	94	1 620	336	253	172
	Rate	130	49	42	59	66	160	59	55	46
10.	No.	513	193	69	169	105	627	287	289	219
	Rate	69	57	69	62	74	62	50	63	58

(546 beds), the City Hospital (447 beds) and the Eastern General Hospital (335 beds). These work in close association with the University of Edinburgh. There are National Health Service and University Departments of Clinical Oncology (radiation oncology and medical oncology) at the Western General Hospital and Royal Infirmary, Edinburgh. Two major pathology departments and a regional laboratory service are located in Edinburgh. For the terminally ill there is a well known hospice service based in Edinburgh, covering a wide area, generously supported by voluntary public effort as well as by the Health Board.

While acute medical and surgical services are provided in hospitals in the Borders area, more specialized care is obtained by referral to Edinburgh. In the heavily populated southern parts of Fife, general surgical, gynaecological and medical care is undertaken in district general hospitals, with reference to Edinburgh for specialized treatment such as radiotherapy.

In the west of Scotland there are a number of major teaching hospitals in association with the University of Glasgow. These are the Royal Infirmary (778 beds), the Western Infirmary (541 beds), the Victoria Infirmary (639 beds), the Southern General Hospital (1168 beds) and Stobhill Hospital (829 beds); all are in Glasgow.

A regional service is provided by the Glasgow Regional Institute of Radiotherapy and Oncology (120 beds), the second largest of its kind in the UK. Other regional services include pathology, paediatrics and plastic surgery, while at the Regional Institute of Neurological Science neurological tumours are treated. Glasgow is also a centre for bone marrow transplantation. A number of institutions are concerned with the treatment and study of cancer. One of these is the Beatson Institute of Cancer Research, named after Sir George Beatson, who first discovered that removal of the ovaries increased the length of survival of breast cancer patients (Beatson, 1896), thus laying the foundation for the modern approach to the treatment of breast cancer. The West of Scotland Cancer Surveillance Unit was established in 1973 and incorporates the cancer registry. Several of these bodies were constituted as the Glasgow Regional Cancer Institutes in 1982. As in the Grampian Health Board area, advice to the Greater Glasgow Health Board is channelled through a cancer advisory group.

Other important developments in recent years have been the establishment of chairs of oncology in Edinburgh (1982) and Glasgow (1974) and a cancer epidemiology unit at the University of Edinburgh, the former funded by the Imperial Cancer Research Fund, the latter two by the Cancer Research Campaign.

So far, this description of cancer care has listed the major institutions involved in cancer care and research in Scotland. However, much of the investigation and medical and surgical treatment of cancer is carried out in district general hospitals. One of the problems in Scotland, as in many regions, is to provide a uniformly high standard of health care in isolated places. Parts of the north of Scotland are particularly affected, not only because of the terrain but also because of the harsh climate. In these circumstances, the smaller hospitals are vital, and a great deal depends on the active family physician and district nursing services, supported by hospital consultants who often travel long distances to conduct clinics. The islands off the west coast present particular problems (Chapter 2). Often, the easiest way to provide specialist care is to transport the patient to suitable hospitals on the mainland. The Air Ambulance Service has been very valuable in providing treatment for patients living in these remote locations.

Finally, a very small number of cancer patients are treated in private hospitals.

CANCER REGISTRATION

The five regional cancer registries which cover the health board areas are listed in Table 4.2.

These registries are run as separate organizations, each with its own director and staff; national coordination is carried out by the Information Services Division of the Common Services Agency of the Scottish Health Service. This division, assisted by a committee on which all registries are represented, is also responsible for collection and dissemination of national data.

In Scotland, the majority of patients with cancer follow a fairly constant pathway towards registration. Patients confirmed or suspected by their GP to be suffering from cancer are almost invariably referred to hospital for further investigation and treatment. Exceptionally, a few patients may not be referred: those whose cancer is too advanced to benefit from specialist

Table 4.2 Health board areas and size of population covered by each cancer registry in Scotland

Cancer registry	Population covered in thousands	Registry location	Health board areas covered
North	218	Inverness	Highland; Western Isles
North-east	504	Aberdeen	Grampian; Orkney, Shetland
East	383	Dundee	Tayside
South-east	1145	Edinburgh	Fife; Lothian; Borders
Western	2785	Glasgow	Greater Glasgow; Lanarkshire; Argyll and Clyde; Ayrshire and Arran; Forth Valley; Dumfries and Galloway

treatment and are mainly elderly. Some of these persons, therefore, escape the registration net, although they may be identified subsequently from the medical certificate of the cause of death. Patients who are treated only in hospital outpatient departments have a better chance of registration, although, in that busy environment, there is still a possibility that registration may not be effected because there is no national in-patient abstract (see below). This can lead to some under-registration of those cancer patients who are investigated and treated mainly as outpatients, e.g., skin cancer patients — a problem common in registries elsewhere. In spite of overall improving efficiency in registration, this could become an increasingly difficult problem as more cancers are investigated and treated without formal in-patient admission. Some cancers of the lymphatic and haematopoietic systems in the elderly now fall into the latter category.

The technique of registration varies between the various regional registries, although all make use of the available sources. The Scottish Hospital Statistics Scheme (SMR1), by which information on every in-patient and day patient discharged from Scottish hospitals is recorded (some 600 000 discharges each year), is used as a primary source by the North and North-east registries. Two of the registries, the East and the South-east, have their own record forms. These are completed by hospital records staff and are submitted to the registries. In the Western registry, data from Greater Glasgow Health Board area are supplemented from SMR1 computer case listings produced locally. Registry staff check all information received from the health board areas, from whatever source, before registering cases.

All the registries obtain information provided by departments of pathology, either directly or indirectly; a further source is the quarterly computer list of deaths bearing a cancer diagnosis that is supplied to each regional registry by the Registrar General for Scotland.

In general, new cases are first recorded on a series of card index files, with careful checking to avoid duplication. Each individual cancer diagnosis is allocated a registration number, enabling second primary malignant cancers to be recorded separately. Individual tumour data are sent by the regional registries on a standard record form (SMR6) to the Information Services Division, which carries out an initial clerical check and codes the occupation of the patient. The data are filed on computer and subjected to validation checks. To assess completeness of registration, checks are made of the expected total annual registrations from each registry, using mortality rates calculated independently from death certificates.

Some regional registries are inaugurating their own computer processing, which will enable data to be submitted to the Information Services Division on magnetic tape.

In the last ten years, each new registration has been flagged in the files of the Scottish National Health Service Central Register (NHSCR), which record every person in the country registered with a GP. Since the NHSCR is notified of every death occurring in Scotland, the date of death of the flagged patients can be communicated to the cancer registration scheme without time-consuming and expensive active follow-up. In this way, up-to-date information on survival is produced. The system is linked to a similar one in England and Wales, so that survival can be recorded of those registered patients who subsequently move between the two countries.

Private medical practice has increased in the UK in recent years. It is difficult to obtain information on its extent, but it is thought to be relatively small in Scotland. Its relevance for cancer registration is that a patient treated privately for cancer is less likely to be included on the cancer registration files. However, notification of some of these cases would reach the registration system *via* pathology laboratory reports, and at death through the Registrar General's lists of cause of death. Since there is no private radiotherapy in Scotland, all patients thus treated should be picked up by the present system.

COMPLETENESS AND RELIABILITY OF SCOTTISH REGISTRATION DATA

In any country there is always the possibility of under-registration. Elderly patients may receive less intensive investigation. Not only can the diagnosis of cancer be missed but the registration system may fail to pick up a diagnosed case. Improvements occur over time in the collection of cases, as well as in the quality of the information collected. These changes are small and are difficult to differentiate from true changes in incidence. It is considered that in Scotland the registration system has been efficient and stable during the period 1975-1980.

There are differences between registries. Large regional registries encounter more difficulties than the smaller ones. This is hardly surprising, since the number of hospitals that supply data varies enormously between the regions. For example, the Western Scotland registry catchment area contains some 180 hospitals treating cancer, whereas the North contains only one main hospital. However, since the mid-1970s, powerful assistance has come to the aid of registration in the form of regular computer listings of

hospital discharge records (the SMR1 discharge record). A second major contribution has come from quarterly computer lists of patients who died with a mention of cancer on the medical certificate of cause of death, supplied by the Registrar General for Scotland (see Death certificate only registration, below).

Deficiency of information can also arise within different parts of a registry region; this possibility is particularly relevant with regard to this atlas. Such patchiness is difficult to detect and to remedy. Hospitals vary in their ability to supply information, due partly to temporary deficiencies in staffing levels, e.g., of records staff.

An attempt was made to assess whether there was likely to be systematic under-registration in one or more districts, by comparing the distribution of the rank order, for persons of each sex, of each cancer site, between the 56 districts. If there were under-registration in certain districts, one might expect the ranks for most cancer sites in such areas to be consistently low; this was not observed. Analysis by Kendall's coefficient of concordance revealed that there was no tendency for any region to have consistently high or low rates for the forms of cancer mapped (males, $W = 0.084$; females, $W = 0.125$).

Elimination of duplicate registrations

Care is taken at the regional registry level to eliminate duplicate registration as far as possible. The record of a patient treated in one region and resident in another is passed to the region of residence for registration. Moreover, once data have been collated and processed by the Central Registry, any duplications that have slipped through can still be detected by a computer data validation check which flags registrations possessing apparently identical information (year of registration; registration number; regional registry code; hospital code).

In preparing data for the atlas, two further checks for duplications were carried out by the Information Services Division and the regional registries. The first was on those patients registered in 1975-1980 who matched, on registry code, surname, birth year, sex and cancer site (to three digits). The second check was on registrations for 1975-1980 for which place of residence was shown on computer analysis to be outside the registry recording the case. These two exercises revealed only 358 true duplicates, out of a total number of registrations in the period 1975-1980 of some 130 000 (0.3%).

Indices of reliability

In the assessment of reliability of registration, three numerical indices are commonly used:

(1) histological verification rate: the proportion of registered cases that had histological verification of diagnosis;

(2) death certificate only rate: the proportion of cases registered for which no information is available other than the statement on the death certificate that the deceased died from or with cancer;

(3) ratio of deaths in period to incidence: the ratio of the number of deaths from a given cancer during a particular period to the number of registered cases of the same cancer in that period.

Histological verification of diagnosis: The proportion of registered cases with histological verification of diagnosis is given by sex and broad age group in Table 4.3. Histological verification of diagnosis implies that a portion of the suspected cancer was removed for microscopic examination. The higher the proportion of histological verification for cancers of deep-seated organs, such as the pancreas and brain, the greater the confidence that a neoplasm existed and that it arose at the stated site. While, for some sites, radiological or biochemical evidence of the presence of a cancer may be as convincing as histological, in the Scottish registries confirmation is based on histology.

A fall in the proportion with histological verification from one time period to another may be due to the emergence of a new diagnostic tool, such as serum α-fetoprotein for primary cancer of the liver. A very high proportion of cases with histological verification does not necessarily mean that figures from the registry in question represent a more reliable estimate of true incidence than those from a registry with a lower proportion. A high proportion may indicate only that the recorded figure is not an overestimation of the true incidence. In some circumstances it may indicate that case-finding was largely limited to notifications received from pathological laboratories.

Problems that arise in interpreting the classification and the coding of cancer (Chapter 5) include handling of multiple cancers in the same patient and of certain benign tumours, and solutions may vary between registries.

Death certificate only registration: The medical certificate of cause of death is used to provide information about persons with cancer who were not registered during their lifetimes. This provides an important check on the completeness of registration. If a high proportion of cases are registered only as a result of death certification, the recorded incidence rates are likely to be too low. Since many cancers are not fatal, it seems likely that in these circumstances the corresponding number of non-fatal cases could have been overlooked.

Cancer registries in most countries obtain batches of death certificates at intervals from national vital statistics offices, and match them against their files. A proportion of cancer cases show no record of a previous registration in the cancer registry. Some registries include all such unmatched deaths in their totals of registered cases of cancer without further enquiry. Others contact the certifying physician and attempt to discover whether the patient was investigated further, or treated, or registered elsewhere at an earlier date. If a case was not registered previously, the registry then exercises its discretion as to whether the underlying evidence for cancer was sufficiently strong to warrant registration. It is known that registry practice varies within Scotland.

Deaths in period: It is axiomatic that if the number of deaths from cancer exceeds the number registered then registration is incomplete, unless the incidence for that site of cancer is declining at a very rapid rate. For some sites such as lung, bone, brain, liver and oesophagus, survival is poor and therefore incidence is likely to approach mortality. For others, such as breast and colon, the survival rate is much higher, and this should be reflected in the mortality:morbidity ratio.

Table 4.3 shows that, as in other countries, in each Scottish registration region the proportion of cases with histological verification tends to fall slightly with advancing age. Further, other indices of reliability tend to be somewhat lower in the West of Scotland than in the other registration districts. It would be unrealistic to expect the figures to be identical, given that the West of Scotland has a much higher level of lung cancer, which is diagnosed frequently by X-ray alone. The information in Table 4.3 for the West of Scotland is little different from that for other large registries in the UK, e.g., Birmingham and Mersey.

Table 4.3 Indices of reliability for Scottish registration data by registry, 1975-1980, for cancers at all sites (ICD 8 140-208), by sex[a]

Age (years)	0-34						35-64						65-74						75+						All ages					
	Male			Female			Male			Female			Male			Female			Male			Female			Male			Female		
	HV	DC	DIP	HV	DC	DIP	HV	DC	DIP	HV	DC	DIP	HV	DC	DIP	HV	DC	DIP	HV	DC	DIP	HV	DC	DIP	HV	DC	DIP	HV	DC	DIP
East	99	-	51	99	-	25	97	*	59	98	*	48	96	1	64	96	*	59	94	1	72	94	1	69	96	1	64	96	1	57
North	97	-	39	99	-	30	83	-	62	91	-	49	77	*	70	84	*	60	74	*	75	79	*	83	79	*	67	86	*	61
North-East	94	-	37	97	*	27	75	*	53	84	*	45	69	1	59	73	1	57	66	2	61	61	3	62	71	1	57	74	1	53
South-East	95	*	32	95	1	25	84	2	56	89	1	48	75	3	64	82	3	58	63	5	69	66	7	65	76	3	62	80	3	55
Western	87	2	45	88	2	38	88	5	66	80	4	57	58	8	74	69	6	65	50	12	75	53	14	73	61	8	71	70	7	63

[a]No cases reported; *, cases reported but percentage less than 0.5%
HV, histological verification (% of all registrations); DC, death certificate only (% of all registrations); DIP, deaths in period (% of all registrations)

Currently, it is not possible to provide the type of information in Table 4.3 for the 56 districts covered in this atlas.

In our opinion, the differences between the regional registries in Scotland are not sufficiently large to negate the comparisons made in this atlas. This is reinforced by comparing the age-specific incidence rates for all forms of cancer in the four cities — Aberdeen, Dundee, Glasgow and Edinburgh — since each is in a different registration region. Despite the variation in registration practice, it appears that the age-specific incidence rates, in both sexes, are virtually identical (Fig. 4.2). The similarity of these curves in the four largest urban populations of Scotland suggests that cancer registration between the regions has an overall net consistency. It is interesting to note (Chapter 7) that elevated rates for Hodgkin's disease were observed in people of both sexes in the contiguous districts of Clydesdale and Tweeddale, districts pertaining to the Western Scotland and the South-East cancer registries, respectively.

In summary, available evidence indicates that the inter-registry and inter-district comparisons made in this atlas are legitimate.

Figure 4.2. Age-specific incidence rates per 100 000 persons from cancer of all forms in:

a)   *Males and females in Scotland*
b)   *Males in the cities of Aberdeen, Dundee, Edinburgh and Glasgow*
c)   *Females in the cities of Aberdeen, Dundee, Edinburgh and Glasgow*

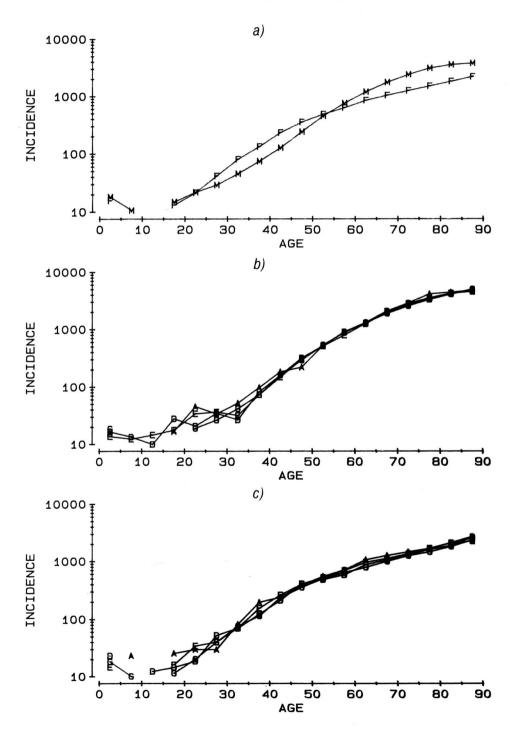

Chapter 5

THE MAPPING OF CANCER

Several aspects of the validity of the incidence data for Scotland were examined in Chapter 4. This chapter includes explanations of some of the terms and conventions used in describing cancer, and of how the information in the atlas is presented. Several of the terms and concepts are explained further in Appendix III.

CANCER

Cancer is a group of diseases which possess a common feature, viz., the uncontrolled growth of the cells that make up the part of the body affected (Cairns, 1977).

The cancers described in this atlas are defined by the 8th revision of the International Classification of Diseases (WHO, 1967) hereafter referred to as ICD8. The ICD8 code numbers for the cancers arising in the various sites (organs) are used in the text, tables and maps.

INCIDENCE

Incidence is the number of newly diagnosed cases of cancer occurring in a given population in a particular time period, usually expressed as a rate per 100 000 population *per annum*. The maps present the incidence of cancer for the 53 local government districts and three Islands areas described in Chapter 6.

PURPOSE AND ADVANTAGES OF
MAPPING INCIDENCE

While the geographical representation of cancer on maps has been recognized as useful in describing the 'cancer scenery' of a particular country (Frentzel-Beyme *et al.*, 1979), the real purpose lies in identifying geographical areas or hypotheses that require more detailed epidemiological study. The maps may also be used as an aid to planning the provision of health services (Boyle *et al.*, 1985a).

A general limitation to the cancer atlases produced to date is that only mortality data have been mapped. Studies that have examined the reliability of cancer death certification have revealed inaccuracies in reporting (Heasman & Lipworth, 1968; Cameron & McCoogan, 1981; Percy *et al.*, 1981), the level of error varying from one cancer site to another; further, the

relation of mortality to incidence varies greatly between sites, depending largely on the death rate due to the tumour. While mortality from cancer of the oesophagus corresponds closely to the incidence, the mortality from lip cancer, for example, would be a gross underestimate of incidence, due to the excellent prognosis associated with such tumours. Many other arguments favour the value of incidence (i.e., occurrence) over mortality (i.e., death) data in representing temporal and spatial patterns of cancer (Boyle, 1985). The material presented here represents the first cancer atlas based on incidence data.

COMPARISON OF INCIDENCE RATES

The validity of comparison of incidence rates between registries and their constituent areas is based on four implicit assumptions:

(1) that the registration process is uniformly complete in the territories covered by the registries, that all items of information collected are accurate and that duplicate registrations have been eliminated;

(2) that only residents of the registration area are included in the numerator of the incidence rate, i.e., cancer patients who are treated within the area, but normally reside outside that area, are excluded from the calculation of the incidence rates and, correspondingly, the residents of the area who are treated outside the area have been included in the numerator;

(3) that the appropriate denominators have been used;

(4) that the data have been classified, coded and processed in the same way.

While there may be minor differences in procedures between the Scottish cancer registries, the existence of the Central Registry and the careful recording of usual address of patients ensure that these assumptions are largely met (Chapter 4).

CHOICE OF AREA SIZE

There are constraints on the choice of areal unit that are outside the control of the cancer mapper; the smallest usable unit is that for which population information is available. For this atlas, the final selection was made by taking the time-period covered into consideration. The intention was to choose the smallest administrative unit that could be expected to provide reliable rates over a period short enough for time trends to be unimportant. Cancer rates in local authority districts and Islands areas of Scotland over the six-year period 1975-1980 were thus thus selected for mapping.

USE OF AGE-STANDARDIZED RATES

As cancer is commoner as people grow older, comparison of cancer levels between areas can be hampered if the age-structures of the populations in the areas differ. To overcome this problem, age-standardization is carried out. The resultant statistic, an age-standardized incidence rate per 100 000 (sometimes called a standardized registration rate), is taken to represent the risk of developing cancer in a particular area. Whether this standardization is accomplished by direct or indirect methods (MacMahon & Pugh, 1968), or by some more sophisticated technique (Breslow & Day, 1975), the result is never perfect.

One advantage of using the direct method is the possibility of standardizing to some universally accepted population. Yet there is a great temptation, when a series of maps is being produced, for a country to standardize by using a local population, as this results in rates close to the crude incidence (see Appendix III). However, this hinders international comparison. The Scottish cancer atlas is based on incidence rates, standardized by the direct method and employing the World Standard Population described and used in successive volumes of *Cancer Incidence in Five Continents* (Doll *et al.*, 1966, 1970; Waterhouse *et al.*, 1976, 1982), since this is the most commonly used reference population.

As the World Standard Population that was used for standardization has a younger age structure than that of the Scottish population, the age-standardized rates are usually lower than the crude (non-standardized) rates. The age-standardized rate can be regarded as giving the magnitude of the risk in each of the districts or countries compared; the crude rate gives the burden of cancer in terms of the number of cases of cancer per 100 000 population in each district or country.

ILLUSTRATING DIFFERENCES
BETWEEN DISTRICTS

The maps indicate the level of incidence in the 53 districts and three Islands areas. To help the reader to distinguish between districts with high and low incidence rates, colour has been used. Some readers might wish to have maps with colours representing the same range of values on each of the maps. However, the incidence rates of cancer in Scottish districts vary from 0.0 to over 130.0 per 100 000, depending on site. Individual cancers may fall within only one segment of this wide spectrum, e.g., lung cancer ranges from 40.2 to 130.6 in men. With a single range of colours to cover all cancers, it would be difficult to separate clearly, by shade and hue, district differences for each cancer site. To obviate this problem, a relative scale using seven classes was employed, viz., the lowest 5% of the district rates, the next 10%, the next 20%, the middle 30%, the next 20%, the next 10%, ending with the highest 5%. These classes were represented by three shades of orange for the higher rates, yellow for the middle of the range and three shades of green for the lower rates.

A more detailed discussion of the mapping techniques employed and of the choice of colours is given in Appendices IIB and IID.

STATISTICAL SIGNIFICANCE

The reader may wish to know whether the differences shown between districts are significant in the statistical sense, i.e., whether or not they are likely to be the result of chance. The districts are clearly of different size, and the larger ones frequently have smaller populations. For the densely populated districts, the standard error of the rates is likely to be small, and, hence, the rate observed is likely to be close to the true rate. Rates in sparsely populated areas with a small total population have a large standard error.

The tables in Chapter 7 indicate whether the level of incidence for a given cancer in a particular district is likely to be significantly different in the statistical sense from that for the rest of Scotland. The statistical significance of differences between two districts can be readily assessed (Appendix III). A significantly high level of a given cancer in a district is not necessarily due to unavoidable exposure to a particular risk, but may be the result of personal habits, diet or at-work exposure of a segment of the population. A district with higher levels of one cancer may have lower levels of another.

It may be asked why the statistical significance of differences in the rates has not been mapped, as in other atlases (e.g., Gardner *et al.*, 1984). The Editors of the Scottish atlas strongly believe that if a choice has to be made between presenting statistical significance and the rates themselves, the latter is preferable as it gives the overall pattern, which is one of the major objectives of mapping, rather than directing attention to those areas where rates are significantly high or low.

A common method of ordering a scale of values involving degrees of statistical significance is as follows:

(1) significantly high values falling in the highest decile;

(2) significantly high values not falling in the highest decile;

(3) values in the highest decile but not significantly different from the average;

(4) values not significantly different from the average;

(5) values significantly lower than the average.

An area in category 3 may have a higher incidence rate than one in category 2, but because it is not significantly higher than the average it is placed in the lower order.

If the 5% significance level is used, then for a large number of districts the rate will be in the intermediate category of 'neither high nor low'. The differences in mortality between districts that fall into this group are then usually disregarded; furthermore, the reader is more likely to see what the mapper wants him to see, as his attention is concentrated only on the high and low rates.

One solution would be to publish rates or significance using a greater number of categories or on a continuum; this would be more feasible when dealing with rates. For example, the high and low rates on the extremes could be represented by black and white, respectively, with intermediate rates represented by some shade of grey, the intensity of blackness being decided by the relative position of the rate for the district in question with respect to the highest and lowest rates observed.

It is possible that use of either the standardized mortality ratio (SMR; a measure of relative risk) or the statistical significance of local deviations from the overall rates could misrepresent the geographical distribution. In the case of the SMR, no account is taken of varying population sizes, so that SMRs, estimated imprecisely on only a few cases, may constitute the high and low extremes of the map, and, hence, its pattern. Mapping significance alone totally ignores the size of the risk being mapped, so that two areas with identical SMRs may be represented quite differently if they are of unequal population size.

It is considered preferable to estimate and present a measure of risk rather than one of significance, and this estimate of risk could be improved if the variation in precision of estimation between districts could be taken into account. More precisely, a problem stems from over-dispersion of the estimates of relative risk in the districts in the presence of a component attributable to Poisson sampling errors within each district. For mapping, it is desirable to have estimates of relative risk the dispersion of which more closely reflects that of the true relative risks.

James and Stein (1961) proposed estimates with this property for the analogous problem of simultaneous estimation of several normal means. Clayton and Kaldor (personal communication) have recently derived empirical Bayes estimates of the true relative risks. A test of the impact of their methodology on changing SMRs to better estimates of the true relative risk is consideration of the incidence of lip cancer among males in the 56 districts of Scotland. Even though data have been aggregated over the six-year period 1975-1980, lip cancer is a reasonably uncommon form of cancer, so that in some sparsely populated regions no case was observed, while in others a high SMR was based on a very small number of observations.

The original SMRs varied between zero and 652. When it was assumed that the true relative risks were independent, identically distributed gamma random variables, the dispersion was from 31 to 422; and when they were assumed to be log-normal the dispersion was from 34 to 394. What was important, however, was that the ranks of the empirical Bayes estimates differed little from the corresponding ranks calculated on the basis of the SMRs. The only exceptions were the very sparsely populated Islands districts where SMRs were based on very low numbers of cases. In these districts, the empirical Bayes estimates were drawn towards the overall mean.

With common forms of cancer, such as lung cancer, no change was found in rank order. In view of the considerable increase in computational complexity and in the burden placed on the reader to interpret such estimates, it is questionable whether such a rigorous, statistical approach to the mapping problem would be justified. However, the results obtained would seem to merit the use of ranking of SMRs, or even directly standardized rates, as a simpler yet accurate approach to obtaining the relative positions of risk in one district compared to another. So, the mapping of rates, with the classes determined by rank order, would not greatly distort the geographic pattern of any but the very rarest forms of cancer (which themselves are unlikely to be mapped).

PATTERNS OF CANCER DISTRIBUTION: AGGREGATIONS AND GRADIENTS

When looking at the spatial distribution of cancer incidence, the observer may see interesting patterns where none exist and, conversely, may fail to observe a grouping. To minimize these risks, a measure has been devised to indicate the likelihood of the pattern in question arising if the distribution of incidence rates truly had no spatial pattern. This measure, designated D, appears in the box at the upper left-hand corner of each map. If the value of D is less than 16, there is less than one chance in 100 that the observed distribution of the incidence rates is random (for further discussion, see Appendix IIC). This index (D) provides a *summary* measure of the probability that the pattern observed over the *whole* map is random. This does not preclude the existence, scattered over the map, of small pockets of contiguous districts with similar rates (e.g., Hodgkin's disease in Chapter 7).

The presence of a group of areas with higher or lower than average cancer incidence, which are contiguous, or near at hand, is always of interest, as this suggests the presence or absence of risk factors common to these areas. Simple inspection of the colour map for lung cancer (Chapter 7), for example, shows a group of districts around the City of Glasgow with high levels of this cancer in males, which look interesting. The map of lip cancer in males shows that the risk falls gradually from north to south (with exceptions such as Wigtown and the south-east coast), a finding that may be worth further study.

In the comments that accompany each map, a statement may be made as to whether that cancer appears to be aggregated. Despite the use of the measure D, such statements are somewhat subjective. In assessing gradients, the editors tended to look at the pattern rather than at the statistical significance of the rates for the adjoining areas.

Chapter 6

THE AREAS MAPPED

This chapter briefly outlines the salient features of the 56 administrative units for which the atlas presents cancer incidence data. These administrative units are contained in nine local government regions and three Islands areas. Most of the regions cover the same territory as the corresponding health board areas — the administrative units of the Scottish Health Service (Chapter 5). The only exception is Strathclyde region which covers four health board areas — Greater Glasgow, Argyll and Clyde, Lanarkshire, Ayrshire and Arran. The local government regions are split further, creating a second administrative tier of 53 local government districts. *The local government district/Islands area is the area/unit of residence selected for analysis in the atlas.* Each has been allotted a number, given in the descriptions below in square brackets, on the reference map (Fig. 6.1) and in the tables in Chapter 7 that give the incidence rates by district.

The populations given for each region, Islands area and district are those 'usually resident' derived from the 1981 census. The descriptions of districts pertain as far as possible to the period 1976-1980.

THE HIGHLAND REGION
(HIGHLAND HEALTH BOARD AREA)
Population, 187 000; area, 2 539 122 hectares

This region, the northern-most part of mainland Scotland, comprises about one-third of Scotland, and is the largest local government unit in western Europe. The whole area, much of it over 300 metres in altitude and sparsely populated, is bisected from south-west to north-east by the Great Glen, which is occupied by a series of lochs, the largest of which is Loch Ness (the home of the 'Loch Ness monster'), linked together by the Caledonian Canal. The greatest concentration of people is found on the relatively flat east coast, where the only large town in the region, Inverness, is situated.

Caithness District [01]
Population, 27 000; persons per hectare, 0.2

Economic activity centres around the towns of Wick and Thurso. Wick's main industries are fishing and the servicing of vessels of the oil industry; while in Thurso the main employment is at the nearby

Dounreay Nuclear Reactor. Caithness is famous for its glass industry.

Sutherland District [02]
Population, 13 000; persons per hectare, 0.02

Industries include farming, crofting, fishing, forestry, craft industries and tourism. At one time, coalmining and brick-making were conducted, now replaced by whisky distilling and wool milling.

Ross and Cromarty District [03]
Population, 45 000; persons per hectare, 0.1

The district contains a major fishing centre (Ullapool), as well as more recent oil-platform and other oil-related construction industries.

Skye and Lochalsh District [04]
Population, 10 000; persons per hectare, 0.04

Over two-thirds of the district's population live on the Island of Skye. The principal industries are farming and tourism, followed by forestry and fishing.

Lochaber District [05]
Population, 19 000; persons per hectare, 0.04

This is a mountainous area, the main occupations being aluminium smelting, forestry and paper-making.

Inverness District [06]
Population, 54 000; persons per hectare, 0.2

The town of Inverness is the main administrative and business centre in the Highlands. Old established occupations are farming, whisky distilling, weaving, food processing and printing with, more recently, developments in the electronics and oil construction industries.

Badenoch and Strathspey District [07]
Population, 9000; persons per hectare, 0.04

This is a mainly mountainous district, with farming in the valleys and with many of the hill slopes given over to forestry.

Nairn District [08]
Population, 10 000; persons per hectare, 0.2

A small prosperous district with noted golfing and resort facilities, where the main industries are farming and forestry.

Figure 6.1. Local Government Districts and Island Areas of Scotland

The names of each local government district and island area are given in Chapter 6 together with the number appearing on this map. They also appear on the tables throughout Chapter 7. Thus 01 gives the location of Caithness; 34 Clydebank and 35 Beardsen and Milngavie (the small districts northwest of the City of Glasgow [33]). For the islands, arrows indicate the relevant district.

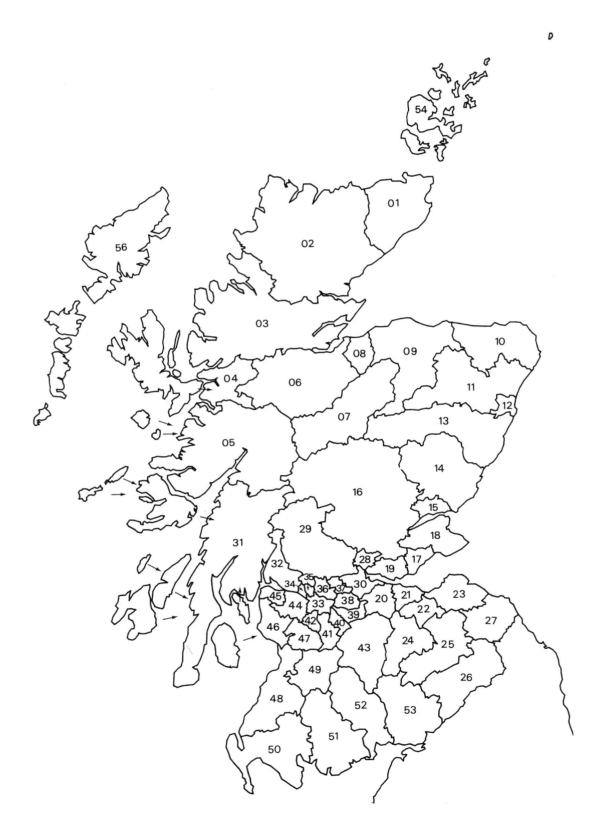
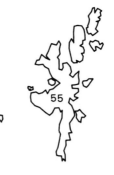

GRAMPIAN REGION
(GRAMPIAN HEALTH BOARD AREA)
Population, 463 000; area, 870 389 hectares

The region, roughly square in shape, is bounded on the east and north by the North Sea. In the south-west are the Cairngorm Mountains with peaks reaching over 1200 metres. Thus, the main access to the area is along the coastal plains to the south-east and north-west. Of the total area, over 60% is agricultural land and 15% productive forest.

Situated on the east coast, the City of Aberdeen is the main centre of population, from which radiate all the main roads serving the rest of the region and those linking the neighbouring regions of Tayside and Highland. Aberdeen also acts as the principal mainland communication point for the Orkney and Shetland Islands. Other centres of population lie on or near the coast.

Moray District [09]
Population, 80 000; persons per hectare, 0.4

Located in the westerly part of the region, where the climate tends to be slightly warmer; the majority of the Scottish whisky distilleries are in this district.

Banff and Buchan District [10]
Population, 80 000; persons per hectare, 0.5

In the north-east of the region, this district includes the towns of Fraserburgh and Peterhead. There are many fishing villages along the coast, with a productive agricultural hinterland.

Gordon District [11]
Population, 62 000; persons per hectare, 0.3

With about 25 km of coastline, the rich agricultural land of the coastal strip of the district gives way towards the west to upland farms.

City of Aberdeen [12]
Population, 200 000; persons per hectare, 10.8

Aberdeen is Scotland's third largest city, dating back to 1179. It is the principal commercial and administrative centre in the north of Scotland, a fishing port and the main European centre for offshore oil exploration. Other industries include engineering, ship-building, food processing, textile production, paper manufacture and chemical production. The site of an ancient university and distinguished research centre, Aberdeen is sometimes called the 'Granite City' because of the high proportion of the buildings built from locally quarried granite. It is said that Aberdeen has the highest level of natural background radiation in the world.

Kincardine and Deeside District [13]
Population, 41 000; persons per hectare, 0.2

This is the region's most sparsely populated district. It includes some dormitory areas for those working in Aberdeen and also some fishing villages. The scenic beauty of the valley of the river Dee and of the mountainous areas is renowned.

TAYSIDE REGION
(TAYSIDE HEALTH BOARD AREA)
Population, 383 000; area, 749 297 hectares

The region extends from the east coast to the Central Highlands belt and straddles the Highland Boundary Fault, approximately two-thirds lying to the north of the fault line. It includes 11% of the agricultural land of Scotland, most of which is to be found to the south and east. The remainder comprises sparsely populated uplands and narrow glens.

Angus District [14]
Population, 91 000; persons per hectare, 0.5

This district is geographically similar to Perth and Kinross [16], with over 75 km of coastline on the North Sea. Agriculture and fishing are traditional occupations, and chemical and oil-related industries have developed recently.

City of Dundee [15]
Population, 178 000; persons per hectare, 7.6

Dundee is the administrative centre of the region. Situated on the north bank of the Tay estuary, the large port and dock installations are important to the offshore oil industry, while the airport also provides servicing facilities. The city has best been known since the late eighteenth century for the manufacture of jute, but other industries include food processing, electronics, other light and heavy engineering, paper making and publishing, and ship-building. University teaching took place in Dundee for many years in association with the University of St Andrews, leading eventually to the establishment of the University of Dundee in 1967.

Perth and Kinross District [16]
Population, 114 000; persons per hectare, 0.2

The district occupies 70% of Tayside; large areas to the north and west are sparsely populated mountains and glens. Rich farmland is found to the south and east, notably in the Carse of Gowrie, which has the largest concentration of raspberry growing in Europe.

Other general crop growing, livestock breeding and rearing are common, while forestry plantations are increasing in the uplands. Perth, 50 km up-river from Dundee, well known as a business centre and for the distilling and distribution of Scotch whisky, has a harbour which can accommodate ships of limited size.

FIFE REGION
(FIFE HEALTH BOARD AREA)
Population, 325 000; area, 130 708 hectares

Fife is a peninsula, bounded on the north and south by the estuaries of the rivers Tay and Forth, respectively, and on the east by the North Sea. A wedge of hilly upland running from the west and north-west constitutes the backbone of the region; it is surrounded by a flatter coastal strip with a more extensive and fertile alluvial plain towards the east. The independent spirit of Fife people may owe something to the former isolation of the region, alleviated first by the building of the railway bridges across the Tay and the Forth in the late nineteenth century and finally by the completion of road bridges in the 1960s.

Coal has been mined in west Fife since the thirteenth century, and the area formerly provided up to one-third of the coal mined in Scotland. However, now many of the coal seams have been worked out. Farming is the staple industry of the eastern half of the region, while shipbuilding and repairing are carried out in small towns on the Forth estuary; the advent of North Sea oil has restored a flicker of life to the latter industries, which are declining.

Kirkcaldy District [17]
Population, 143 000; persons per hectare, 5.8

In the centre of the region, the main town, Kirkcaldy, is well known for linoleum manufacture, while the new town of Glenrothes to the north is a rapidly developing electronics manufacturing centre (Chapter 3).

North-east Fife District [18]
Population, 60 000; persons per hectare, 0.8

On the east of the region, the district is predominantly a rich agricultural area. Cropping and beefstock are the dominant types of farming. Fishing is carried on from a number of small towns on the south-east coast. The principal town is St Andrews, the seat of Scotland's oldest university and the home of golf.

Dunfermline District [19]
Population, 122 000; persons per hectare, 4.1

Located in the west of the region, Dunfermline, like the neighbouring district of Kirkcaldy, has never become heavily industrialized. The principal industry, now much reduced, has been coalmining. The main town, Dunfermline, was important in the past for linen manufacture. A large naval dockyard at Rosyth provides considerable employment in the area.

LOTHIAN REGION
(LOTHIAN HEALTH BOARD AREA)
Population, 723 000; area, 175 518 hectares

Lothian is the second largest of the regions in terms of population. It is a predominantly lowland area, bounded to the south and west by hills ranging from 200-500 metres high. The northern and eastern limits are the coastline formed by the estuary of the river Forth and the North Sea.

West Lothian District [20]
Population, 137 000; persons per hectare, 3.2

In this district, to the west of Edinburgh, coal has been mined extensively. There are prosperous farms, with emphasis on pigs, poultry and market gardening. Industrial development, also, has taken place in recent years, particularly at the New Town of Livingston (Chapter 3), to which many Glasgow people have migrated.

City of Edinburgh [21]
Population, 425 000; persons per hectare, 16.3

Strategically situated on the east coast route between England and the north of Scotland, Edinburgh has been the accepted capital since about 1500. It remained the largest Scottish city until the early nineteenth century when it was overtaken by Glasgow. The differences in the characters of the two cities are quite marked, not least in their occupational activities. Edinburgh is a main centre of service and administration, while Glasgow is dominated by heavy industry. Nevertheless, Edinburgh still outranks Dundee and Aberdeen as an industrial town, with some 46 000 persons employed in 1981 in manufacture — one-fifth of the city's workforce. In the 1970s, manufacturing employment comprised three main groups; first, food, drink and tobacco manufactures; second, engineering, electrical goods manufacture and electronics industries; third, paper making, printing and publishing. Leith, the port of Edinburgh, contributes considerably to this manufacturing output. For over 200 years, Edinburgh has been a centre of medicine, law, banking, insurance and investment. It is now second only to London in the UK as a tourist centre and is the main communications centre for road, rail and air for the eastern side of the country.

The city's housing stock varies from fine Georgian houses and late nineteenth century apartment blocks in the centre, to large developments of private villas and local authority housing erected over the last 50 years in the outskirts. During this period, the problems of high-density living and lack of social amenities of the city centre have been largely solved, only to emerge, to a certain extent, in some of the suburban areas.

Midlothian District [22]
Population, 82 000; persons per hectare, 2.3

This district is located to the south-east of Edinburgh. Coal has been mined here for many years, leading to the development of a number of small towns, but the industry is now in decline. The land is extensively farmed.

East Lothian District [23]
Population, 79 000; persons per hectare, 1.1

This district east of Edinburgh is primarily a rich agricultural area. The greatest concentration of market gardening in Scotland is located on the Lothian coastlands, yielding two-thirds of the country's market garden crops.

BORDERS REGION
(BORDERS HEALTH BOARD AREA)
Population, 97 000; area, 467 158 hectares

Comprising the eastern part of the Southern Uplands, the countryside of this region is predominantly hill and moorland, the small lowland areas being confined to river valleys and a coastal strip on the eastern flank. The main river system, the Tweed, runs from west to east between the Cheviot Hills to the south and the Lammermuir Hills to the north.

The main industry is sheep farming, but towards the lower land in the east, in the Berwickshire District, hill-grazing gradually gives way to grassland and crops. The woollen industry was established in the Tweed Valley in the eighteenth century and led to the development of a number of wool-working towns, but

their prosperity, particularly those engaged in tweed manufacture, has since fluctuated considerably.

The constituent districts are generally similar in nature.

Tweeddale District [24]
Population, 14 000; persons per hectare, 0.2

This district covers the west of the Borders region. The main town, Peebles, specializes in tweed manufacture. The Forestry Commission has planted extensive forests in the area.

Ettrick and Lauderdale District [25]
Population, 31 000; persons per hectare, 0.2

In this district at the centre of the Borders region, the principal town, Galashiels, specializes in manufacture of woollen cloth. Neighbouring Selkirk has woollen mills, also.

Roxburgh District [26]
Population, 35 000; persons per hectare, 0.2

In this district to the south of the Borders region, the main towns are Hawick, Jedburgh and Kelso. In Hawick, knitwear of all kinds is manufactured. In Jedburgh rayon has been added to wool manufacture, while Kelso functions mainly as a market town and a centre of livestock sales.

Berwickshire District [27]
Population, 18 000; persons per hectare, 0.2

The North Sea forms the eastern border of this district, which includes two small towns, Eyemouth and Duns. The rich lowland farms of this eastern district contrast with those on the comparable west coast district of Wigtown. A much higher proportion of land in the east is under tillage: barley, oats, fodder crops and potatoes are grown, in contrast to Wigtown's highly developed dairy farming. The differences are due to the greater rainfall in the west.

CENTRAL REGION
(FORTH VALLEY HEALTH BOARD AREA)
Population, 268 000; area, 263 147 hectares

The Central region lies between the Clyde and Forth estuaries, linking the heavily industrialized west and the government and professional centre in the east. It stretches from Loch Lomond and the picturesque Trossachs in the west across fertile plains to the industrial east around the petrochemical port of Grangemouth. The eastern part of the region contains 90% of its population and all the urban and industrial development. The major towns of Stirling, Falkirk, Grangemouth and Alloa are located there. Farming is an important industry in the Forth Valley, and coalmining, long established in the region, remains significant.

Clackmannan District [28]
Population, 47 000; persons per hectare, 2.9

Covering the north-east of the Central region, this district is largely unspoiled countryside, with hills, rivers and glens. Alloa is the largest town. The bulk of employment is in textile manufacture, engineering, distilling, brewing and glass manufacture. Agriculture is economically important.

Stirling District [29]
Population, 77 000; persons per hectare, 0.4

Located to the north and west of the region, much of this district is impressive scenically. The main town is Stirling, an ancient fortress and capital of Scotland. The University of Stirling was established in 1967. Major activities are tourism, forestry and agriculture.

Falkirk District [30]
Population, 144 000; persons per hectare, 4.8

The largest district in terms of population and the most industrially developed, Falkirk is in the south-east of the Central region, lying strategically across the narrow waist of Scotland. The district became the centre of the Industrial Revolution, which reached Scotland with the founding, in 1759, of the famous Carron Company, the major supplier of cannon for the Napoleonic Wars. The older dependence upon coal and iron has now given way to high technology industry. On the Forth, the port of Grangemouth has expanded fast in comparatively recent years, acquiring chemical engineering, light manufacturing and a large oil refinery.

STRATHCLYDE REGION
(HEALTH BOARD AREAS - GREATER GLASGOW, LANARKSHIRE, ARGYLL AND CLYDE, AYRSHIRE AND ARRAN)
Population, 2 375 000; area, 1 353 698 hectares

By far the largest of the regions in area and population, and the most diverse, Strathclyde contains a massive industrial conurbation, several inland lochs, mountains and pleasant undulating countryside and straddles the western extremities of Scotland's great natural divisions — the Highlands, the Midland Valley and the Southern Uplands.

In the Highland area, covering the Argyll and Bute District and the northern part of Dumbarton District, high mountains alternate with deep glens hollowed out in ancient times by the action of ice, while to the west the hills and valleys became islands and sea lochs.

The western end of the great Midland Valley of Scotland contained much of the coal and all of the iron-ore deposits that were the basis for the great industrial growth of the Clydeside conurbation. The soil is much richer than in the Highland area and supports agriculture alongside the industrial towns.

The Southern Uplands, comprising high moorland and hills with thinner soil and vegetation, include parts of the districts of Kyle and Carrick, Cumnock and Doon Valley, and Clydesdale. The higher parts are sheep-farming country.

The people of Strathclyde are employed to a lesser extent in the primary sector of agriculture, forestry and mining than in other parts of Scotland, and a correspondingly higher proportion (37%) work in manufacturing. All Scotland's steel-making plants are located within the Strathclyde region, employing some 25 000 people.

The River Clyde has a long ship-building tradition, but this activity is now much reduced. In addition to naval vessels, current major construction includes oil-drilling rigs and drillships. There is now a movement away from heavy towards light engineering, an expanding sector being the electronics industry. Most of Scotland's overseas trade is shipped through the Clyde ports — Glasgow, Gourock, Greenock, Ardrossan and Finnart (where oil is imported). There are international airports at Glasgow and Prestwick.

Since 1945, the development of new towns (Chapter 3) has been a major instrument in helping to relieve Scotland's overcrowded cities and in encouraging the dispersal of its industry to new locations. In Strathclyde, these have been established in existing villages or communities at Cumbernauld, Irvine and East Kilbride.

Argyll and Bute District [31]
Population, 64 000; persons per hectare, 0.1

Located to the north-west of the Strathclyde region, this district comprises a mainland of mountains, moors and glens, while offshore lie the pleasant islands of the Inner Hebrides. There is a flourishing tourist industry, and the traditional activities of agriculture, forestry and whisky distilling are still carried out.

Dumbarton District [32]

Population 77 000; persons per hectare, 1.6

Dumbarton, the main town of Dumbarton District, lies on the River Clyde, to the west of Glasgow, and was formerly a centre for ship-building and marine engineering. Non-ferrous metals are worked, also. Other industry includes the distilling, blending and bottling of whisky. To the north of the town is open hilly country.

The City of Glasgow [33]

Population, 761 000; persons per hectare, 38.2

Glasgow, the third largest city in the UK, has a history of more than 800 years, and even more ancient historical roots as a settlement. In the eighteenth century, the city began its great expansion, brought about initially by the trans-Atlantic tobacco trade, then by the cotton textile industry, followed by ship-building and heavy engineering during the nineteenth century. Glasgow is the commercial focus of Scotland and is a major centre for accountancy, law, medicine and the arts.

The nineteenth century industrial development drew many people into the city, particularly from the Highlands, bringing the population to over one million by 1914, many of whom lived in congested tenements. Before the Second World War, Glasgow had a reputation for having a high proportion of slum dwellings, overcrowding, poor environmental and social conditions and poverty (Berry, 1984). Since then, many of the poorer tenements have been torn down and some of the remainder have been restored, blending well with the elegant Victorian office blocks, stately Georgian and Victorian residences and public parks created from the past prosperity. Unfortunately, much of the former congested and deprived living conditions of the core of the city have been transferred to the newer housing conglomerations that have sprung up in the suburbs.

Although reduced by recession, the main manufacturing concerns today are with steel, ships, marine and aero engines, automotive engineering, printing and publishing. However, the major share in the city's economy is taken up by service industries such as professional and scientific services, distribution, transport and communication and public and private administration. By 1981, service industries provided employment for some 320 000 persons, as compared with 122 000 in manufacturing.

Clydebank District [34]

Population, 52 000; persons per hectare, 14.6

Situated to the west of Glasgow, Clydebank district is based on the town of Clydebank, which has a higher proportion of workers in metal trades than any other Scottish town, concentrating on ship-building and marine engineering and, more recently, oil-rig construction and sewing-machine manufacture.

Bearsden and Milngavie District [35]

Population, 39 000; persons per hectare, 10.8

This is a high-amenity residential area serving the City of Glasgow; it is situated to the north of the city.

Strathkelvin District [36]

Population, 87 000; persons per hectare, 5.3

Strathkelvin District is also north of Glasgow, but contains a mixture of industrial activity and open farming.

Cumbernauld and Kilsyth District [37]

Population, 62 000; persons per hectare, 6.5

Forming part of the eastern boundary of Strathclyde region, this district was formerly a coal-mining area. Industry is now mixed, mainly centred around the new town of Cumbernauld and the old established burgh of Kilsyth (Chapter 3).

Monklands District [38]

Population, 110 000; persons per hectare, 6.7

This industrial district east of Glasgow includes the towns of Coatbridge and Airdrie, in which steel manufacture is complemented by a variety of other manufacturing industries. Although the dominant character of the district is industrial, it contains a large rural area.

Motherwell District [39]

Population, 149 000; persons per hectare, 8.7

This district south-east of Glasgow is an industrial area, concentrating mainly on steel manufacture and, more recently, electronics. The main towns are Motherwell and Wishaw.

Hamilton District [40]

Population, 108 000; persons per hectare, 8.2

This district, also, forms part of the heavily industrialized region south-east of Glasgow. The main town of Hamilton, once an important coalmining centre, now supports electronics, printing, machine-tool manufacturing and aluminium refining.

East Kilbride District [41]
Population, 83 000; persons per hectare, 2.9

Some 25 km south of Glasgow, East Kilbride District is largely agricultural and residential. East Kilbride new town (Chapter 3), founded 20 years ago, is perhaps the most successful of all the Scottish new towns, having attracted a large and varied range of manufacturing and service industries and research centres.

Eastwood District [42]
Population, 53 000; persons per hectare, 4.6

Eastwood District is south-west of Glasgow, and contains extensive areas of high-amenity housing where many of the region's managerial and professional population live.

Clydesdale District [43]
Population, 57 000; persons per hectare, 0.4

Clydesdale District is south of Glasgow, and is mainly a farming area; the principal town, Lanark, is a major cattle and sheep marketing centre.

Renfrew District [44]
Population, 205 000; persons per hectare, 6.7

To the west of Glasgow and south of the Clyde, Renfrew District is largely industrial. Paisley, the main town, has a long tradition of cotton manufacture. Other industries include engineering and whisky distilling.

Inverclyde District [45]
Population, 100 000; persons per hectare, 6.3

This district in the west of Strathclyde region, on the Clyde estuary, is largely industrial, with large port facilities. Port Glasgow and Greenock are outports of Glasgow, famous for ship-building and heavy engineering and, more recently, the manufacture of oil-drilling ships and an electronics industry. Greenock also has a long-established sugar-refining industry.

Cunninghame District [46]
Population, 136 000; persons per hectare, 1.6

This district comprises rich farming land, with a tradition of high-quality textile and woollen goods manufacture. The main urban area is Irvine, a new town (Chapter 3). Hunterston on the coast has an atomic power station which began producing electricity in 1964. A newly created iron-ore import terminal is located nearby. Offshore lies Arran, a mountainous island with fertile valleys, the main industries being sheep and cattle grazing, fishing and tourism.

Kilmarnock & Loudon District [47]
Population, 82 000; persons per hectare, 2.2

This district contains mixed farming and industry. More than half the population lives in the town of Kilmarnock, known for its whisky, carpets, engineering products and manufacture of fine textiles.

Kyle and Carrick District [48]
Population, 112 000; persons per hectare, 0.9

This southern part of Strathclyde region is traditionally agricultural, dairy farming being the most important sector. The largest town, Ayr, is a market and administrative centre and a holiday resort. Other activities include engineering and fishing.

Cumnock & Doon Valley District [49]
Population, 45 000; persons per hectare, 0.6

In this district on the southern boundary of the Strathclyde region, farming is the main activity, but some industry is developing in the Cumnock area.

DUMFRIES AND GALLOWAY REGION (DUMFRIES AND GALLOWAY HEALTH BOARD AREA)
Population, 142 000; area, 637 006 hectares

This region in the south-west corner of Scotland forms part of the Southern Uplands but has a long coastline on the Solway Firth. The countryside is varied — rolling farmland, moor, forested hills and a many lochs. Dairy farming is highly developed in the valleys and coastal lowlands, the region supplying about one-third of the total Scottish milk output. Current market instability is leading to some change towards beef production. In the more exposed locations, the native Galloway beef-stock cattle are kept, while in the uplands sheep farming predominates. Forestry has expanded in the last 20 years; in 1970 over 14% of the region was under woodland; by 1980, 18%. Tourism has been important for many years, and fishing has always made a contribution to the economy.

Wigtown District [50]
Population, 30 000; persons per hectare, 0.2

Wigtown District is in the extreme west of Dumfries and Galloway Region. The main town and port is Stranraer, and major employment is in dairy farming, 90% of the farmland being in grass.

Stewartry District [51]

Population 22 000; persons per hectare, 0.1

To the east of Wigtown District is Stewartry District, where the main towns are Kirkcudbright and Castle Douglas. Dairy and sheep farming, tourism and fishing are the principal occupations.

Nithsdale District [52]

Population, 55 000; persons per hectare, 0.4

Further east, again, of the Stewartry District, is Nithsdale District. The main town, Dumfries, is the largest in the Dumfries and Galloway Region. One-quarter of its workforce is in manufacture, the largest share being taken by engineering, but textile manufacture is prominent. Chemicals, including synthetic fibres and rubber, are also manufactured. The district has a very large dairy farming industry and extensive forestry. An atomic power station on the coast became operational in 1959.

Annandale and Eskdale District [53]

Population, 35 000; persons per hectare, 0.2

This district is east of Nithsdale and includes the Langholm area, which is noted for high-quality knitwear and textiles. Some industry (pharmaceuticals and engineering) is located around the town of Annan.

ORKNEY ISLANDS ISLANDS AREA [54]
Population, 18 000; persons per hectare, 0.2

The Orkney Islands lie just off the north-eastern coast of Scotland and resemble closely the neighbouring mainland of Caithness [01]. The surface is low, with few hills, and, although the climate is mild for that latitude, the islands are constantly windswept.

As 45% of the land area is suitable for crops and grass, the islands have a flourishing agricultural industry — mainly dairy and poultry farming. The main impact from the North Sea oil industry has been the establishment of an oil terminal on the island of Flotta; this increased the population of the islands by 11% between 1971 and 1981, reversing a previous trend of depopulation.

SHETLAND ISLANDS ISLANDS AREA [55]
Population, 23 000; persons per hectare, 0.2

About 80 km to the north-east of Orkney lie the Shetland Islands, about 100 in number, of which 27 are inhabited. The landscape tends to be bleak, rising to 150 metres above sea level in only a few places. Although the climate is humid and comparatively mild, severe storms are frequent. The Islands have a close affinity to Norway, many of the islanders having Norwegian ancestors. The effect of oil development, including supply of services to the offshore oil platforms and the associated onshore bases, has been greater in Shetland than anywhere else in Scotland. It increased the population by 29% between 1971 and 1981, mainly in the younger age groups, following a steady decline during the previous part of this century. Since over 90% of the land in Shetland is classified as rough grazing, agriculture does not take the important part in the economy that it does in Orkney.

THE WESTERN ISLES ISLANDS AREA
(WESTERN ISLES HEALTH BOARD AREA) [56]
Population, 31 000; persons per hectare, 0.1

The Western Isles or Outer Hebrides, many of which are uninhabited, form a linear archipelago extending south-west to north-east, some 60 km off the north-west mainland. The largest islands are Lewis and Harris, North Uist and Barra. They lie in the path of the almost ceaseless gales of the North Atlantic which, to a greater or lesser extent, affect living conditions — from the cultivation of food to the maintainance of communications with the rest of the country. In the past, physical communications were a serious problem, the only contact being by sea, but the development of local air services in recent years has helped. The main occupations of the Western Isles are fishing and farming, mainly sheep with some cattle. Many islanders combine fishing with crofting.

Chapter 7

THE SCOTTISH CANCER MAPS

PRESENTATION OF THE MAPS

For each cancer site considered, the material is presented in a uniform fashion, in groups of eight pages.

First page. The title gives the cancer site and its code number in the 8th revision of the International Classification of Diseases (WHO, 1967).

Second and third pages. Following a brief summary, the cancer in question is described in greater detail. After the headings 'Males' and 'Females', two percentages appear in brackets. These are, in sequence: (1) the percentage of all cancers in Scotland due to that particular cancer; and (2) the chance of developing this cancer between birth and 75 years of age (see Cumulative risk in Appendix III for further information).

In the following narrative, the incidence rates in Scotland as a whole are compared, for each sex, with those in cancer registries in England and Wales and other parts of the world. In presenting information on cancer incidence from other parts of the world, abbreviated registry titles are used, e.g., 'Birmingham' denotes 'Birmingham and West Midlands Cancer Registry', 'Detroit' the white population of that city, 'Alameda' the 'Alameda County (California) Cancer Registry'. (See Appendix IV for complete list.) The average annual age-standardized registration rates per 100 000 are given, each rate standardized to the World Standard Population to allow valid comparisons (Appendix III).

Next, the incidence rates by local government district for males, females and the four Scottish cities are presented. This is followed by an assessment of the degree of clustering of districts with similar levels of the form of cancer being discussed (Appendix IIC).

A discussion of what is known about the risk factors for this cancer follows and, when possible, some assessment of how these relate to the Scottish findings. Suggestions of possible reasons for observed

differences may be made, but many of these are merely speculative and do not represent proven facts.

Fourth and fifth pages. These contain the maps, males on the left, females on the right (except for sex-specific sites, which are paired). At the top left-hand corner is a smoothed histogram showing the range of incidence of the cancer, which corresponds to a particular colour on the vertical axis. The frequency with which this range of incidence was found among the 56 local government districts and Islands areas is reflected by the horizontal axis. The rate for Scotland as a whole is marked by a horizontal line with a value at the right-hand margin. These maps are on a relative scale (see Appendix IIA for further details).

Sixth and seventh pages. These black-and-white maps (males on the left, females on the right) present the incidence of cancer on an absolute scale (Appendix IIA). The actual numbers of cases of the specific form of cancer are presented for each district, followed by the crude rate, the age-standardized rate, the standard error of the age-standardized rate and an assessment of whether the district rate is significantly different from the rate observed in the rest of Scotland (symbols +, ++, —, and —— denote rates that are significantly higher and lower at the 5% and 1% level, respectively; see Appendix III). Several district titles are abbreviated because space is limited. The full title, with district number, is given in Chapter 6. Below these statistics, the rates in the four Scottish cities (Aberdeen, Dundee, Edinburgh and Glasgow) are presented as histograms.

Eighth page. This page contains three diagrams. The top figure presents the age-specific incidence rates for this form of cancer in people of each sex; the middle figure presents the age-specific rates found in males in each of the four cities (Aberdeen, Dundee, Edinburgh and Glasgow); and the bottom figure presents the corresponding age-specific rates for females.

PRINCIPAL SITES

CANCER OF THE LIP
(ICD8 140)

Cancer of the vermilion border of the lip arises in a relatively small anatomical region between the mouth and the hair-bearing skin of the face; this rubric specifically excludes cancer of the skin of the lip. The incidence of lip cancer in Scotland, as elsewhere, is approximately 10 times higher in males than in females, although the incidence has been falling in people of each sex since 1960.

Males (0.8%; 0.28%)

The incidence of cancer of lip in Scotland is similar to that found in many other countries (Waterhouse *et al.*, 1982). The rate in males (2.7) is close to that of Sweden (2.8), but a little lower than those of Norway (3.7), Denmark (4.4) and Finland (5.3). The Scottish male incidence is well below that of Newfoundland (22.8), Romania, County Cluj (11.9), Utah (10.3) or South Australia (10.1).

Within Scotland there is wide variation. In males, the lowest incidence occurs in Tweeddale (0.0), Annandale and Eskdale (0.0), Strathkelvin (0.4), Eastwood (0.6), Bearsden and Milngavie (0.7) and the highest rates in Skye and Lochalsh (17.0), Banff and Buchan (11.3), Sutherland (9.6), Caithness (9.5), Ross and Cromarty (9.2) and Berwickshire (8.7). The highest incidence occurs in rural areas in the Highland and Grampian regions in the north and the Border regions in the south, with the rates higher in the north. The high-incidence areas are all coastal or rural in character.

Females (0.1%; 0.02%)

The female Scottish incidence (0.2) is considerably lower than that in males, as in almost every country (Waterhouse *et al.*, 1982); it is identical with the incidence of Sweden and Norway, and close to that of Denmark (0.3) and Finland (0.4), while considerably lower than the highest rates, which are found in Brazil (1.2) and in Newfoundland (1.1).

In females, lip cancer again occurs mainly in rural areas of the north and south — Lochaber (1.9), Inverness (1.2) and Wigtown (0.8), although it is difficult to discern any pattern, as there were many districts where no case occurred.

Scottish cities

The highest male incidence rate was reported from the most northern city, Aberdeen (3.5), with the rates in the cities of the south — Edinburgh (1.0), Dundee (0.9) and Glasgow (0.8) — being considerably lower. For females, the rate in Aberdeen (0.5) was slightly higher than that in Edinburgh (0.3), Glasgow (0.1) or Dundee (0.0).

Spatial aggregation

The strongest aggregation of districts with similar rates was found for male lip cancer (D = 10.88, $p < 0.0001$). Weaker, although still statistically significant aggregation was found for females (D = 16.50, $p = 0.013$), probably because of many zero rates.

Comment

Examination of geographical patterns in the incidence of lip cancer has shown that the highest rates are in Newfoundland (Waterhouse *et al.*, 1982), although the levels in Saskatchewan, Manitoba and Alberta are also higher than elsewhere. Rates in these prairie provinces of Canada have fallen gradually over the period spanned by the four volumes of *Cancer Incidence in Five Continents* (Doll *et al.*, 1966, 1970; Waterhouse *et al.*, 1976, 1982), while the rate in Newfoundland has decreased only latterly (Boyle *et al.*, 1983a). The incidence of lip cancer has been falling also in Finland (Lindqvist, 1979), Norway and Yugoslavia (Stukonis, 1982), Czechoslovakia (Svejda & Kosut, 1971) and Connecticut (USA) (Shedd *et al.*, 1970).

Descriptive epidemiological studies have revealed three characteristics of lip cancer. First, it affects mainly men; second, it is far more common in rural than in urban areas, and, third, the tumour almost always arises on the lower lip (e.g., 97.3% in Northern Ireland) (Lynch, 1967).

As regards etiology, cancer of the lip appears to share risk factors with both the mouth (e.g., cigarette-smoking) and facial skin (e.g., sunlight exposure). Many studies have demonstrated a significantly increased risk of lip cancer with pipe-smoking (Broders, 1920; Ebenius, 1943; Hamalainen, 1955), tobacco chewing (Ebenius, 1943) and smoking generally (Hamalainen, 1955; Williams & Horm, 1977).

Cancer of the lip is rare, however, in negroes, among whom even pipe-smokers seldom develop lip cancer (Ackerman & del Regato, 1962). It has been suggested that lip cancer incidence decreases with latitude (Wynder *et al.*, 1957; Keller, 1970). In the USA, mortality was higher in southern states than northern ones (Mason *et al.*, 1975), although a recent analysis of data from the US Third National Cancer

Survey showed no association between lip cancer incidence rates and latitude (Szpak *et al.*, 1977). Increased exposure to sunlight has been implicated in the excess occurrence of lip cancers seen among rural populations, producing high rates in specific occupations such as fishing (Spitzer *et al.*, 1975) and in farmers and other outdoor workers (Clemmesen, 1965). Increased rates among fishermen in western Scotland have been reported previously, by the former Director of the Western Scotland Cancer Registry (Haddow, 1968).

Exposure to sunlight appears to be the single most important risk factor in lip cancer, with a smaller, independent contribution from smoking. Other etiological associations proposed are with trauma (Ebenius, 1943) and herpes simplex virus (Shillitoe & Silverman, 1979).

The association with exposure to sun could explain the geographical pattern of lip cancer seen in Scotland. Almost exclusively, rates are highest in areas where substantial proportions of the population are employed in farming, forestry and fishing — all activities involving outdoor work (Fig. 3.6).

Lip — Males

D=10.88, p<0.0001

Lip — Females

D=16.50, p=0.013

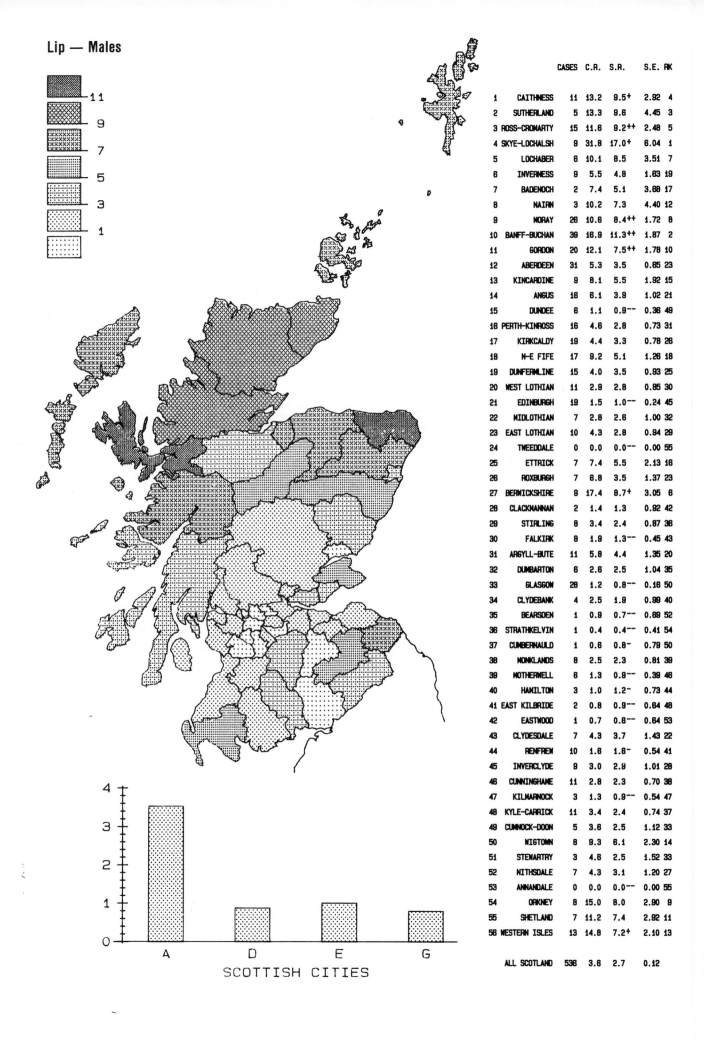

Lip — Males

		CASES	C.R.	S.R.	S.E.	RK
1	CAITHNESS	11	13.2	9.5+	2.92	4
2	SUTHERLAND	5	13.3	9.6	4.45	3
3	ROSS-CROMARTY	15	11.6	9.2++	2.48	5
4	SKYE-LOCHALSH	9	31.8	17.0+	6.04	1
5	LOCHABER	6	10.1	8.5	3.51	7
6	INVERNESS	9	5.5	4.6	1.63	19
7	BADENOCH	2	7.4	5.1	3.68	17
8	NAIRN	3	10.2	7.3	4.40	12
9	MORAY	26	10.6	8.4++	1.72	8
10	BANFF-BUCHAN	39	16.9	11.3++	1.87	2
11	GORDON	20	12.1	7.5++	1.78	10
12	ABERDEEN	31	5.3	3.5	0.65	23
13	KINCARDINE	9	8.1	5.5	1.92	15
14	ANGUS	16	6.1	3.9	1.02	21
15	DUNDEE	6	1.1	0.9--	0.36	49
16	PERTH-KINROSS	16	4.6	2.8	0.73	31
17	KIRKCALDY	19	4.4	3.3	0.78	26
18	N-E FIFE	17	9.2	5.1	1.26	18
19	DUNFERMLINE	15	4.0	3.5	0.93	25
20	WEST LOTHIAN	11	2.9	2.8	0.85	30
21	EDINBURGH	19	1.5	1.0--	0.24	45
22	MIDLOTHIAN	7	2.8	2.6	1.00	32
23	EAST LOTHIAN	10	4.3	2.8	0.94	29
24	TWEEDDALE	0	0.0	0.0--	0.00	55
25	ETTRICK	7	7.4	5.5	2.13	16
26	ROXBURGH	7	6.8	3.5	1.37	23
27	BERWICKSHIRE	9	17.4	8.7+	3.05	6
28	CLACKMANNAN	2	1.4	1.3	0.92	42
29	STIRLING	8	3.4	2.4	0.87	36
30	FALKIRK	8	1.9	1.3--	0.45	43
31	ARGYLL-BUTE	11	5.8	4.4	1.35	20
32	DUMBARTON	6	2.6	2.5	1.04	35
33	GLASGOW	26	1.2	0.8--	0.16	50
34	CLYDEBANK	4	2.5	1.9	0.99	40
35	BEARSDEN	1	0.9	0.7--	0.69	52
36	STRATHKELVIN	1	0.4	0.4--	0.41	54
37	CUMBERNAULD	1	0.6	0.8-	0.79	50
38	MONKLANDS	8	2.5	2.3	0.81	39
39	MOTHERWELL	6	1.3	0.9--	0.39	46
40	HAMILTON	3	1.0	1.2-	0.73	44
41	EAST KILBRIDE	2	0.8	0.9--	0.64	48
42	EASTWOOD	1	0.7	0.6-	0.64	53
43	CLYDESDALE	7	4.3	3.7	1.43	22
44	RENFREW	10	1.6	1.6-	0.54	41
45	INVERCLYDE	9	3.0	2.9	1.01	28
46	CUNNINGHAME	11	2.8	2.3	0.70	38
47	KILMARNOCK	3	1.3	0.9--	0.54	47
48	KYLE-CARRICK	11	3.4	2.4	0.74	37
49	CUMNOCK-DOON	5	3.6	2.5	1.12	33
50	WIGTOWN	8	9.3	6.1	2.30	14
51	STEWARTRY	3	4.6	2.5	1.52	33
52	NITHSDALE	7	4.3	3.1	1.20	27
53	ANNANDALE	0	0.0	0.0--	0.00	55
54	ORKNEY	8	15.0	8.0	2.90	9
55	SHETLAND	7	11.2	7.4	2.92	11
56	WESTERN ISLES	13	14.8	7.2+	2.10	13
	ALL SCOTLAND	536	3.8	2.7	0.12	

SCOTTISH CITIES

Lip — Females

		CASES	C.R.	S.R.		S.E.	RK
1	CAITHNESS	0	0.0	0.0	—	0.00	32
2	SUTHERLAND	0	0.0	0.0	—	0.00	32
3	ROSS-CROMARTY	1	0.8	0.4		0.39	8
4	SKYE-LOCHALSH	0	0.0	0.0	—	0.00	32
5	LOCHABER	2	3.3	1.9		1.34	1
6	INVERNESS	4	2.4	1.2		0.82	2
7	BADENOCH	0	0.0	0.0	—	0.00	32
8	NAIRN	0	0.0	0.0	—	0.00	32
9	MORAY	2	0.8	0.4		0.34	7
10	BANFF-BUCHAN	1	0.4	0.2		0.18	22
11	GORDON	0	0.0	0.0	—	0.00	32
12	ABERDEEN	7	1.0	0.5		0.20	6
13	KINCARDINE	0	0.0	0.0	—	0.00	32
14	ANGUS	2	0.7	0.1		0.10	29
15	DUNDEE	0	0.0	0.0	—	0.00	32
16	PERTH-KINROSS	2	0.5	0.3		0.21	15
17	KIRKCALDY	4	0.9	0.4		0.18	10
18	N-E FIFE	2	1.0	0.3		0.21	14
19	DUNFERMLINE	2	0.5	0.3		0.18	17
20	WEST LOTHIAN	1	0.3	0.2		0.17	26
21	EDINBURGH	8	0.5	0.3		0.12	12
22	MIDLOTHIAN	1	0.4	0.2		0.18	22
23	EAST LOTHIAN	1	0.4	0.2		0.22	20
24	TWEEDDALE	1	2.2	0.7		0.71	4
25	ETTRICK	1	1.0	0.2		0.18	24
26	ROXBURGH	0	0.0	0.0	—	0.00	32
27	BERWICKSHIRE	0	0.0	0.0	—	0.00	32
28	CLACKMANNAN	1	0.7	0.4		0.39	6
29	STIRLING	1	0.4	0.3		0.30	13
30	FALKIRK	2	0.5	0.3		0.20	16
31	ARGYLL-BUTE	0	0.0	0.0	—	0.00	32
32	DUMBARTON	0	0.0	0.0	—	0.00	32
33	GLASGOW	5	0.2	0.1		0.04	30
34	CLYDEBANK	0	0.0	0.0	—	0.00	32
35	BEARSDEN	0	0.0	0.0	—	0.00	32
36	STRATHKELVIN	1	0.4	0.2		0.24	18
37	CUMBERNAULD	0	0.0	0.0	—	0.00	32
38	MONKLANDS	0	0.0	0.0	—	0.00	32
39	MOTHERWELL	1	0.2	0.2		0.17	26
40	HAMILTON	2	0.6	0.5		0.35	5
41	EAST KILBRIDE	0	0.0	0.0	—	0.00	32
42	EASTWOOD	0	0.0	0.0	—	0.00	32
43	CLYDESDALE	0	0.0	0.0	—	0.00	32
44	RENFREW	1	0.2	0.1		0.05	31
45	INVERCLYDE	2	0.6	0.2		0.17	18
46	CUNNINGHAME	1	0.2	0.2		0.15	26
47	KILMARNOCK	1	0.4	0.2		0.18	24
48	KYLE-CARRICK	1	0.3	0.2		0.20	21
49	CUMNOCK-DOON	0	0.0	0.0	—	0.00	32
50	WIGTOWN	1	1.1	0.8		0.79	3
51	STEWARTRY	0	0.0	0.0	—	0.00	32
52	NITHSDALE	0	0.0	0.0	—	0.00	32
53	ANNANDALE	0	0.0	0.0	—	0.00	32
54	ORKNEY	0	0.0	0.0	—	0.00	32
55	SHETLAND	0	0.0	0.0	—	0.00	32
56	WESTERN ISLES	2	2.2	0.4		0.25	11
	ALL SCOTLAND	64	0.4	0.2		0.03	

0.55
0.45
0.35
0.25
0.15
0.05

SCOTTISH CITIES

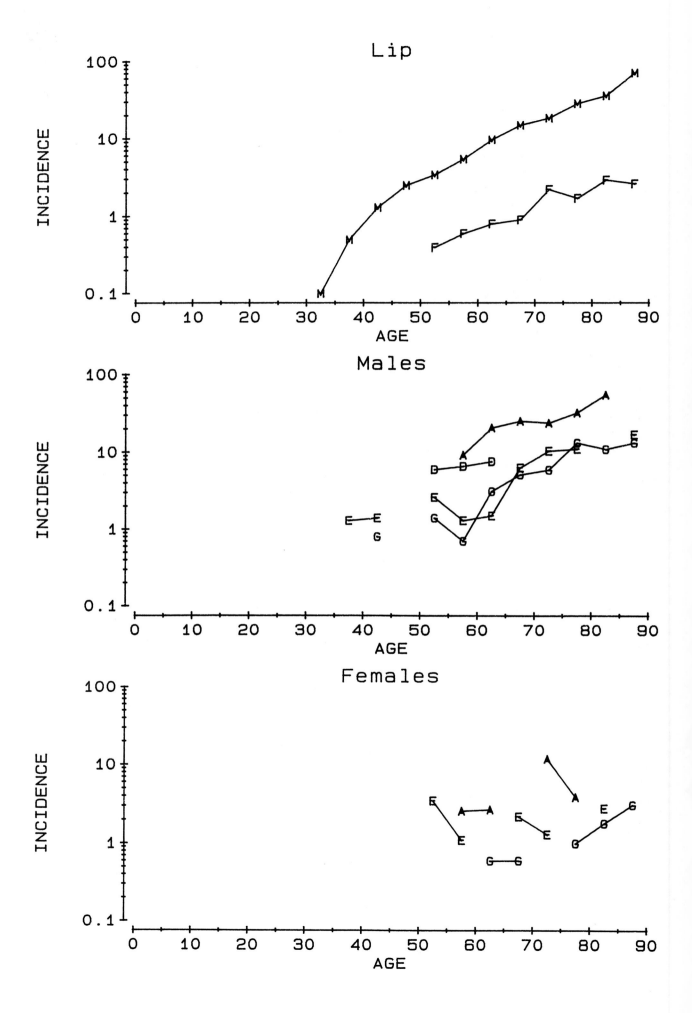

CANCER OF THE ORAL CAVITY AND PHARYNX
(ICD8 141, 143-145, 146, 148)

Cancers of the tongue, gums, roof and floor of the mouth, the inner aspects of the cheeks, oropharynx and hypopharynx have been grouped together under the label 'oral cavity and pharynx'. These sites are fairly uncommon and, for this reason, not presented individually. Further, there is difficulty in deciding the exact site of origin of a significant proportion of these tumours. A certain number of cancers in this region that were labelled 'pharynx' were not included in the figures given below, in order to permit international comparisons of incidence. Inclusion of these cancers would increase rates by about 8% for both sexes.

Males (1.3%; 0.43%)

The incidence of these forms of cancer is similar throughout Britain and northern Europe, being considerably higher in Latin regions of Europe (Geneva, 18.2; Bas-Rhin, 38.8), parts of North America (Connecticut, 11.9; New Orleans blacks, 19.3) and Bombay (28.7).

The all-Scotland rate is 3.9, with considerable regional variation; the extremes occur in Nairn (0.0) and Badenoch and Strathspey (11.2). Apart from Nairn, lower rates are found in Caithness (0.8), Lochaber (1.2), the Orkney Islands (1.3) and Cumnock and Doon Valley (1.5). The districts with the higher incidence rates are Badenoch and Strathspey (11.2), Skye and Lochalsh (8.0), Monklands (7.7), Tweeddale (7.4) and Argyll and Bute (6.8).

There appears to be no striking pattern, although the higher and lower rates are found in predominantly rural areas, in both the north and south of the country.

Females (0.8%; 0.21%)

The all-Scottish rate in females (1.9) is about one-half that in males; incidence ranges from 0.2 in Roxburgh to 3.5 in the Western Isles.

The lower female rates are found in Roxburgh (0.2), Cumnock and Doon Valley (0.5), Dunfermline (0.7), Cumbernauld and Kilsyth (0.8) and Tweeddale (0.8). The higher rates are found in the Western Isles (3.5), West Lothian (3.5), Ettrick (2.9), Inverness (2.8) and the Orkney Islands (2.8).

There is a suggestion for higher rates to be found in northern rural areas and lower rates in the southern part of the country. Small pockets of higher incidence are found to the south-east and south-west of Glasgow City and to the west of Edinburgh.

Scottish cities

The highest male rate is found in Edinburgh (5.3), followed by Glasgow (4.3), Dundee (4.2) and Aberdeen (2.2). Among females, the highest rates are recorded in Aberdeen (2.6), Glasgow (1.8), Edinburgh (1.9) and Dundee (1.4).

Spatial aggregation

There was no evidence of spatial aggregation, either for males ($D = 20.03$, $p = 0.833$) or for females ($D = 19.56$, $p = 0.694$).

Comment

The risk factors for the different sites of cancer in this grouping are discussed individually, although they possess some similarities; for the most part, the etiology of these forms of cancer is poorly understood.

It has been known for many years that mortality from tongue cancer has been elevated among people in trades or professions with easy access to alcohol, such as publicans and draymen (Young & Russell, 1926). The mechanisms for such an association remain unclear. Tongue cancer was once held to be associated with syphilis, but it was demonstrated subsequently that the increased risk was due to the treatment rather than the disease; when antibiotics replaced arsenic as the preferred mode of therapy, the association between an elevated risk of tongue cancer and a positive Wasserman reaction disappeared.

The link between cigarette smoking and mouth cancer (ICD8 143, 144, 145) has been well established (Doll & Peto, 1976; US Surgeon General, 1982). Alcohol consumption acts synergistically with cigarette-smoking in increasing risk (Rothman & Keller, 1972), while a high dietary intake of vitamin A reduces the risk of this cancer (Graham et al., 1977).

Cancers of the oropharynx and hypopharynx are also influenced by tobacco and alcohol consumption, and are held to explain the high rate of incidence in Geneva and Bas-Rhin. On the Indian sub-continent the habit of chewing the betel quid, with or without tobacco, increases the risk of mouth cancer. If the chewer also smokes, the risk seems to be displaced, resulting in increased levels of cancers of the oropharynx and hypopharynx (Sanghvi et al., 1955).

Reductions in alcohol and tobacco consumption would result in a fall in these, as well as in lung and other cancers. Should the practice of chewing tobacco sachets spread to this country, as seems likely, this will in all probability result in an increase in the number of oral cancers — as in the United States. Although there appears to be little pattern to the occurrence of these forms of cancer in the local authority districts of Scotland, the high rate in males in Badenoch and Strathspey (albeit based on four cases) is worth further study, as these men may follow a particular occupation.

Oral Cavity and Pharynx — Males

D=20.03, p=0.833

Oral Cavity and Pharynx — Females

D=19.56, p=0.694

Oral Cavity and Pharynx — Males

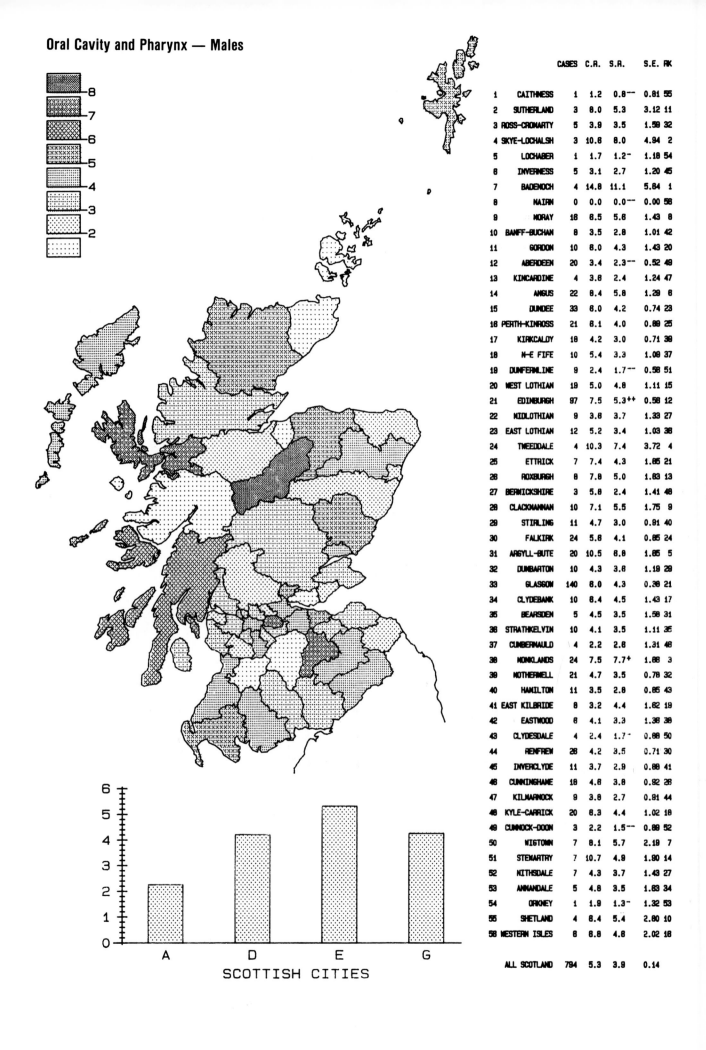

SCOTTISH CITIES

		CASES	C.R.	S.R.	S.E.	RK
1	CAITHNESS	1	1.2	0.8--	0.81	55
2	SUTHERLAND	3	8.0	5.3	3.12	11
3	ROSS-CROMARTY	5	3.9	3.5	1.59	32
4	SKYE-LOCHALSH	3	10.8	8.0	4.94	2
5	LOCHABER	1	1.7	1.2-	1.18	54
6	INVERNESS	5	3.1	2.7	1.20	45
7	BADENOCH	4	14.8	11.1	5.84	1
8	NAIRN	0	0.0	0.0--	0.00	56
9	MORAY	16	6.5	5.6	1.43	6
10	BANFF-BUCHAN	8	3.5	2.8	1.01	42
11	GORDON	10	6.0	4.3	1.43	20
12	ABERDEEN	20	3.4	2.3--	0.52	49
13	KINCARDINE	4	3.8	2.4	1.24	47
14	ANGUS	22	8.4	5.8	1.29	6
15	DUNDEE	33	6.0	4.2	0.74	23
16	PERTH-KINROSS	21	6.1	4.0	0.89	25
17	KIRKCALDY	18	4.2	3.0	0.71	39
18	N-E FIFE	10	5.4	3.3	1.09	37
19	DUNFERMLINE	9	2.4	1.7--	0.58	51
20	WEST LOTHIAN	19	5.0	4.8	1.11	15
21	EDINBURGH	97	7.5	5.3++	0.58	12
22	MIDLOTHIAN	9	3.6	3.7	1.33	27
23	EAST LOTHIAN	12	5.2	3.4	1.03	36
24	TWEEDDALE	4	10.3	7.4	3.72	4
25	ETTRICK	7	7.4	4.3	1.65	21
26	ROXBURGH	6	7.8	5.0	1.83	13
27	BERWICKSHIRE	3	5.8	2.4	1.41	48
28	CLACKMANNAN	10	7.1	5.5	1.75	9
29	STIRLING	11	4.7	3.0	0.91	40
30	FALKIRK	24	5.8	4.1	0.85	24
31	ARGYLL-BUTE	20	10.5	6.8	1.65	5
32	DUMBARTON	10	4.3	3.8	1.19	29
33	GLASGOW	140	6.0	4.3	0.36	21
34	CLYDEBANK	10	6.4	4.5	1.43	17
35	BEARSDEN	5	4.5	3.5	1.56	31
36	STRATHKELVIN	10	4.1	3.5	1.11	35
37	CUMBERNAULD	4	2.2	2.6	1.31	46
38	MONKLANDS	24	7.5	7.7+	1.68	3
39	MOTHERWELL	21	4.7	3.5	0.78	32
40	HAMILTON	11	3.5	2.8	0.85	43
41	EAST KILBRIDE	8	3.2	4.4	1.62	19
42	EASTWOOD	6	4.1	3.3	1.36	38
43	CLYDESDALE	4	2.4	1.7-	0.86	50
44	RENFREW	26	4.2	3.5	0.71	30
45	INVERCLYDE	11	3.7	2.9	0.88	41
46	CUNNINGHAME	18	4.6	3.6	0.92	28
47	KILMARNOCK	9	3.6	2.7	0.91	44
48	KYLE-CARRICK	20	6.3	4.4	1.02	18
49	CUMNOCK-DOON	3	2.2	1.5--	0.89	52
50	WIGTOWN	7	8.1	5.7	2.19	7
51	STEWARTRY	7	10.7	4.9	1.90	14
52	NITHSDALE	7	4.3	3.7	1.43	27
53	ANNANDALE	5	4.8	3.5	1.63	34
54	ORKNEY	1	1.9	1.3-	1.32	53
55	SHETLAND	4	6.4	5.4	2.80	10
56	WESTERN ISLES	6	6.6	4.6	2.02	16
	ALL SCOTLAND	794	5.3	3.9	0.14	

Oral Cavity and Pharynx — Females

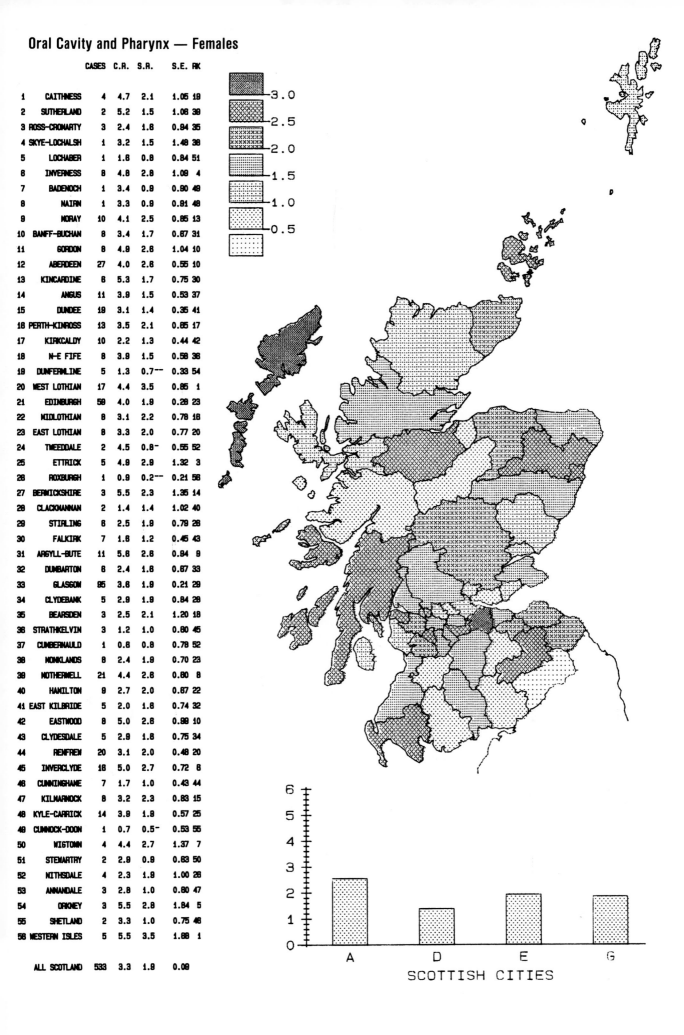

		CASES	C.R.	S.R.	S.E.	RK
1	CAITHNESS	4	4.7	2.1	1.05	19
2	SUTHERLAND	2	5.2	1.5	1.06	39
3	ROSS-CROMARTY	3	2.4	1.6	0.94	35
4	SKYE-LOCHALSH	1	3.2	1.5	1.48	38
5	LOCHABER	1	1.6	0.8	0.84	51
6	INVERNESS	8	4.6	2.8	1.09	4
7	BADENOCH	1	3.4	0.9	0.90	49
8	NAIRN	1	3.3	0.9	0.91	48
9	MORAY	10	4.1	2.5	0.85	13
10	BANFF-BUCHAN	8	3.4	1.7	0.67	31
11	GORDON	8	4.8	2.6	1.04	10
12	ABERDEEN	27	4.0	2.6	0.55	10
13	KINCARDINE	8	5.3	1.7	0.75	30
14	ANGUS	11	3.9	1.5	0.53	37
15	DUNDEE	19	3.1	1.4	0.35	41
16	PERTH-KINROSS	13	3.5	2.1	0.65	17
17	KIRKCALDY	10	2.2	1.3	0.44	42
18	N-E FIFE	8	3.9	1.5	0.58	36
19	DUNFERMLINE	5	1.3	0.7--	0.33	54
20	WEST LOTHIAN	17	4.4	3.5	0.85	1
21	EDINBURGH	58	4.0	1.9	0.28	23
22	MIDLOTHIAN	8	3.1	2.2	0.78	16
23	EAST LOTHIAN	8	3.3	2.0	0.77	20
24	TWEEDDALE	2	4.5	0.8-	0.55	52
25	ETTRICK	5	4.9	2.9	1.32	3
26	ROXBURGH	1	0.9	0.2--	0.21	56
27	BERWICKSHIRE	3	5.5	2.3	1.35	14
28	CLACKMANNAN	2	1.4	1.4	1.02	40
29	STIRLING	8	2.5	1.8	0.79	26
30	FALKIRK	7	1.6	1.2	0.45	43
31	ARGYLL-BUTE	11	5.6	2.8	0.94	9
32	DUMBARTON	8	2.4	1.6	0.67	33
33	GLASGOW	95	3.6	1.9	0.21	29
34	CLYDEBANK	5	2.9	1.9	0.84	28
35	BEARSDEN	3	2.5	2.1	1.20	18
36	STRATHKELVIN	3	1.2	1.0	0.60	45
37	CUMBERNAULD	1	0.6	0.6	0.78	52
38	MONKLANDS	8	2.4	1.9	0.70	23
39	MOTHERWELL	21	4.4	2.6	0.60	8
40	HAMILTON	9	2.7	2.0	0.67	22
41	EAST KILBRIDE	5	2.0	1.6	0.74	32
42	EASTWOOD	8	5.0	2.6	0.99	10
43	CLYDESDALE	5	2.9	1.8	0.75	34
44	RENFREW	20	3.1	2.0	0.48	20
45	INVERCLYDE	18	5.0	2.7	0.72	6
46	CUNNINGHAME	7	1.7	1.0	0.43	44
47	KILMARNOCK	8	3.2	2.3	0.83	15
48	KYLE-CARRICK	14	3.9	1.9	0.57	25
49	CUMNOCK-DOON	1	0.7	0.5-	0.53	55
50	WIGTOWN	4	4.4	2.7	1.37	7
51	STEWARTRY	2	2.8	0.9	0.63	50
52	NITHSDALE	4	2.3	1.9	1.00	26
53	ANNANDALE	3	2.8	1.0	0.60	47
54	ORKNEY	3	5.5	2.8	1.64	5
55	SHETLAND	2	3.3	1.0	0.75	46
56	WESTERN ISLES	5	5.5	3.5	1.88	1
	ALL SCOTLAND	533	3.3	1.9	0.09	

SCOTTISH CITIES

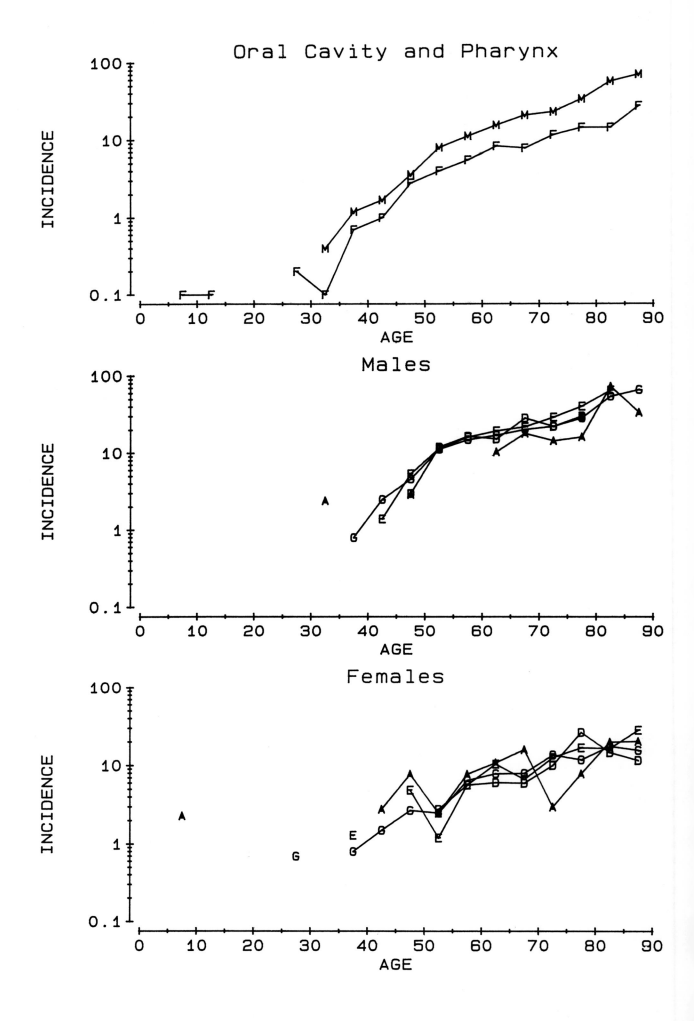

CANCER OF THE OESOPHAGUS
(ICD8 150)

The incidence of cancer of the oesophagus (or gullet) is increasing dramatically in Scotland; since 1970 it has almost doubled in people of each sex. The increase is particularly evident in younger (35-64) age groups.

Males (2.5%; 0.89%)

The incidence of oesophageal cancer in Scottish males (7.5) is higher than that found in other British regions, e.g., Mersey (7.0), North-west (6.5), Birmingham (5.5) and Trent (5.1). Incidence is also lower than the Scottish figures in Finland (4.0), Denmark (3.0), Sweden (2.9) and Norway (2.7), and is generally lower in most other regions of Europe except for the two regions of France for which there are incidence data — Doubs (13.0) and Bas-Rhin (17.0). The incidence in North America is also lower than that in Scotland, e.g., Connecticut (5.3), Ontario (4.2) and Alberta (2.6).

Within Scotland, the male rate varies between 10.4 and 3.8. High rates are recorded in Inverness (10.4), Angus (10.3), Nithsdale (10.3), Clackmannan (10.3) and the Shetland Islands (10.1); and low rates in Tweeddale (3.8), Berwickshire (4.1), Roxburgh (4.2), Eastwood (4.4), and Kincardine and Deeside (4.7).

Both high and low incidence rates occur in rural areas. Areas with high rates are concentrated in the Grampian region, the western Highland region and the Firth of Clyde. Smaller pockets of high incidence occur in the eastern Tayside region and the Shetland Islands.

Females (2.2%; 0.48%)

The incidence rate in Scottish females (4.2) is substantially lower than that in males (7.5) but, as for males, the incidence is higher than in any region of England [e.g., Trent (2.7), Birmingham (2.9) and Oxford (2.2)], Europe and North America. The only rates that are currently higher are those reported from Bombay (10.7) and Poona (10.4) (both regions of India), the Netherlands Antilles (8.7), Shanghai (China) (8.0), Hong Kong (5.5), Puerto Rico (5.2) and Detroit blacks (4.3). The incidence among Scottish females may be viewed, therefore, as high [although very low by Iranian standards (Joint Iran-IARC Study Group, 1977)]. The highest rates in the Scottish districts appear in Annandale and Eskdale (6.8), Caithness (6.4), the Western Isles (6.2) and East Kilbride (6.1); areas of low incidence include the Orkney Islands (0.4), the Shetland Islands (0.8), Sutherland (0.9), Skye and Lochalsh (1.6) and Nairn (1.7).

High levels of incidence of oesophageal cancer in women are apparent in the southern half of Scotland, notably in the eastern borders, south Ayrshire, Clydesdale and south Lothian. The rural regions showed areas of both high and low incidence.

Scottish cities

The incidence in the cities is close to the national rate for people of both sexes. In males, the rates in Aberdeen (8.7), Glasgow (8.6) and Dundee (8.5) are almost identical, that in Edinburgh (6.6) being slightly lower. In females, the rate in each of the cities was close to the all-Scottish female rate (4.2): Dundee (4.4), Edinburgh (4.2), Glasgow (4.0) and Aberdeen (3.9).

Spatial aggregation

Evidence of spatial aggregation of districts with similar rates of oesophageal cancer was weak in males ($D = 17.35$, $p = 0.066$) and even weaker in females ($D = 17.63$, $p = 0.102$).

Comment

The high risk of oesophageal cancer in those who drink and smoke heavily has been known for many years (Young & Russell, 1926; Mosbech & Videback, 1955), and the association has been confirmed in a number of countries (Steiner, 1956; Schwartz et al., 1962; Day & Muñoz, 1982). The available evidence taken as a whole suggests that wine and beer consumption do not increase risk unduly, whereas high consumption of alcohol that includes a substantial proportion of spirits increases the risk very substantially. Diets very low in fresh fruit and vegetables may render the oesophagus more susceptible to cancer-causing agents (Craver, 1932; Tuyns, 1970; Joint Iran-IARC Study Group, 1977; Cook-Mozaffari et al., 1979).

These investigations must lead to a consideration of the role played by whisky, diet and cigarette smoking in producing oesophageal cancer in Scotland. There is little direct information on the regional drinking habits of Scottish people. The main whisky-distilling areas are, however, situated in the northern part of the Grampian region and in the Highland region (Chapters 3 and 4), where the incidence of the disease is high. This association does not necessarily imply cause and effect, but rather a need for further study. Consumption of green vegetables and fruits is probably relatively low in these areas, as the volume grown locally on farms is much lower than in the south-east of the country, and lower than in the country as a whole (Table 3.5). While it is possible that local people grow their own vegetables, reliable information on this point would require special study. In Iran and China, where oesophageal cancer is 20 times more prevalent than in Scotland, vitamin deficiencies are fairly common. It is interesting that there are a number of districts with low rates of oesophageal cancer in males in the south-east, where much of Scotland's home-produced vegetable crops originate.

No regional data are available regarding cigarette consumption. The overall cigarette consumption is thought to be high; in a small sample survey, the average weekly expenditure on tobacco in Scotland in 1981 was nearly 30% higher than that for Great Britain as a whole.

For oesophageal cancer, therefore, several factors may serve to increase risk — cigarette smoking, consumption of strong alcoholic drinks and a diet low in the protective substances believed to occur in fresh fruit and vegetables. Studies to show whether these factors are, indeed, responsible for oesophageal cancer in Scotland should be carried out with some priority, in view of the rapidly increasing incidence in younger people, the extraordinary level of this cancer in females in Scotland and the fatal course of this malignant disease.

Oesophagus — Males

10.5
9.7
8.8
7.4   —— 7.5
6.0
5.2
4.4
3.5

D=17.35, p=0.066

Oesophagus — Females

D=17.63, p=0.102

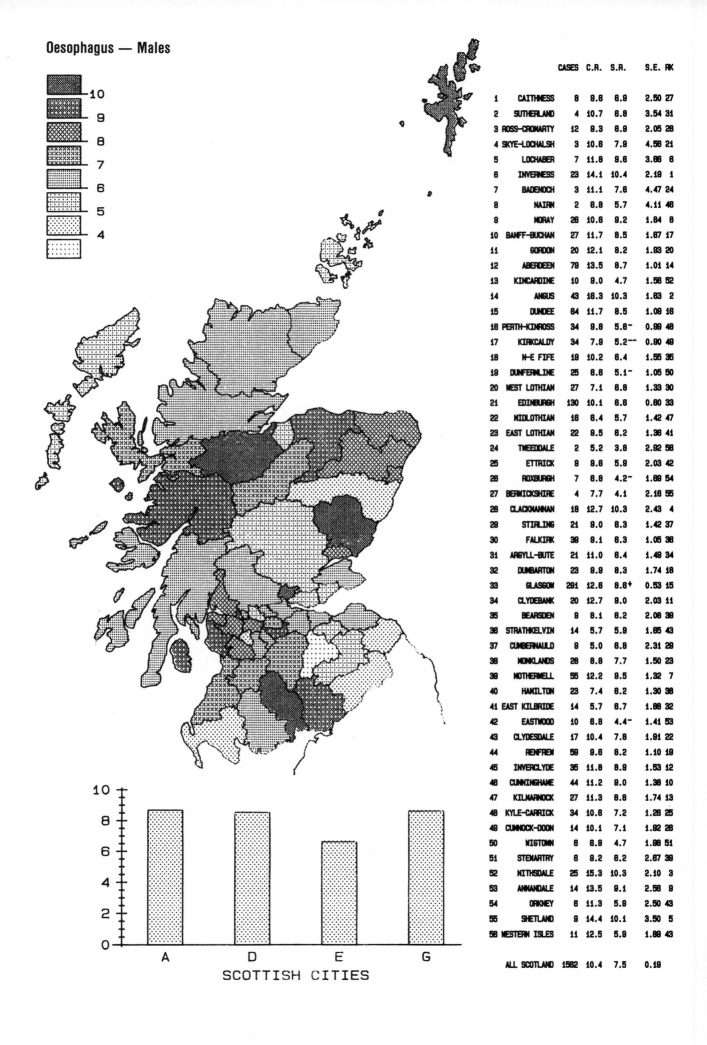

Oesophagus — Males

		CASES	C.R.	S.R.	S.E.	RK
1	CAITHNESS	8	9.6	6.9	2.50	27
2	SUTHERLAND	4	10.7	6.8	3.54	31
3	ROSS-CROMARTY	12	9.3	6.9	2.05	28
4	SKYE-LOCHALSH	3	10.6	7.9	4.56	21
5	LOCHABER	7	11.8	9.6	3.66	6
6	INVERNESS	23	14.1	10.4	2.19	1
7	BADENOCH	3	11.1	7.6	4.47	24
8	NAIRN	2	6.6	5.7	4.11	46
9	MORAY	26	10.6	9.2	1.84	8
10	BANFF-BUCHAN	27	11.7	8.5	1.67	17
11	GORDON	20	12.1	8.2	1.83	20
12	ABERDEEN	79	13.5	8.7	1.01	14
13	KINCARDINE	10	9.0	4.7	1.56	52
14	ANGUS	43	18.3	10.3	1.63	2
15	DUNDEE	64	11.7	8.5	1.09	16
16	PERTH-KINROSS	34	9.8	5.6⁻	0.99	48
17	KIRKCALDY	34	7.9	5.2⁻⁻	0.90	49
18	N-E FIFE	19	10.2	6.4	1.55	35
19	DUNFERMLINE	25	8.6	5.1⁻	1.05	50
20	WEST LOTHIAN	27	7.1	6.6	1.33	30
21	EDINBURGH	130	10.1	6.6	0.60	33
22	MIDLOTHIAN	16	6.4	5.7	1.42	47
23	EAST LOTHIAN	22	9.5	6.2	1.38	41
24	TWEEDDALE	2	5.2	3.8	2.92	56
25	ETTRICK	9	9.6	5.9	2.03	42
26	ROXBURGH	7	6.6	4.2⁻	1.69	54
27	BERWICKSHIRE	4	7.7	4.1	2.18	55
28	CLACKMANNAN	18	12.7	10.3	2.43	4
29	STIRLING	21	9.0	6.3	1.42	37
30	FALKIRK	38	9.1	6.3	1.05	36
31	ARGYLL-BUTE	21	11.0	6.4	1.49	34
32	DUMBARTON	23	9.9	6.3	1.74	16
33	GLASGOW	291	12.6	8.6⁺	0.53	15
34	CLYDEBANK	20	12.7	9.0	2.03	11
35	BEARSDEN	9	8.1	6.2	2.08	39
36	STRATHKELVIN	14	5.7	5.9	1.65	43
37	CUMBERNAULD	9	5.0	6.8	2.31	29
38	MONKLANDS	26	8.6	7.7	1.50	23
39	MOTHERWELL	55	12.2	8.5	1.32	7
40	HAMILTON	23	7.4	6.2	1.30	38
41	EAST KILBRIDE	14	5.7	6.7	1.88	32
42	EASTWOOD	10	8.6	4.4⁻	1.41	53
43	CLYDESDALE	17	10.4	7.6	1.91	22
44	RENFREW	59	9.6	6.2	1.10	19
45	INVERCLYDE	35	11.8	8.9	1.53	12
46	CUNNINGHAME	44	11.2	8.0	1.38	10
47	KILMARNOCK	27	11.3	8.6	1.74	13
48	KYLE-CARRICK	34	10.6	7.2	1.26	25
49	CUMNOCK-DOON	14	10.1	7.1	1.92	26
50	WIGTOWN	6	8.9	4.7	1.96	51
51	STEWARTRY	6	8.2	6.2	2.67	39
52	NITHSDALE	25	15.3	10.3	2.10	3
53	ANNANDALE	14	13.5	9.1	2.58	9
54	ORKNEY	6	11.3	5.9	2.50	43
55	SHETLAND	9	14.4	10.1	3.50	5
56	WESTERN ISLES	11	12.5	5.9	1.89	43
	ALL SCOTLAND	1582	10.4	7.5	0.19	

SCOTTISH CITIES

Oesophagus — Females

		CASES	C.R.	S.R.	S.E.	RK
1	CAITHNESS	9	10.6	6.4	2.27	2
2	SUTHERLAND	1	2.6	0.9--	0.85	54
3	ROSS-CROMARTY	8	6.3	4.2	1.59	24
4	SKYE-LOCHALSH	2	6.4	1.6-	1.18	53
5	LOCHABER	3	4.9	3.0	1.78	47
6	INVERNESS	12	7.1	3.4	1.06	39
7	BADENOCH	3	10.1	3.6	2.24	38
8	NAIRN	1	3.3	1.7	1.68	52
9	MORAY	27	11.1	5.5	1.20	9
10	BANFF-BUCHAN	17	7.2	3.2	0.85	44
11	GORDON	10	6.1	2.2--	0.72	50
12	ABERDEEN	61	9.1	3.9	0.54	32
13	KINCARDINE	9	7.9	4.2	1.53	27
14	ANGUS	41	14.6	5.6	1.01	7
15	DUNDEE	58	9.6	4.4	0.64	20
16	PERTH-KINROSS	35	9.5	4.4	0.85	23
17	KIRKCALDY	34	7.4	3.9	0.70	31
18	N-E FIFE	23	11.2	4.5	1.06	17
19	DUNFERMLINE	17	4.5	2.4--	0.60	49
20	WEST LOTHIAN	28	7.3	5.5	1.06	10
21	EDINBURGH	141	9.5	4.2	0.40	26
22	MIDLOTHIAN	14	5.4	3.8	1.04	33
23	EAST LOTHIAN	23	9.5	4.2	0.95	25
24	TWEEDDALE	6	13.5	3.7	1.59	37
25	ETTRICK	7	6.9	2.2-	0.88	51
26	ROXBURGH	18	16.2	4.5	1.13	16
27	BERWICKSHIRE	7	12.6	5.1	2.39	14
28	CLACKMANNAN	13	8.9	4.4	1.31	19
29	STIRLING	17	7.0	3.8	1.00	34
30	FALKIRK	34	7.8	4.1	0.73	29
31	ARGYLL-BUTE	24	12.1	5.3	1.23	12
32	DUMBARTON	15	6.0	3.3	0.89	43
33	GLASGOW	238	9.0	4.0	0.28	30
34	CLYDEBANK	10	5.9	3.2	1.03	44
35	BEARSDEN	8	6.7	3.3	1.26	41
36	STRATHKELVIN	13	5.1	3.4	0.99	39
37	CUMBERNAULD	4	2.2	2.6	1.33	46
38	MONKLANDS	27	8.0	5.7	1.14	6
39	MOTHERWELL	43	8.0	5.4	0.85	11
40	HAMILTON	19	5.7	3.7	0.88	35
41	EAST KILBRIDE	19	7.5	6.1	1.43	4
42	EASTWOOD	11	6.9	3.3	1.16	41
43	CLYDESDALE	18	10.5	5.6	1.39	8
44	RENFREW	52	7.9	4.7	0.69	15
45	INVERCLYDE	31	8.6	5.3	1.02	13
46	CUNNINGHAME	24	5.8	3.1	0.69	46
47	KILMARNOCK	20	7.9	4.1	0.96	28
48	KYLE-CARRICK	36	10.1	4.5	0.81	18
49	CUMNOCK-DOON	11	7.7	4.4	1.37	21
50	WIGTOWN	10	10.9	4.4	1.49	22
51	STEWARTRY	8	8.7	3.7	1.88	35
52	NITHSDALE	23	13.3	5.7	1.26	5
53	ANNANDALE	14	13.1	6.6	2.00	1
54	ORKNEY	1	1.6	0.4--	0.38	56
55	SHETLAND	2	3.3	0.8--	0.57	55
56	WESTERN ISLES	10	11.1	6.2	2.31	3
	ALL SCOTLAND	1388	8.5	4.2	0.12	

SCOTTISH CITIES

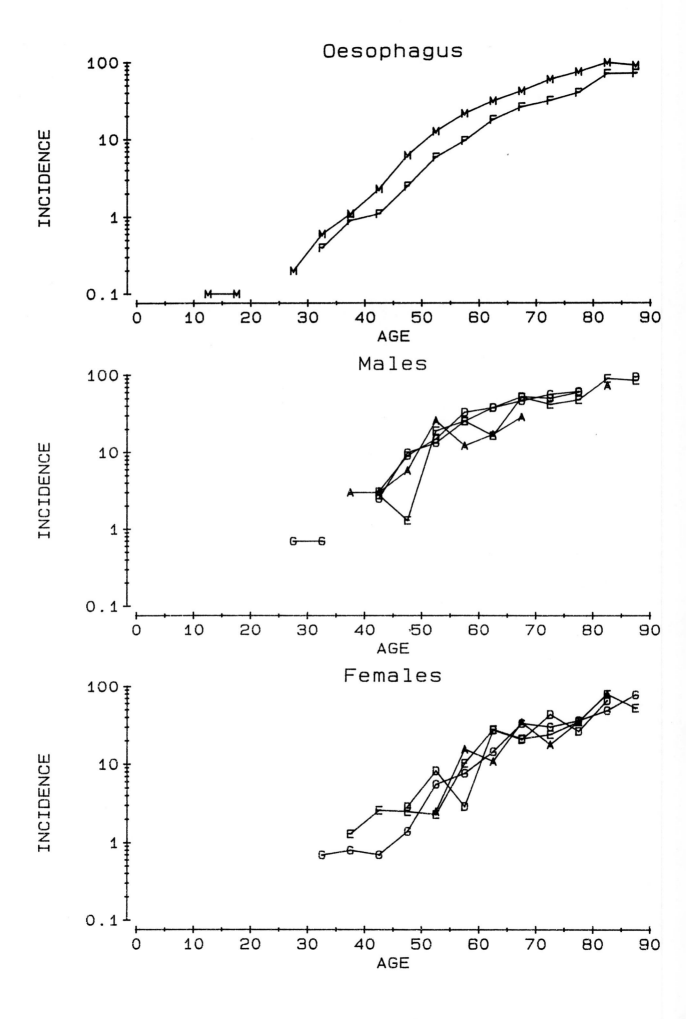

CANCER OF THE STOMACH
(ICD8 151)

As in most Western countries, the incidence of stomach cancer is decreasing in Scotland. However, stomach cancer is currently the third commonest site of cancer in Scottish males and the fourth in Scottish females.

Males (6.8%; 2.39%)

The average incidence (20.7) in Scottish males is comparable with that of neighbouring countries and registries, e.g., Denmark (17.1), Sweden (18.3), Norway (20.6) and Birmingham (22.1), but lower than that in Finland (29.3) and many times lower than that in Nagasaki, Japan (100.2).

Within Scotland, the highest incidence levels are found in Dundee (29.1), Monklands (28.0), Inverclyde (26.6), Motherwell (26.3) and Clydesdale (25.3), while the lowest rates are found in Badenoch and Strathspey (5.2), the Shetland Islands (11.1), Roxburgh (11.6), Caithness (11.7) and Gordon (11.5).

The highest incidence levels in males thus occur in the industrial areas surrounding Glasgow and in Dundee, while the lowest are found in the sparsely populated areas of the north and in one rural area in the south, Roxburgh.

Females (5.4%; 1.12%)

The average incidence rate of stomach cancer in females in Scotland (10.0) is, as elsewhere, approximately one-half the rate reported in males (20.7) in the same population. The rate among Scottish females is similar to that found in most populations of Britain, western Europe and North America, with a similar racial-ethnic structure. Incidence rates are, however, much higher in eastern Europe (e.g., Kraków in Poland (15.6)) and Japan (Nagasaki 51.0 and Osaka 38.5).

Areas of highest incidence are Cumbernauld and Kilsyth (15.4), Motherwell (14.0), Clydebank (13.5), Inverclyde (13.1) and East Kilbride (12.4); areas of low incidence are Badenoch and Strathspey (1.6), Sutherland (2.9), Tweeddale (4.0), Gordon (4.2) and Nairn (4.8). The geographical pattern of higher urban incidence is similar to that in males, i.e., high rates are found in and around cities with heavy industry.

Scottish cities

Male incidence in each of the cities exceeds the all-Scottish male incidence rate (20.7); Dundee (29.1) has the highest incidence in Scotland, and Glasgow (23.3) the seventh highest, with Edinburgh (20.8) and Aberdeen (20.7) having slightly lower rates.

Female incidence in each city is higher than the national female rate (10.0); that of Glasgow (11.2) is followed by those of Dundee (11.0), Aberdeen (10.7) and Edinburgh (9.9). A high incidence is thus consistently present in people of both sexes in Scottish cities, Aberdeen and Edinburgh having slightly lower rates than Glasgow and Dundee.

Spatial aggregation

There is visual and statistical evidence of clustering of districts with similar rates of stomach cancer in both males (D = 15.98, $p = 0.004$) and females (D = 15.38, $p < 0.001$) in Scotland.

Comment

The main feature of the geographic pattern of stomach cancer in people of each sex is the consistent finding of low rates in the rural north and south of Scotland and concentrations of high rates in and around the major industrial centres of Glasgow (notably Monklands, Motherwell and Clydesdale) and Dundee.

Fifty years ago, stomach cancer was the leading cause of death from cancer in males in the USA (Levin et al., 1974). Since then, mortality and incidence have fallen virtually everywhere, but nonetheless this cancer remains the fourth commonest fatal cancer in US males today. Stomach cancer death rates have halved in Scotland since 1911. The reasons for this global decline are not known (Nomura, 1982), although it is suspected that it is caused by wider availability of fresh fruit and vegetables and better food preservation. The risk of stomach cancer is consistently reported to be higher among members of the lower socio-economic classes (Cohart, 1954; Torgersen & Petersen, 1956; Sigurjonsson, 1967; Hirayama, 1971), a finding which also held true 70 years ago (Logan, 1982).

Alcohol seems to have little influence on stomach cancer risk (Nomura, 1982), but a small excess risk has been noted among smokers (Dorn, 1959; Doll & Peto, 1976; Hirayama, 1977). Large studies from different countries have shown that more persons with stomach cancer are likely to be of blood group A than would be expected. This risk may apply to the intestinal histologic sub-type (Correa et al., 1973).

In recent years, a good deal of attention has been paid to items of diet that protect against stomach cancer. Japanese farmers, among whom this disease is very common, reduced their risk by frequent consumption of lettuce and celery (Haenszel *et al.*, 1976), and these findings were repeated in other populations in the USA (Graham *et al.*, 1972) and Norway (Bjelke, 1978). Consumption of fresh fruit, tomatoes and milk also lowers the risk. The decrease observed in most of the world is thus consistent with a greater availability of dairy products, fresh fruit and vegetables throughout the year. It is known that the Scots do not eat as much fruit as inhabitants of some other countries, e.g., some 30% less than in southern England (Table 3.3). However, no data are available to document possible regional differences within Scotland.

Another suggestion is that a high intake of grilled meat and of smoked and salted foods is related to a high stomach cancer level. There is no information on the Scottish consumption of such foods, but the Scots buy 50% more pre-cooked and tinned meat than do residents of south-east England (Table 3.3). Further investigation appears to be worthwhile.

The occupational mortality report for Scotland for 1969-1973 (Kemp & Bledin, 1981) showed that workers exposed to chemicals and metals have a higher than average risk of stomach cancer, a finding similar to that in England and Wales (Registrar General for England and Wales, 1978). Industries of this kind are situated in the central areas of the country. Such a finding does not prove that the excess risk is due to exposure at the workplace — e.g., these workers may take less fresh fruit and vegetables than others — but such a conclusion should be reached only after excluding any possible occupational risk factor. A series of epidemiological studies of stomach cancer in Scotland would be justified, to determine whether the risk and the protective factors are the same as elsewhere.

Stomach — Males

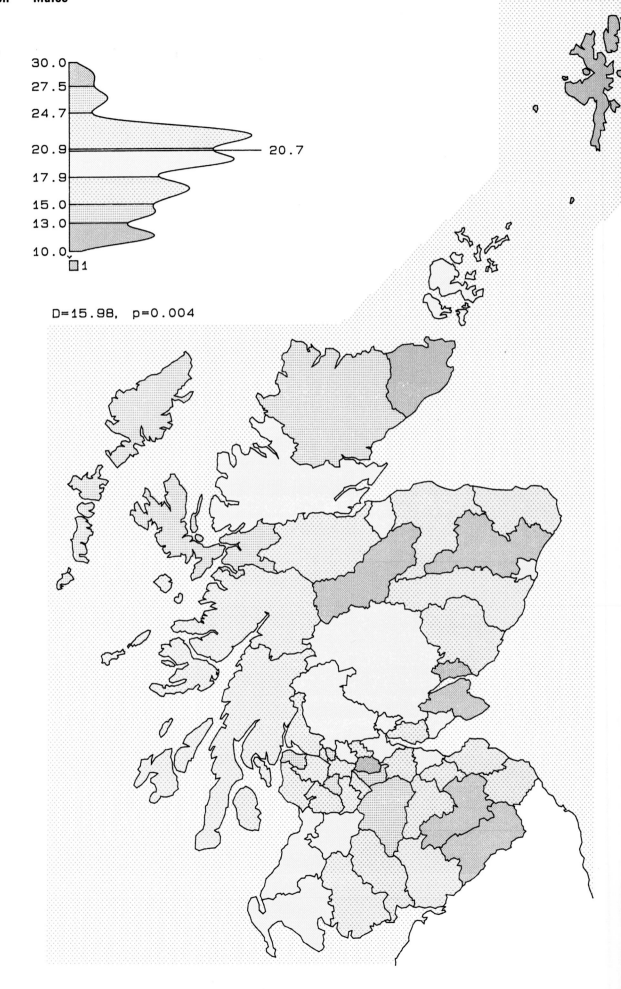

D=15.98, p=0.004

Stomach — Females

D=15.38, p=0.001

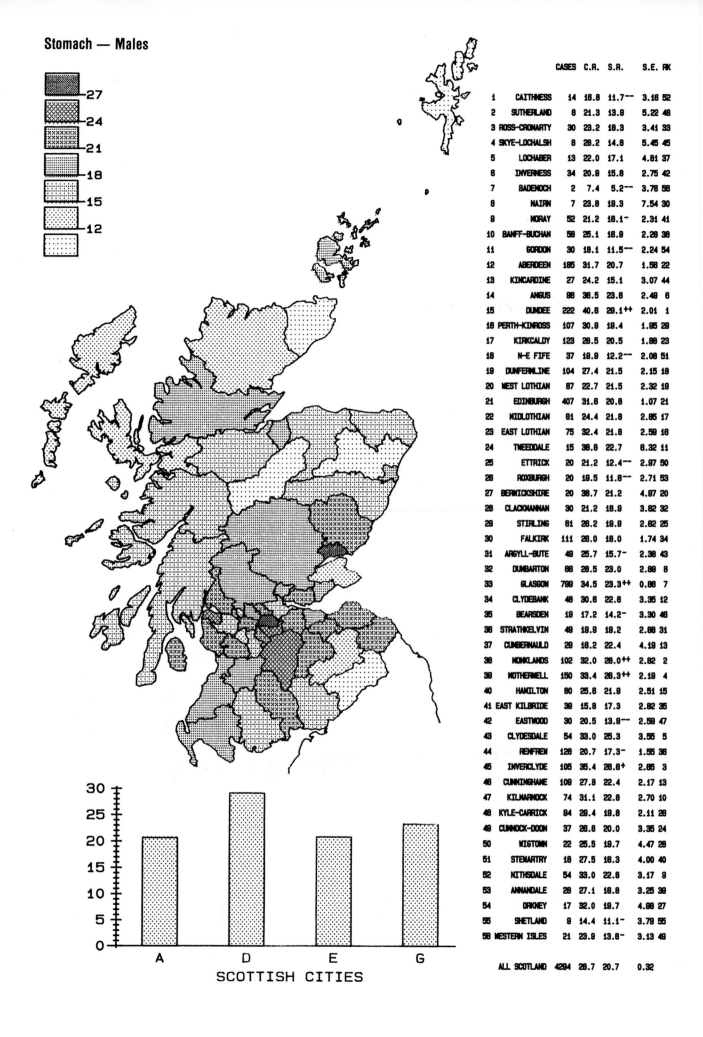

Stomach — Males

		CASES	C.R.	S.R.	S.E.	RK
1	CAITHNESS	14	18.8	11.7--	3.16	52
2	SUTHERLAND	8	21.3	13.9	5.22	46
3	ROSS-CROMARTY	30	23.2	18.3	3.41	33
4	SKYE-LOCHALSH	8	28.2	14.8	5.45	45
5	LOCHABER	13	22.0	17.1	4.81	37
6	INVERNESS	34	20.9	15.8	2.75	42
7	BADENOCH	2	7.4	5.2--	3.78	56
8	NAIRN	7	23.8	19.3	7.54	30
9	MORAY	52	21.2	16.1-	2.31	41
10	BANFF-BUCHAN	58	25.1	16.9	2.28	36
11	GORDON	30	18.1	11.5--	2.24	54
12	ABERDEEN	185	31.7	20.7	1.58	22
13	KINCARDINE	27	24.2	15.1	3.07	44
14	ANGUS	96	36.5	23.6	2.49	6
15	DUNDEE	222	40.6	29.1++	2.01	1
16	PERTH-KINROSS	107	30.9	18.4	1.95	29
17	KIRKCALDY	123	28.5	20.5	1.88	23
18	N-E FIFE	37	19.8	12.2--	2.08	51
19	DUNFERMLINE	104	27.4	21.5	2.15	18
20	WEST LOTHIAN	87	22.7	21.5	2.32	19
21	EDINBURGH	407	31.6	20.8	1.07	21
22	MIDLOTHIAN	61	24.4	21.8	2.85	17
23	EAST LOTHIAN	75	32.4	21.8	2.59	16
24	TWEEDDALE	15	36.8	22.7	6.32	11
25	ETTRICK	20	21.2	12.4--	2.97	50
26	ROXBURGH	20	18.5	11.8--	2.71	53
27	BERWICKSHIRE	20	38.7	21.2	4.97	20
28	CLACKMANNAN	30	21.2	18.9	3.62	32
29	STIRLING	61	26.2	19.9	2.62	25
30	FALKIRK	111	26.0	18.0	1.74	34
31	ARGYLL-BUTE	49	25.7	15.7-	2.38	43
32	DUMBARTON	66	26.5	23.0	2.89	8
33	GLASGOW	799	34.5	23.3++	0.86	7
34	CLYDEBANK	46	30.6	22.8	3.35	12
35	BEARSDEN	19	17.2	14.2-	3.30	46
36	STRATHKELVIN	49	19.9	19.2	2.86	31
37	CUMBERNAULD	29	18.2	22.4	4.19	13
38	MONKLANDS	102	32.0	26.0++	2.82	2
39	MOTHERWELL	150	33.4	26.3++	2.19	4
40	HAMILTON	80	25.6	21.9	2.51	15
41	EAST KILBRIDE	39	15.6	17.3	2.82	35
42	EASTWOOD	30	20.5	13.9--	2.59	47
43	CLYDESDALE	54	33.0	25.3	3.55	5
44	RENFREW	126	20.7	17.3-	1.55	38
45	INVERCLYDE	105	35.4	26.6+	2.65	3
46	CUNNINGHAME	109	27.8	22.4	2.17	13
47	KILMARNOCK	74	31.1	22.8	2.70	10
48	KYLE-CARRICK	94	29.4	19.6	2.11	28
49	CUMNOCK-DOON	37	26.8	20.0	3.35	24
50	WIGTOWN	22	25.5	19.7	4.47	26
51	STEWARTRY	18	27.5	16.3	4.00	40
52	NITHSDALE	54	33.0	22.8	3.17	9
53	ANNANDALE	28	27.1	16.8	3.25	39
54	ORKNEY	17	32.0	19.7	4.98	27
55	SHETLAND	9	14.4	11.1-	3.79	55
56	WESTERN ISLES	21	23.9	13.6-	3.13	49
	ALL SCOTLAND	4294	28.7	20.7	0.32	

SCOTTISH CITIES

Stomach — Females

		CASES	C.R.	S.R.	S.E.	RK
1	CAITHNESS	8	7.1	5.2⁻	2.17	50
2	SUTHERLAND	3	7.8	2.9⁻⁻	1.72	55
3	ROSS-CROMARTY	18	14.1	8.0	2.05	40
4	SKYE-LOCHALSH	5	18.0	5.3	2.49	49
5	LOCHABER	10	18.4	9.5	3.32	30
6	INVERNESS	30	17.8	8.9	1.75	34
7	BADENOCH	2	8.7	1.8⁻⁻	1.13	56
8	NAIRN	3	9.8	4.8	2.77	52
9	MORAY	38	15.8	7.8	1.33	42
10	BANFF-BUCHAN	30	12.7	6.1⁻	1.25	44
11	GORDON	19	11.5	4.2⁻⁻	1.02	53
12	ABERDEEN	154	22.9	10.7	0.98	16
13	KINCARDINE	16	14.1	6.4⁻	1.82	43
14	ANGUS	78	27.9	11.2	1.43	13
15	DUNDEE	149	24.6	11.0	1.01	14
16	PERTH-KINROSS	71	19.3	8.2	1.12	39
17	KIRKCALDY	88	19.3	10.0	1.14	23
18	N-E FIFE	47	22.9	9.1	1.51	32
19	DUNFERMLINE	72	19.2	10.1	1.26	20
20	WEST LOTHIAN	82	16.1	11.7	1.58	8
21	EDINBURGH	383	24.6	9.9	0.58	25
22	MIDLOTHIAN	44	17.0	11.4	1.77	9
23	EAST LOTHIAN	58	24.0	10.1	1.42	20
24	TWEEDDALE	8	18.0	4.0⁻⁻	1.47	54
25	ETTRICK	18	17.8	5.7⁻⁻	1.46	46
26	ROXBURGH	29	26.1	12.3	2.67	6
27	BERWICKSHIRE	9	16.5	4.9⁻⁻	1.73	51
28	CLACKMANNAN	25	17.1	8.8	1.84	35
29	STIRLING	50	20.6	11.4	1.70	9
30	FALKIRK	84	19.3	10.4	1.19	18
31	ARGYLL-BUTE	51	25.8	10.9	1.75	15
32	DUMBARTON	38	14.5	7.8	1.37	41
33	GLASGOW	697	26.4	11.2⁺⁺	0.47	12
34	CLYDEBANK	43	25.4	13.5	2.15	3
35	BEARSDEN	22	18.3	8.7	2.18	27
36	STRATHKELVIN	42	16.5	10.3	1.65	19
37	CUMBERNAULD	29	16.2	15.4	2.94	1
38	MONKLANDS	80	17.8	11.3	1.51	11
39	MOTHERWELL	118	24.6	14.1⁺⁺	1.38	2
40	HAMILTON	54	16.3	8.7	1.37	28
41	EAST KILBRIDE	40	15.8	12.4	2.03	5
42	EASTWOOD	34	21.2	8.1	1.74	31
43	CLYDESDALE	41	24.0	12.3	2.11	7
44	RENFREW	103	15.7	8.8	0.89	37
45	INVERCLYDE	81	25.1	13.1⁺	1.60	4
46	CUNNINGHAME	72	17.4	8.7	1.09	36
47	KILMARNOCK	52	20.5	8.9	1.46	24
48	KYLE-CARRICK	71	20.0	8.3	1.12	38
49	CUMNOCK-DOON	23	16.1	10.5	2.28	17
50	WIGTOWN	21	22.9	9.8	2.18	29
51	STEWARTRY	15	21.8	8.0	2.84	33
52	NITHSDALE	35	20.3	8.8	1.79	26
53	ANNANDALE	15	14.0	8.0⁻	1.75	45
54	ORKNEY	7	12.8	5.4⁻	2.19	48
55	SHETLAND	11	18.0	10.0	3.49	22
56	WESTERN ISLES	10	11.1	5.5⁻	2.03	47
	ALL SCOTLAND	3373	20.9	10.0	0.19	

Legend: 14, 12, 10, 8, 6, 4

SCOTTISH CITIES: A, D, E, G

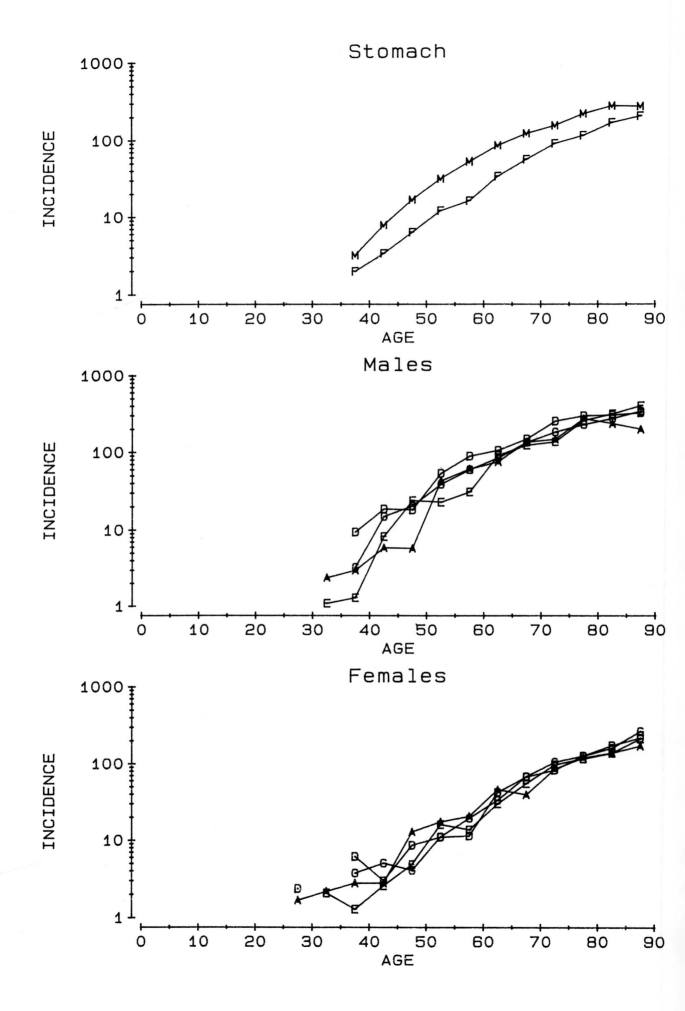

CANCER OF THE LARGE BOWEL
(ICD8 153, 154)

The large bowel is one of the three major sites of malignant disease in Scotland, the other two being the lung in men and the breast in women. Cancer of the large bowel includes cancers of the colon (ICD 153) and the rectum (ICD 154). The incidence of both colonic and rectal cancer is increasing in people of each sex in Scotland. The two sites have been combined in this section of the atlas, as the anatomical boundary between them is not always easy to determine. Colonic and rectal cancer are, nonetheless, presented separately later.

Males (11.2%; 3.83%)

The male incidence rate (34.2) is higher than that recorded elsewhere in Great Britain: Birmingham (33.0), North-western (32.0), Mersey (31.2), Trent (30.9), Oxford (30.7) and South Thames (28.0). The incidence is similar to that found in Denmark (36.0), but considerably higher than that in Finland (17.0), Norway (26.0) or Sweden (27.2). Incidence rates are generally higher in regions of North America, e.g., Newfoundland (34.9), Manitoba (38.8), Ontario (41.0), Iowa (41.2), Detroit (42.5) and Connecticut (50.0).

Within Scotland, the highest rate is found in Tweeddale (55.3) followed by Badenoch and Strathspey (53.9), Nairn (51.9), Skye and Lochalsh (50.1) and Ross and Cromarty (47.3); the lowest rates are recorded in the Orkney islands (21.2), the Western Isles (24.1), Annandale and Eskdale (24.8), Dunfermline (25.0) and Clydesdale (27.5). The outstanding feature of the distribution of large-bowel cancer among males in Scotland is the aggregation of rural areas with a similar incidence of colorectal cancer. Apart from the very highest level in Tweeddale (55.3), most of the high levels are found in contiguous areas in the north of Scotland: Badenoch and Strathspey (53.9), Nairn (51.9), Skye and Lochalsh (50.1), Ross and Cromarty (47.3), Inverness (44.6), Aberdeen City (44.1), Angus (40.9), Kincardine (40.9), Banff and Buchan (39.2), Gordon (39.2), Sutherland (38.7) and Perth and Kinross (38.4).

Females (13.7%; 3.09%)

The Scottish female incidence rate of colorectal cancer (27.5) is higher than any incidence rate recorded in regions of England and Wales: Birmingham (25.0), Mersey (24.7), Oxford (24.5), North-western (24.4), South Thames (23.5) and Trent (23.3). The Scottish rate is lower than in Denmark

(29.2), but higher than that in Norway (22.6), Sweden (22.3) and Finland (16.0). Incidence is higher in Ontario (35.9), Manitoba (32.8), Newfoundland (28.2), Connecticut (37.5), Iowa (35.2) and Detroit (30.2).

In Scotland, the lowest rates are found in the Orkney Islands (8.9), Tweeddale (14.3), Berwickshire (18.7), Clydesdale (19.1) and Stewartry (21.8); the highest rates in Ross and Cromarty (40.2), Kincardine and Deeside (39.6), Skye and Lochalsh (38.6), Badenoch and Strathspey (38.1) and Nairn (37.5). As in males, the outstanding feature of the map for females is the aggregation of districts with high rates in the north of Scotland. As well as those just mentioned, high rates were found in Moray (36.0), Aberdeen City (35.3), Caithness (34.8), Lochaber (34.2), Angus (32.1) and Banff and Buchan (29.7).

Scottish cities

The highest male rate is in Aberdeen City (44.1), followed by Glasgow (36.3), Dundee (36.0) and Edinburgh (33.4). For females, those in Aberdeen (35.3) have a higher rate than those in Dundee (28.0), Edinburgh (27.3) and Glasgow (26.0).

Spatial aggregation

There is very strong evidence of clustering of districts with similar rates of large-bowel cancer in both males (D = 14.28, $p < 0.0001$) and females (D = 14.93, $p < 0.001$). For colonic cancer, the aggregation is stronger among males (D = 14.63, $p < 0.001$) than among females (D = 17.49, $p = 0.083$); conversely, for rectal cancer, the aggregation is stronger for females (D = 14.70, $p < 0.001$) than for males (D = 16.02, $p = 0.004$).

Comment

Together with cancer of the breast and lung, colorectal cancer accounts for over 50% of all cancer deaths in Western countries. Colorectal cancer is typically frequent in North America and Europe, rare in Asia and particularly uncommon in sub-Saharan Africa. Internationally, colonic cancer incidence varies 60-fold between areas, with the highest (Connecticut, USA) and lowest (Dakar, Senegal) incidence rates recorded in the mid-1970s. Within the continent of Europe, variation in the incidence of colonic cancer in males between cancer registry regions is four-fold (with the highest incidence in the Scotland North Registry).

Within the UK, colorectal cancer incidence is higher in the five cancer registry regions of Scotland than among the six regions in England and Wales with comparable published incidence data (Waterhouse *et al.*, 1982), due mainly to colon cancer. Mortality from colorectal cancer has been higher, in people of each sex, in Scotland than in England and Wales by a relatively constant margin since data of sufficient detail were first published in 1911 (Boyle *et al.*, 1985a). In Scotland, there is a three-fold variation in the incidence of colorectal cancer between districts of highest and lowest incidence. The north and the south of the country are clearly demarcated, incidence rates being one-third higher in the north than in the south. Moreover, there is marked aggregation of landward areas with high, intermediate and low rates (Boyle *et al.*, 1985a).

The signal discrepancy of the rates in males (highest in Scotland) and females (second lowest) in Tweeddale is worthy of further study; it is apparently due to a very high male rate in a part of the country where the incidence rates are generally low. Investigation could be made of some special exposure of males — occupational, recreational or dietary in origin. The finding may also be due to chance.

In people of younger ages, the age-specific incidence of large-bowel cancer is higher in women than in men. However, after the age of 55, there appears to be a crossover, and at each later age the incidence is higher in men. This could indicate that as they get older, men become increasingly exposed to factors that initiate (or more likely promote) cancer of the large bowel (Boyle *et al.*, 1985a).

Evidence from descriptive epidemiological studies (Boyle *et al.*, 1985a), migrant studies (Haenszel, 1982) and studies of religious groups such as Seventh Day Adventists (Phillips *et al.*, 1980) and Mormons (Enstrom, 1980) suggests that environmental factors are of particular importance in the etiology of these cancers. Evidence from these studies, substantiated by experimental findings, indicates that dietary factors are likely to be responsible for most colorectal cancers. Briefly, diets with a large component of meat and fat have been associated with high risk, and diets with high levels of fibre and cruciferous vegetables (such as cabbage) with low risk (Zaridze, 1983). Although several lines of evidence suggest an association between colonic cancer risk and a high-fat diet, the fat appears to promote rather than initiate large-bowel carcinogenesis (Nigro *et al.*, 1975; Bull *et al.*, 1979). Little is known about initiators, some of which may be related to mode of preparation of food, e.g., grilling. Differences in exposure to initiators, along with differences in exposure to protective factors such as dietary fibre, ascorbic acid, retinoids and selenium, may explain some of the geographical variations in the incidence of this cancer.

The range of variation present in Scotland, together with the clear aggregation of areas of low and high incidence, presents a unique setting for detailed epidemiological studies of colorectal cancer (Boyle & Zaridze, 1984; Boyle *et al.*, 1984).

Large Bowel — Males

D=14.28, p<0.0001

Large Bowel — Females

D=14.93, p<0.001

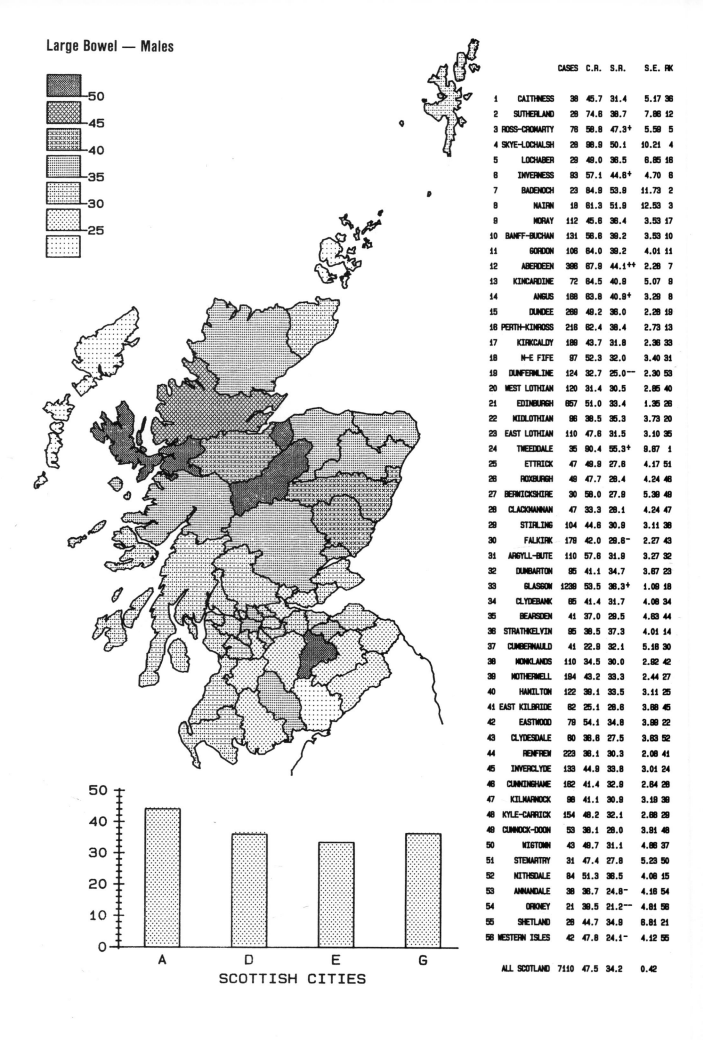

Large Bowel — Males

		CASES	C.R.	S.R.	S.E.	RK
1	CAITHNESS	38	45.7	31.4	5.17	36
2	SUTHERLAND	28	74.6	38.7	7.86	12
3	ROSS-CROMARTY	76	58.8	47.3+	5.59	5
4	SKYE-LOCHALSH	28	98.9	50.1	10.21	4
5	LOCHABER	29	49.0	36.5	6.85	16
6	INVERNESS	93	57.1	44.6+	4.70	6
7	BADENOCH	23	84.9	53.9	11.73	2
8	NAIRN	18	61.3	51.9	12.53	3
9	MORAY	112	45.6	36.4	3.53	17
10	BANFF-BUCHAN	131	56.6	39.2	3.53	10
11	GORDON	106	64.0	39.2	4.01	11
12	ABERDEEN	396	67.9	44.1++	2.26	7
13	KINCARDINE	72	64.5	40.9	5.07	9
14	ANGUS	168	63.8	40.9+	3.29	8
15	DUNDEE	289	49.2	36.0	2.26	19
16	PERTH-KINROSS	216	62.4	38.4	2.73	13
17	KIRKCALDY	189	43.7	31.8	2.36	33
18	N-E FIFE	97	52.3	32.0	3.40	31
19	DUNFERMLINE	124	32.7	25.0--	2.30	53
20	WEST LOTHIAN	120	31.4	30.5	2.85	40
21	EDINBURGH	657	51.0	33.4	1.35	26
22	MIDLOTHIAN	96	38.5	35.3	3.73	20
23	EAST LOTHIAN	110	47.6	31.5	3.10	35
24	TWEEDDALE	35	90.4	55.3+	9.87	1
25	ETTRICK	47	49.9	27.6	4.17	51
26	ROXBURGH	49	47.7	28.4	4.24	46
27	BERWICKSHIRE	30	58.0	27.9	5.39	49
28	CLACKMANNAN	47	33.3	28.1	4.24	47
29	STIRLING	104	44.6	30.9	3.11	38
30	FALKIRK	179	42.0	29.6-	2.27	43
31	ARGYLL-BUTE	110	57.6	31.9	3.27	32
32	DUMBARTON	95	41.1	34.7	3.67	23
33	GLASGOW	1239	53.5	36.3+	1.09	18
34	CLYDEBANK	65	41.4	31.7	4.08	34
35	BEARSDEN	41	37.0	29.5	4.63	44
36	STRATHKELVIN	95	36.5	37.3	4.01	14
37	CUMBERNAULD	41	22.9	32.1	5.16	30
38	MONKLANDS	110	34.5	30.0	2.92	42
39	MOTHERWELL	184	43.2	33.3	2.44	27
40	HAMILTON	122	39.1	33.5	3.11	25
41	EAST KILBRIDE	62	25.1	28.6	3.68	45
42	EASTWOOD	79	54.1	34.8	3.99	22
43	CLYDESDALE	60	36.6	27.5	3.63	52
44	RENFREW	223	38.1	30.3	2.06	41
45	INVERCLYDE	133	44.9	33.6	3.01	24
46	CUNNINGHAME	162	41.4	32.9	2.64	28
47	KILMARNOCK	98	41.1	30.9	3.19	39
48	KYLE-CARRICK	154	46.2	32.1	2.68	29
49	CUMNOCK-DOON	53	38.1	28.0	3.91	48
50	WIGTOWN	43	49.7	31.1	4.86	37
51	STEWARTRY	31	47.4	27.8	5.23	50
52	NITHSDALE	84	51.3	36.5	4.08	15
53	ANNANDALE	38	36.7	24.8-	4.16	54
54	ORKNEY	21	38.5	21.2--	4.81	56
55	SHETLAND	26	44.7	34.9	6.81	21
56	WESTERN ISLES	42	47.8	24.1-	4.12	55
	ALL SCOTLAND	7110	47.5	34.2	0.42	

SCOTTISH CITIES

Large Bowel — Females

		CASES	C.R.	S.R.		S.E.	RK
1	CAITHNESS	50	58.9	34.8		5.25	9
2	SUTHERLAND	19	49.5	24.3		6.38	44
3	ROSS-CROMARTY	85	66.6	40.2 ++		4.71	1
4	SKYE-LOCHALSH	25	79.8	38.6		9.52	3
5	LOCHABER	32	52.8	34.2		6.42	10
6	INVERNESS	80	47.5	28.8		3.27	35
7	BADENOCH	21	70.8	38.1		9.18	4
8	NAIRN	17	55.8	37.5		9.85	5
9	MORAY	180	65.7	36.0 ++		3.09	6
10	BANFF-BUCHAN	133	56.5	29.7		2.81	16
11	GORDON	101	61.3	29.5		3.35	18
12	ABERDEEN	491	73.1	35.3 ++		1.78	7
13	KINCARDINE	86	77.6	39.6 ++		4.86	2
14	ANGUS	205	73.2	32.1		2.53	11
15	DUNDEE	352	58.1	28.0		1.68	27
16	PERTH-KINROSS	236	64.2	28.9		2.12	22
17	KIRKCALDY	239	51.8	28.4		1.96	26
18	N-E FIFE	140	66.1	29.5		2.89	17
19	DUNFERMLINE	155	41.4	23.1 -		1.99	48
20	WEST LOTHIAN	153	39.6	26.8		2.40	25
21	EDINBURGH	818	62.2	27.4		1.02	29
22	MIDLOTHIAN	97	37.6	27.2		2.85	32
23	EAST LOTHIAN	146	60.4	28.7		2.58	23
24	TWEEDDALE	21	47.2	14.3 --		3.59	55
25	ETTRICK	67	65.6	28.7		4.03	24
26	ROXBURGH	80	71.9	27.3		3.46	30
27	BERWICKSHIRE	25	45.8	18.7 -		4.18	54
28	CLACKMANNAN	78	53.3	32.0		3.81	12
29	STIRLING	121	49.9	29.4		2.88	19
30	FALKIRK	213	49.0	27.3		1.97	30
31	ARGYLL-BUTE	108	54.6	23.7		2.63	46
32	DUMBARTON	127	51.1	30.3		2.88	15
33	GLASGOW	1485	58.7	26.0 --		0.75	36
34	CLYDEBANK	88	51.9	31.2		3.46	13
35	BEARSDEN	66	55.0	30.9		4.11	14
36	STRATHKELVIN	96	37.7	26.3		2.77	37
37	CUMBERNAULD	46	26.8	27.2		4.01	32
38	MONKLANDS	124	36.7	24.7		2.32	43
39	MOTHERWELL	185	36.8	22.9 --		1.76	49
40	HAMILTON	121	36.5	23.5		2.23	47
41	EAST KILBRIDE	80	31.7	26.5		3.04	38
42	EASTWOOD	81	50.5	22.4		2.71	50
43	CLYDESDALE	66	36.6	19.1 -		2.53	53
44	RENFREW	274	41.9	24.8		1.58	42
45	INVERCLYDE	161	49.8	25.8		2.15	40
46	CUNNINGHAME	182	43.9	23.8		1.90	45
47	KILMARNOCK	121	47.7	25.9		2.50	39
48	KYLE-CARRICK	227	63.9	28.9		2.09	21
49	CUMNOCK-DOON	59	41.3	27.1		3.71	34
50	WIGTOWN	43	46.8	22.1		3.65	51
51	STEWARTRY	33	47.9	21.8		4.21	52
52	NITHSDALE	121	70.2	35.1 +		3.48	8
53	ANNANDALE	61	58.9	27.8		3.78	28
54	ORKNEY	13	23.8	8.9 --		2.63	56
55	SHETLAND	36	62.2	29.3		5.47	20
56	WESTERN ISLES	55	60.9	25.5		4.03	41
	ALL SCOTLAND	8622	53.5	27.5		0.32	

SCOTTISH CITIES

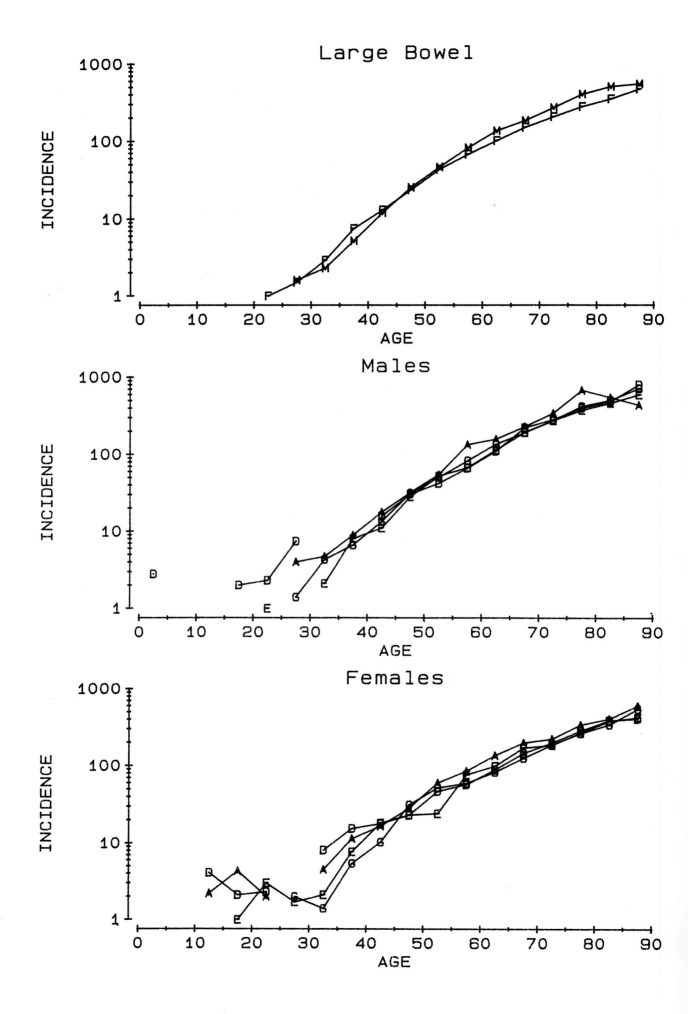

CANCER OF THE PANCREAS
(ICD8 157)

Pancreatic cancer is an increasingly common form of cancer, for which survival five years from initial diagnosis is exceedingly rare. The incidence (averaged over all age groups) has been increasing in Scotland, and the increase is most dramatic among younger (35-64) age groups.

Males (2.9%; 1.03%)

In comparison to rates in other countries, the incidence of cancer of the pancreas in Scottish males (8.7) is in the middle to upper range. Several registries record high incidence among coloured people, including New Zealand Maoris (14.8) and the black population of the San Francisco Bay Area (18.3). Few white populations have a higher incidence than that in Scotland, e.g., Atlanta (11.3) and Ontario (11.0). The Scottish male incidence is similar to that in northern European countries such as Finland (9.7), Denmark (9.5), Sweden (9.1) and Norway (8.2) and marginally higher than that in English registries, North-western (8.5), Oxford (8.4), Trent (8.2), Birmingham (8.1) and Mersey (7.6).

In Scotland, the highest male rates are found in Clackmannan (13.0), Moray (12.0), Inverclyde (12.0), East Lothian (11.4) and Angus (10.9), the lowest rates in Sutherland (3.0), Gordon (3.1), Skye and Lochalsh (3.8), Caithness (4.3) and the Shetland Islands (5.0).

Females (3.0%; 0.67%)

The Scottish female incidence (5.8) is lower than that in males, but again lies in the middle to upper international range. The incidence among female coloured populations is generally higher, e.g., Alameda blacks (9.9), New Zealand Maoris (10.2). Northern European countries show similar rates, e.g., Finland (6.3), Denmark (6.2), Sweden (6.0) and Norway (5.2), as do the English registries of Oxford (5.7), North-western (5.1), Mersey (5.0), Birmingham (4.8) and Trent (4.5).

In Scotland, the highest female rates are found in Ettrick and Lauderdale (8.0), East Lothian (7.6), Kilmarnock and Loudon (7.6), Strathkelvin (7.4) and East Kilbride (7.3); the lowest are in Badenoch and Strathspey (0.0), Lochaber (1.9), Skye and Lochalsh (2.0), Ross and Cromarty (2.4) and Berwickshire (2.7).

Incidence in females has the same general distribution as that of males, with higher levels concentrated in the north-eastern and eastern areas, and areas of high incidence stretching across the central low-lands from East Lothian to Strathkelvin and Inverclyde. Areas with a higher incidence for both males and females include Angus, East Lothian, Ettrick and Lauderdale, Strathkelvin and Inverclyde.

Scottish cities

In the cities, the male incidence rate is higher than the all-Scottish male rate (8.7), with only small inter-city differences: Edinburgh (9.7), Aberdeen (9.6), Dundee (9.5) and Glasgow (9.3). The female incidence in cities was a little closer to the average (5.8), with Edinburgh (6.6) and Glasgow (6.1) having slightly higher rates than Aberdeen (5.8) and Dundee (5.4)

Special aggregation

Aggregation of districts with similar rates of pancreatic cancer is moderate among males (D = 16.43, $p = 0.011$), but weaker among females (D = 17.41, $p = 0.073$).

Comment

Several interesting features can be noted in the geographic distribution of cancer of the pancreas in Scotland. Consistently low rates are reported in people of each sex in Western Highland districts, and high rates are found in and around the four Scottish cities (except for the female rate in Aberdeen). The high rates found in males in both urban and rural districts of south-east Scotland, and the belt of high incidence rates found in females in southern central Scotland present further epidemiological challenges.

Higher urban frequencies of pancreatic cancer have been reported from a number of countries such as Denmark, Norway and Finland (Waterhouse et al., 1982). These findings are consistent with the high rates of pancreatic cancer found in the well-populated central belt and the east coast of Scotland.

An excess risk of pancreatic cancer has been described among smokers of cigarettes, and possibly among cigar smokers (Hammond & Horn, 1958; Cederlof et al., 1975; Doll & Peto, 1976). Lower levels of pancreatic cancer have been reported among Seventh Day Adventists (Phillips et al., 1980) and Mormons (Enstrom, 1980) — both groups of individuals who rarely smoke.

Coffee drinking has been claimed to increase risk (MacMahon et al., 1981) although there is some disagreement on this finding (Goldstein, 1982). Pancreatic cancer is reported more frequently in countries with a high average consumption of fat, oil, sugar, eggs and milk (Armstrong & Doll, 1975). Excess alcohol intake has also been associated with an increased risk (Blot *et al.*, 1978).

In summary, little is known about the causes of pancreatic cancer and there is, as yet, no explanation for the regional differences in Scotland. This numerically important cancer, which is increasing in Scotland, carries a very poor prognosis. It seems important, therefore, to utilize the variations within Scotland to improve our understanding of the etiology of this cancer and thereby increase prospects for prevention.

Pancreas — Males

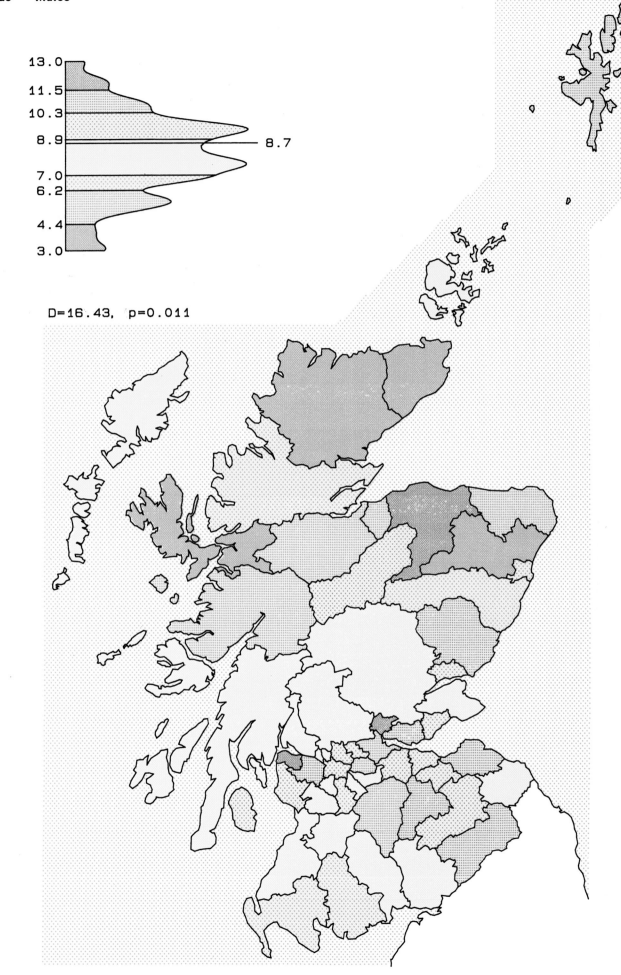

13.0
11.5
10.3
8.9 — 8.7
7.0
6.2
4.4
3.0

D=16.43, p=0.011

Pancreas — Females

D=17.41, p=0.073

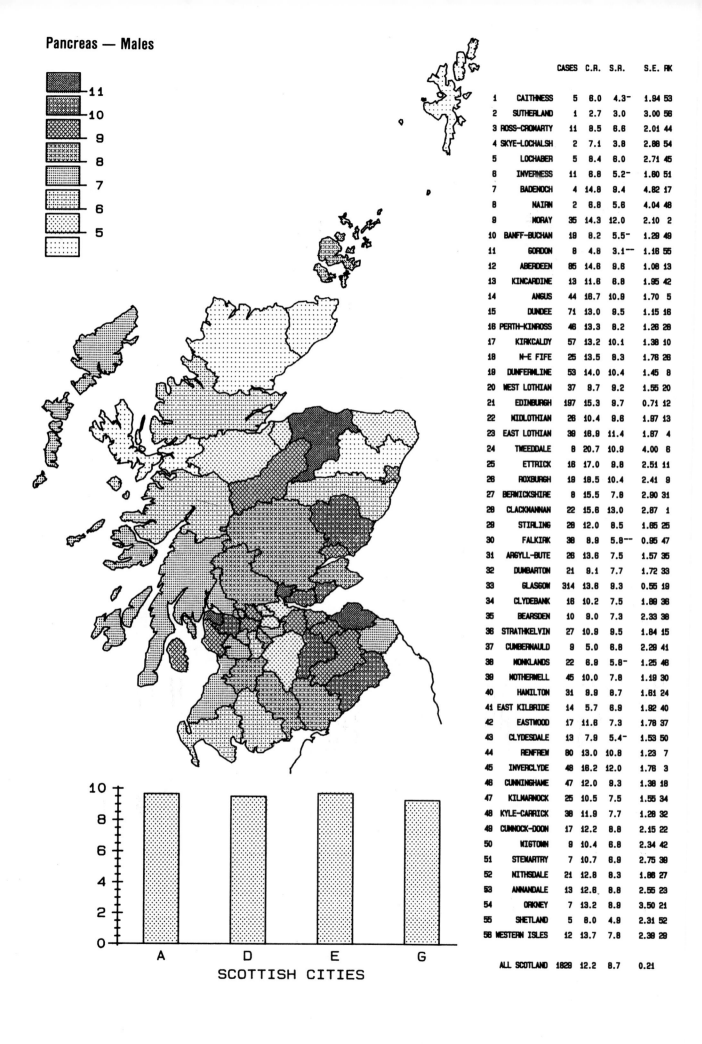

Pancreas — Males

Pancreas — Females

		CASES	C.R.	S.R.	S.E.	RK
1	CAITHNESS	5	5.9	3.9	1.85	45
2	SUTHERLAND	5	13.0	5.3	2.71	33
3	ROSS-CROMARTY	5	3.8	2.4--	1.18	53
4	SKYE-LOCHALSH	2	8.4	2.0--	1.38	54
5	LOCHABER	3	4.9	1.9--	1.14	55
6	INVERNESS	11	8.5	3.9	1.20	43
7	BADENOCH	0	0.0	0.0--	0.00	56
8	NAIRN	4	13.1	5.2	3.02	34
9	MORAY	29	11.9	6.4	1.30	14
10	BANFF-BUCHAN	33	14.0	6.8	1.30	9
11	GORDON	13	7.9	3.7	1.12	47
12	ABERDEEN	94	14.0	5.8	0.68	26
13	KINCARDINE	15	13.2	4.7	1.28	38
14	ANGUS	50	17.9	7.2	1.12	7
15	DUNDEE	75	12.4	5.4	0.70	30
16	PERTH-KINROSS	52	14.1	5.4	0.82	31
17	KIRKCALDY	50	10.8	6.1	0.91	21
18	N-E FIFE	31	15.1	6.3	1.35	16
19	DUNFERMLINE	41	11.0	5.8	0.95	27
20	WEST LOTHIAN	34	8.8	6.9	1.23	6
21	EDINBURGH	224	15.2	6.6	0.50	11
22	MIDLOTHIAN	15	5.8	3.9	1.04	44
23	EAST LOTHIAN	35	14.5	7.8	1.39	2
24	TWEEDDALE	7	15.7	6.2	2.85	18
25	ETTRICK	17	16.7	8.0	2.05	1
26	ROXBURGH	16	14.4	5.3	1.47	32
27	BERWICKSHIRE	4	7.3	2.7-	1.41	52
28	CLACKMANNAN	11	7.5	4.8	1.50	37
29	STIRLING	27	11.1	5.9	1.22	24
30	FALKIRK	47	10.8	6.8	1.01	10
31	ARGYLL-BUTE	26	13.1	5.7	1.32	28
32	DUMBARTON	26	10.5	6.1	1.25	21
33	GLASGOW	350	13.3	6.1	0.36	20
34	CLYDEBANK	21	12.4	6.3	1.42	15
35	BEARSDEN	5	4.2	3.0-	1.38	49
36	STRATHKELVIN	27	10.8	7.4	1.45	4
37	CUMBERNAULD	10	5.8	5.0	1.81	36
38	MONKLANDS	22	8.5	4.3	0.94	41
39	MOTHERWELL	38	7.9	4.3-	0.73	40
40	HAMILTON	23	6.9	4.8	1.00	39
41	EAST KILBRIDE	22	8.7	7.3	1.81	5
42	EASTWOOD	20	12.5	5.8	1.38	29
43	CLYDESDALE	21	12.3	6.5	1.58	13
44	RENFREW	54	8.3	5.1	0.73	35
45	INVERCLYDE	46	14.2	7.2	1.11	6
46	CUNNINGHAME	31	7.5	4.2-	0.79	42
47	KILMARNOCK	37	14.6	7.6	1.32	3
48	KYLE-CARRICK	54	15.2	6.5	0.95	12
49	CUMNOCK-DOON	14	9.8	6.2	1.70	18
50	WIGTOWN	7	7.6	3.3	1.37	46
51	STEWARTRY	6	8.7	3.8	1.79	48
52	NITHSDALE	23	13.3	5.9	1.35	25
53	ANNANDALE	14	13.1	6.0	1.69	23
54	ORKNEY	8	14.7	6.2	2.38	17
55	SHETLAND	3	4.9	3.0	2.08	50
56	WESTERN ISLES	7	7.8	2.8-	1.35	51
	ALL SCOTLAND	1870	11.8	5.8	0.14	

SCOTTISH CITIES

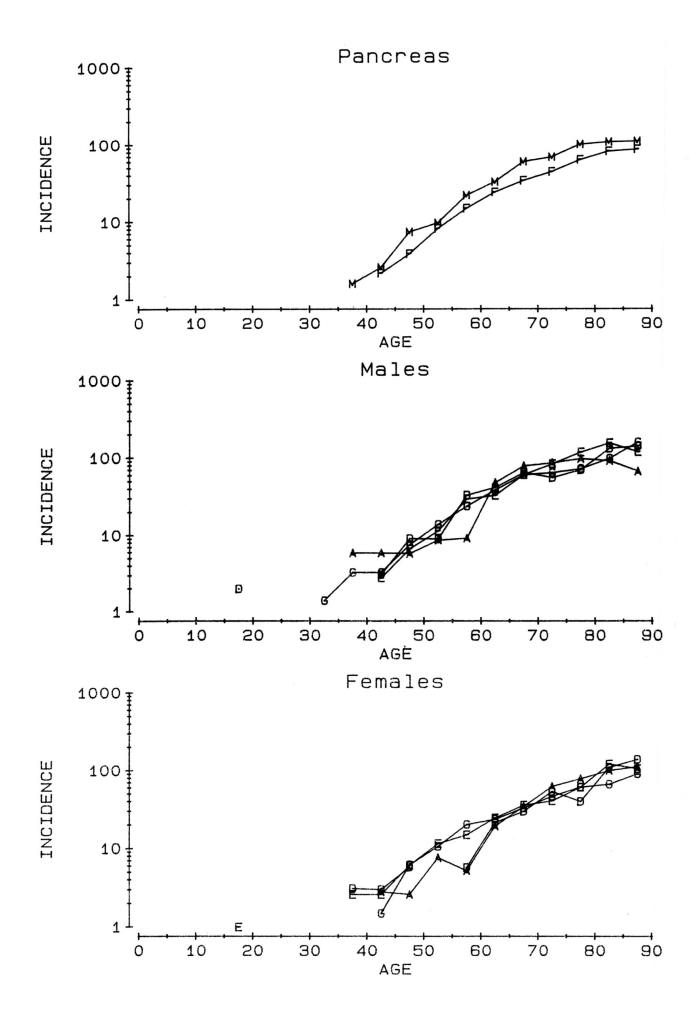

CANCER OF THE LARYNX
(ICD8 161)

Cancer of the larynx arises on either the vocal cords or the surrounding structures. However, the exact anatomical site of origin of many laryngeal cancers is difficult to establish, and hence the larynx is here considered as a whole. The incidence is increasing in people of each sex in Scotland, but this form of cancer remains four to five times commoner in males than in females.

Males (1.3%; 0.51%)

The all-Scottish rate (4.1) is very similar to that recorded in regions of England such as Mersey (4.4), Birmingham (4.0), Oxford (3.9) and Trent (3.7). Rates in Denmark (4.2) are fairly similar, rates in Finland (5.5), higher, and rates in Norway (2.8) and Sweden (2.8), lower. The incidence of laryngeal cancer in males is highest in Spain, e.g., Navarra (15.0) and Zaragoza (12.0), in Geneva (10.1) and in France [Doubs (17.6) and Bas-Rhin (11.2)]. Incidence rates in Canadian provinces such as Alberta (4.0), Newfoundland (4.0) and Manitoba (4.2) are similar to the Scottish rates, while the incidence is much higher in US regions such as the San Francisco Bay Area (7.2), Los Angeles (7.4), Detroit (7.7) and Connecticut (8.2).

No case were found, during the period studied, in two districts: the Orkney Islands and Nairn. The next lowest rates are found in Caithness (0.8) and Eastwood (0.9); the highest rates occur in Berwickshire (8.6), Dumbarton (7.7), Tweeddale (7.1), Cumbernauld and Kilsyth (6.7) and Aberdeen (5.7).

There does not appear to be any pattern to the geographical distribution of laryngeal cancer among males in Scotland and no urban-rural or coastal-inland split.

Females (0.4%; 0.12%)

The female rate (1.0) for Scotland is generally higher than that reported among other female populations of Britain, such as Birmingham (0.5), Oxford (0.6), Trent (0.7) and Mersey (1.1). Rates in Denmark (0.7), Finland (0.3), Sweden (0.3) and Norway (0.2) are lower. Whereas several European centres have high rates of laryngeal cancer among males, female rates in these areas, viz., Doubs (0.9), Geneva (0.8), Navarra (0.2), Zaragoza (0.2) and Bas-Rhin (0.1), are similar to the rates from the northern European countries. Further, rates in Canadian and US populations are low also, e.g., Alberta (0.4), Newfoundland (0.4), Manitoba (0.7), Connecticut (1.2), Detroit (1.3), Los Angeles (1.4), and the San Francisco Bay Area (1.5).

In Scotland, no case was found in 15 districts: Skye and Lochalsh, Lochaber, Inverness, Badenoch and Strathspey, the Western Isles, Sutherland, Kincardine and Deeside, the Orkney Islands, Caithness, Tweeddale, Stewartry, Strathkelvin, Berwickshire, Annandale and Eskdale and Cumnock and Doon Valley. There is little variation. The highest rates are found in Nairn (2.5), Ettrick and Lauderdale (1.7), Stirling (1.7) and Glasgow (1.7). The large number of districts with no case reported in females limits the search for a pattern.

Scottish cities

In the cities, the male rates are similar in Aberdeen (5.7), Glasgow (5.3), Dundee (5.0) and Edinburgh (4.6). The female rates are lower than those of males but similar in Glasgow (1.7), Dundee (1.5), Aberdeen (1.2) and Edinburgh (0.8). In each city except Edinburgh (for females), the rates are higher than the all-Scottish rates.

Spatial aggregation

The lack of any discernible pattern of clustering of laryngeal cancer is reinforced by the absence of statistical evidence for both males (D = 18.83, $p = 0.426$) and females (D = 17.92, $p = 0.158$).

Comment

The high rates of laryngeal cancer in people of each sex in all four cities and in the region adjacent to Glasgow were not sufficient to produce a significant degree of spatial aggregation. This lack of a pattern is in contrast to the strong aggregation of lung cancer in the central belt (see below). The incidence of laryngeal cancer has been increasing in people of each sex in Scotland, although mortality rates have remained constant. The disease is four to five times commoner in males than in females, as in most parts of the world.

The main risk factors have been clearly shown to be cigarette smoking (US Surgeon General, 1982) and alcohol consumption (Wynder & Stellman, 1977); the latter study showed an interaction between these habits and cancer risk. A large body of evidence demonstrates the association between increased risk of laryngeal cancer and cigarette consumption, and the risk among heavy smokers is generally estimated to be 30 times that of non-smokers. Although most laryngeal cancer is due to consumption of alcohol and tobacco (Stell & McGill, 1973), nickel refining (Pederson et al., 1973) and certain other specific occupations (Austin, 1982) have been shown to be associated with an excess risk of laryngeal cancer. However, these enquiries did not take the smoking and drinking habits of the workers into consideration.

Larynx — Males

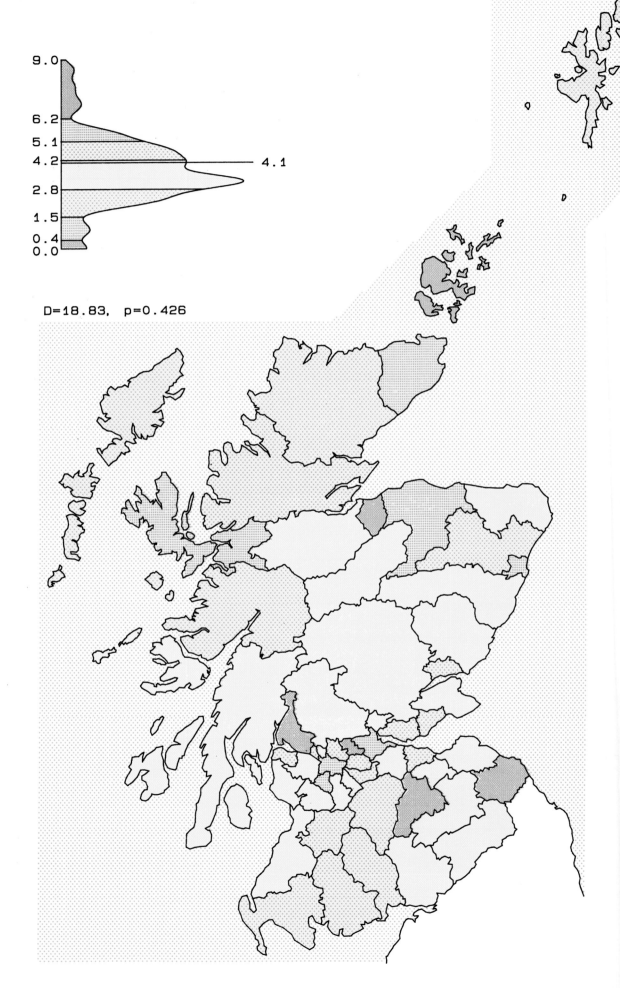

D=18.83, p=0.426

Larynx — Females

D=17.93, p=0.158

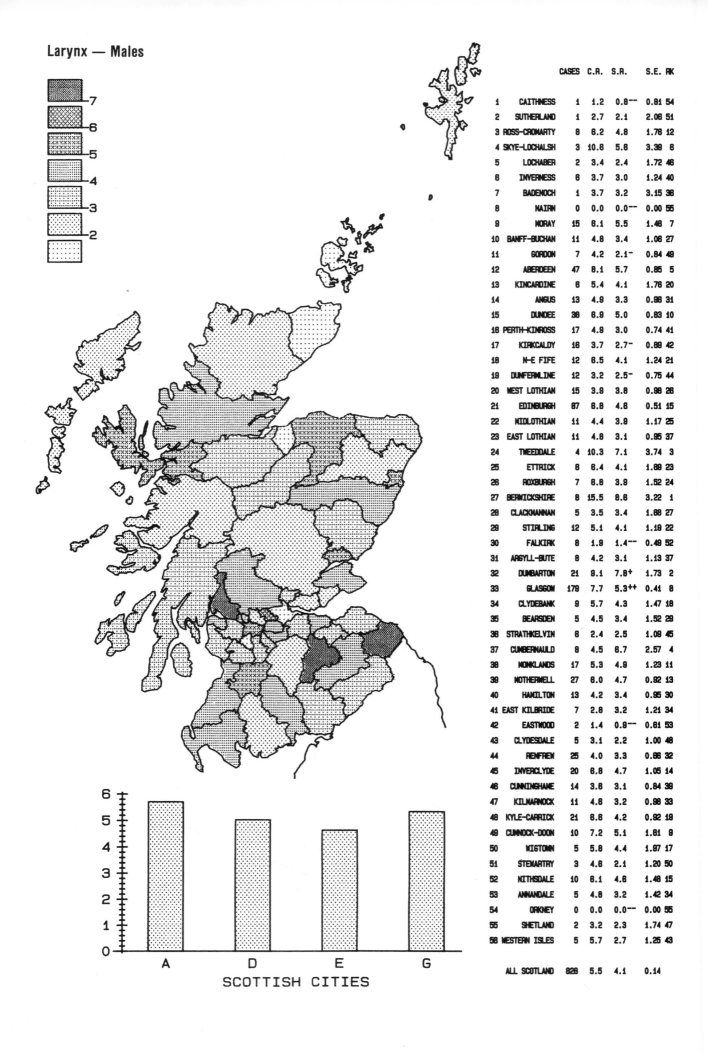

Larynx — Males

		CASES	C.R.	S.R.	S.E.	RK
1	CAITHNESS	1	1.2	0.8--	0.81	54
2	SUTHERLAND	1	2.7	2.1	2.08	51
3	ROSS-CROMARTY	8	6.2	4.8	1.78	12
4	SKYE-LOCHALSH	3	10.6	5.6	3.39	6
5	LOCHABER	2	3.4	2.4	1.72	46
6	INVERNESS	8	3.7	3.0	1.24	40
7	BADENOCH	1	3.7	3.2	3.15	36
8	NAIRN	0	0.0	0.0--	0.00	55
9	MORAY	15	6.1	5.5	1.46	7
10	BANFF-BUCHAN	11	4.8	3.4	1.06	27
11	GORDON	7	4.2	2.1-	0.84	49
12	ABERDEEN	47	6.1	5.7	0.85	5
13	KINCARDINE	8	5.4	4.1	1.76	20
14	ANGUS	13	4.9	3.3	0.98	31
15	DUNDEE	38	6.9	5.0	0.83	10
16	PERTH-KINROSS	17	4.9	3.0	0.74	41
17	KIRKCALDY	16	3.7	2.7-	0.69	42
18	N-E FIFE	12	6.5	4.1	1.24	21
19	DUNFERMLINE	12	3.2	2.5-	0.75	44
20	WEST LOTHIAN	15	3.9	3.6	0.98	26
21	EDINBURGH	87	6.6	4.6	0.51	15
22	MIDLOTHIAN	11	4.4	3.9	1.17	25
23	EAST LOTHIAN	11	4.8	3.1	0.95	37
24	TWEEDDALE	4	10.3	7.1	3.74	3
25	ETTRICK	6	6.4	4.1	1.69	23
26	ROXBURGH	7	6.6	3.9	1.52	24
27	BERWICKSHIRE	8	15.5	8.6	3.22	1
28	CLACKMANNAN	5	3.5	3.4	1.66	27
29	STIRLING	12	5.1	4.1	1.19	22
30	FALKIRK	8	1.9	1.4--	0.49	52
31	ARGYLL-BUTE	8	4.2	3.1	1.13	37
32	DUMBARTON	21	9.1	7.6+	1.73	2
33	GLASGOW	179	7.7	5.3++	0.41	8
34	CLYDEBANK	9	5.7	4.3	1.47	18
35	BEARSDEN	5	4.5	3.4	1.52	29
36	STRATHKELVIN	8	2.4	2.5	1.09	45
37	CUMBERNAULD	8	4.5	6.7	2.57	4
38	MONKLANDS	17	5.3	4.9	1.23	11
39	MOTHERWELL	27	6.0	4.7	0.92	13
40	HAMILTON	13	4.2	3.4	0.95	30
41	EAST KILBRIDE	7	2.8	3.2	1.21	34
42	EASTWOOD	2	1.4	0.9--	0.61	53
43	CLYDESDALE	5	3.1	2.2	1.00	48
44	RENFREW	25	4.0	3.3	0.66	32
45	INVERCLYDE	20	6.6	4.7	1.05	14
46	CUNNINGHAME	14	3.6	3.1	0.84	39
47	KILMARNOCK	11	4.6	3.2	0.98	33
48	KYLE-CARRICK	21	6.6	4.2	0.92	19
49	CUMNOCK-DOON	10	7.2	5.1	1.61	9
50	WIGTOWN	5	5.8	4.4	1.97	17
51	STEWARTRY	3	4.6	2.1	1.20	50
52	NITHSDALE	10	6.1	4.6	1.46	15
53	ANNANDALE	5	4.8	3.2	1.42	34
54	ORKNEY	0	0.0	0.0--	0.00	55
55	SHETLAND	2	3.2	2.3	1.74	47
56	WESTERN ISLES	5	5.7	2.7	1.25	43
	ALL SCOTLAND	828	5.5	4.1	0.14	

SCOTTISH CITIES

Larynx — Females

		CASES	C.R.	S.R.	S.E.	RK
1	CAITHNESS	0	0.0	0.0--	0.00	42
2	SUTHERLAND	0	0.0	0.0--	0.00	42
3	ROSS-CROMARTY	1	0.8	0.9	0.85	22
4	SKYE-LOCHALSH	0	0.0	0.0--	0.00	42
5	LOCHABER	0	0.0	0.0--	0.00	42
6	INVERNESS	0	0.0	0.0--	0.00	42
7	BADENOCH	0	0.0	0.0--	0.00	42
8	NAIRN	1	3.3	2.5	2.50	1
9	MORAY	3	1.2	1.1	0.78	14
10	BANFF-BUCHAN	2	0.8	0.6	0.47	31
11	GORDON	2	1.2	0.9	0.68	20
12	ABERDEEN	12	1.6	1.2	0.35	13
13	KINCARDINE	0	0.0	0.0--	0.00	42
14	ANGUS	5	1.6	1.1	0.52	15
15	DUNDEE	13	2.1	1.5	0.44	5
16	PERTH-KINROSS	7	1.8	1.3	0.50	11
17	KIRKCALDY	9	2.0	1.3	0.45	10
18	N-E FIFE	3	1.5	1.2	0.73	12
19	DUNFERMLINE	3	0.8	0.5	0.32	34
20	WEST LOTHIAN	3	0.8	0.7	0.39	29
21	EDINBURGH	24	1.6	0.8	0.18	23
22	MIDLOTHIAN	1	0.4	0.3-	0.29	39
23	EAST LOTHIAN	4	1.7	1.0	0.50	18
24	TWEEDDALE	0	0.0	0.0--	0.00	42
25	ETTRICK	3	2.8	1.7	1.00	2
26	ROXBURGH	2	1.6	1.0	0.74	17
27	BERWICKSHIRE	0	0.0	0.0--	0.00	42
28	CLACKMANNAN	2	1.4	0.8	0.61	24
29	STIRLING	5	2.1	1.7	0.77	3
30	FALKIRK	6	1.4	0.8	0.35	24
31	ARGYLL-BUTE	3	1.5	1.0	0.59	19
32	DUMBARTON	1	0.4	0.2--	0.20	40
33	GLASGOW	74	2.8	1.7++	0.21	4
34	CLYDEBANK	3	1.8	1.1	0.63	16
35	BEARSDEN	1	0.8	0.3	0.33	38
36	STRATHKELVIN	0	0.0	0.0--	0.00	42
37	CUMBERNAULD	1	0.8	0.6	0.63	30
38	MONKLANDS	7	2.1	1.5	0.57	6
39	MOTHERWELL	3	0.8	0.4-	0.24	37
40	HAMILTON	5	1.5	1.4	0.62	7
41	EAST KILBRIDE	3	1.2	1.4	0.79	7
42	EASTWOOD	1	0.8	0.2--	0.15	41
43	CLYDESDALE	2	1.2	0.8	0.59	28
44	RENFREW	4	0.8	0.5-	0.25	35
45	INVERCLYDE	4	1.2	0.8	0.42	27
46	CUNNINGHAME	5	1.2	0.7	0.35	26
47	KILMARNOCK	3	1.2	0.8	0.34	33
48	KYLE-CARRICK	6	1.7	0.9	0.39	21
49	CUMNOCK-DOON	0	0.0	0.0--	0.00	42
50	WIGTOWN	1	1.1	0.6	0.59	31
51	STEWARTRY	0	0.0	0.0--	0.00	42
52	NITHSDALE	1	0.8	0.4	0.42	36
53	ANNANDALE	0	0.0	0.0--	0.00	42
54	ORKNEY	0	0.0	0.0--	0.00	42
55	SHETLAND	1	1.8	1.4	1.35	8
56	WESTERN ISLES	0	0.0	0.0--	0.00	42
	ALL SCOTLAND	240	1.5	1.0	0.07	

Legend:
- 1.4
- 1.2
- 1.0
- 0.8
- 0.6
- 0.4
- 0.2

SCOTTISH CITIES (A, D, E, G)

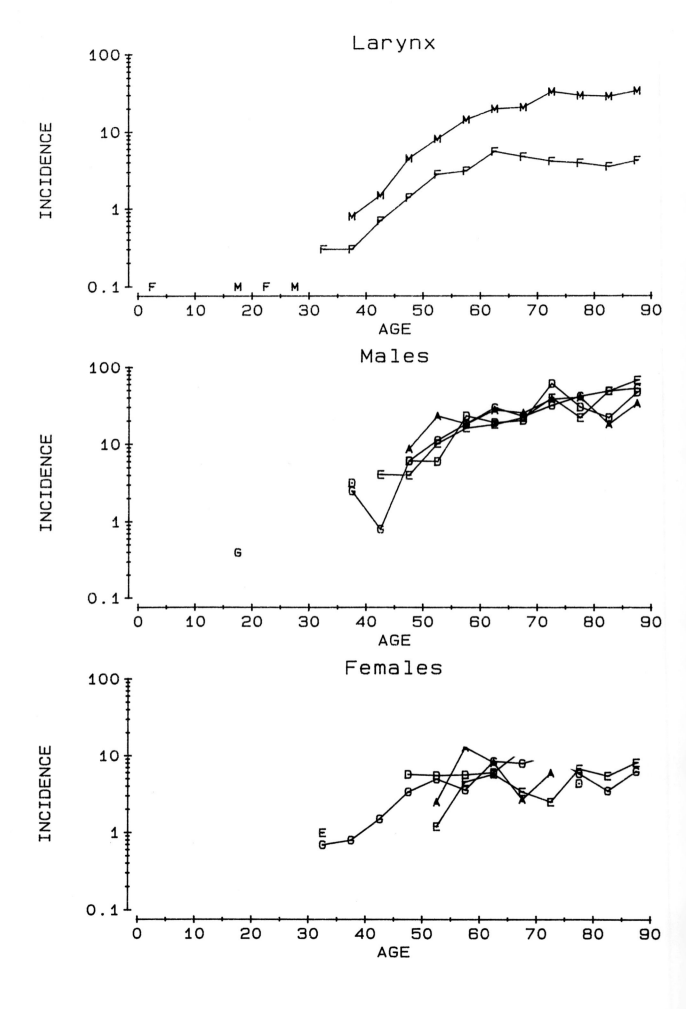

CANCER OF THE TRACHEA, BRONCHUS AND LUNG (ICD8 162)

Cancer of the trachea, bronchus and lung (hereafter termed 'lung cancer' for simplicity) is tragically common in Scotland, the incidence among males in the west of Scotland being higher than that recorded in any other white population (Waterhouse *et al.*, 1982). The average incidence among males (91.4) is higher than that among females (23.1); in people of both sexes the incidence has been increasing for 20 years.

Males (30.4%; 11.82%)

The all-Scottish male rate (91.4) is higher than that recorded in any other region of Britain; the next highest rates are found in Mersey (88.3) and Northwest (86.0). Scottish rates are higher than those recorded in the Nordic countries and elsewhere in Europe, e.g., Finland (74.4), Denmark (51.8), Norway (25.4), Sweden (23.8), Warsaw (68.4), Geneva (69.4) and Zaragoza (30.9). The incidence in Scotland is much higher than that recorded in North American white populations, such as those in Connecticut (60.9), Iowa (58.5), Ontario (53.3) and Alberta (36.0). The male Scottish rate is surpassed only by that of blacks in New Orleans (107.2), New Zealand Maoris (105.7) and native Hawaiians (96.2).

Among males, there is a three- to four-fold variation in incidence rates between the local authority districts. The lowest rate occurs in the Orkney Islands (40.2), followed closely by Skye and Lochalsh (40.7), Caithness (42.8), the Shetland Islands (46.1) and the Western Isles (46.3). The highest rate is found in Glasgow City (130.6) — considerably higher than the next ranking Inverclyde (109.9), Cumbernauld and Kilsyth (109.1), Clydebank (108.0), East Kilbride (104.3), Monklands (103.9), Edinburgh City (103.2), Dundee City (100.0) and Aberdeen City (98.5). The geographic pattern demonstrates clustering of high incidence rates in the region surrounding Glasgow, the remainder of the central area of Scotland, on Tayside and in Aberdeen City. It is also apparent that rates in the Isles and in the exclusively rural areas of north and northwest Scotland are particularly low.

Females (9.8%; 2.88%)

The Scottish female rate (23.1) is also extraordinarily high in national, continental and global terms. Among white populations, the Scottish rate is approached or passed only by rates in the sparsely populated Northwest territories and Yukon (34.7), Alameda County (24.1), New Orleans (23.0), Los Angeles (22.4), Seattle (22.2) and Mersey (19.4). Rates are higher among Hawaiians in Hawaii (40.6), New Zealand Maoris (48.8), Chinese in the San Francisco Bay Area (25.1), Chinese in Hong Kong (23.4) and New Orleans blacks (23.2).

In females, there is a six- to seven-fold variation in incidence. The lowest rate is seen in the Shetland Islands (5.6), followed by the Western Isles (7.0), Skye and Lochalsh (10.3), Sutherland (11.6) and Lochaber (11.6); the highest rate is recorded in Glasgow City (33.3), followed by Badenoch and Strathspey (31.8), Tweeddale (29.1), Dumbarton (28.4), Inverclyde (27.5) and Aberdeen City (26.4). The geographic pattern among females is quite similar to that in males, with high rates in the central belt, Tayside and Aberdeen. The pattern is not so clear-cut as that found in males, given the high rates reported from Tweeddale and, in particular, from Badenoch and Strathspey.

Scottish cities

Incidence rates in the four cities are uniformly high. In males, the highest rate is reported from Glasgow (130.6), followed by Edinburgh (103.2), Dundee (100.0) and Aberdeen (98.5). The rates in the cities other than Glasgow are remarkably similar; although lower than in Glasgow, they are nonetheless among the highest currently recorded (Waterhouse *et al.*, 1982).

For females, Glasgow (33.3) reports the highest lung cancer rate among the Scottish cities, followed by Aberdeen (26.4), Edinburgh (25.9) and Dundee (24.9). Again, the rates from Aberdeen, Edinburgh and Dundee are remarkably similar to one another, and, although considerably lower than in Glasgow, they are among the highest recorded in women for lung cancer.

Spatial aggregation

The strong visual impression of geographic clusters of high and low incidence districts is reflected statistically. The aggregation appears stronger among males ($D = 13.00$, $p < 0.0001$) than among females ($D = 15.73$, $p = 0.002$).

Comment

In broad terms, low rates of lung cancer are reported in rural areas and high rates in urban areas, defining these in terms of population density (Fig. 3.1). This pattern is stronger among males, the highest rates being found in the four cities and the central belt. The highest rate is reported from Glasgow, and the rates in contiguous areas are also extremely high. The Glasgow City rate (130.6) is much higher than the highest recorded in international publications — that among blacks in New Orleans (107.2) (Waterhouse *et al.*, 1982). If the level of lung cancer in Glasgow were lowered to that of New Orleans blacks, the number of new cases of lung cancer occurring among males in Glasgow City (male population 400 000) would be reduced by over 100 each year; abolition of cigarette smoking (see below) would reduce the number of premature deaths from this type of cancer by 440 each year.

The predominantly urban pattern observed in females is perturbed by the high rates reported from Tweeddale and, particularly, from Badenoch and Strathspey. This latter result is entirely unexpected. Examination of mortality rates from this district over the same period reveals highly variable standardized mortality ratios, ranging from 41 to 163 in individual years. While the finding of such a high rate should not be dismissed, it must be interpreted with great caution in view of the small size of the population and the low rates observed among the contiguous areas.

The high rates of lung cancer found in densely populated urban areas should not be interpreted to imply an association between urban living *per se* and an increasing risk of lung cancer. More likely, it reflects the pattern of cigarette usage found in large industrial conurbations. Cigarette smoking is perhaps the best understood, major, human cancer hazard. The evidence of a causal association between cigarette smoking and lung cancer is overwhelming and has recently been reviewed thoroughly (US Surgeon General, 1982; IARC, 1985). Most lung cancer, perhaps as much as 85% of the total incidence, is caused by smoking cigarettes and could be avoided by not smoking.

High levels of lung cancer have been reported in specific groups of workmen. The occupational mortality report for Scotland covering the five-year period around the 1971 census found *mortality* from lung cancer to be high in males in many of the occupations in Occupation Order V — furnace, forge foundry and rolling mill workers — and Occupation Order VII — engineering and allied trade workers (Kemp & Bledin, 1981). These occupations have been practised for over one hundred years in central Scotland, particularly in Glasgow and surrounding areas, which have now been shown to have the highest lung cancer risks. This report (Kemp & Bledin, 1981) also high-lighted excess rates in Scottish butchers, found similarly in England and Wales and other countries (Registrar General for England and Wales, 1978).

This excess, of course, could be caused mainly by high levels of cigarette consumption in these occupations, but it could also be due, to a greater or lesser extent, to the inhalation of noxious substances or dust produced at the place of work. This idea is reinforced by the finding of levels of lung cancer among these occupational groups that are considerably higher than those in the same social-class grouping taken as a whole. Several studies have shown that atmospheric pollution is not likely to increase lung cancer risk other than in smokers. There are, of course, other aesthetic and health reasons for reducing such pollution to a minimum. In view of the extraordinarily high levels of lung cancer in people of each sex in Glasgow, the recent establishment of an aggressive compaign to make Glasgow a non-smoking city by the year 2000 is to be thoroughly applauded.

Trachea, Bronchus and Lung — Males

□ 1

110
105

96
89 ———————————————— 91.4

65
59

48
40

D=13.00, p<0.0001

Trachea, Bronchus and Lung — Females

35
31
25
23.1
20
16
12
10
2

D=15.73, p=0.002

Trachea, Bronchus and Lung — Males

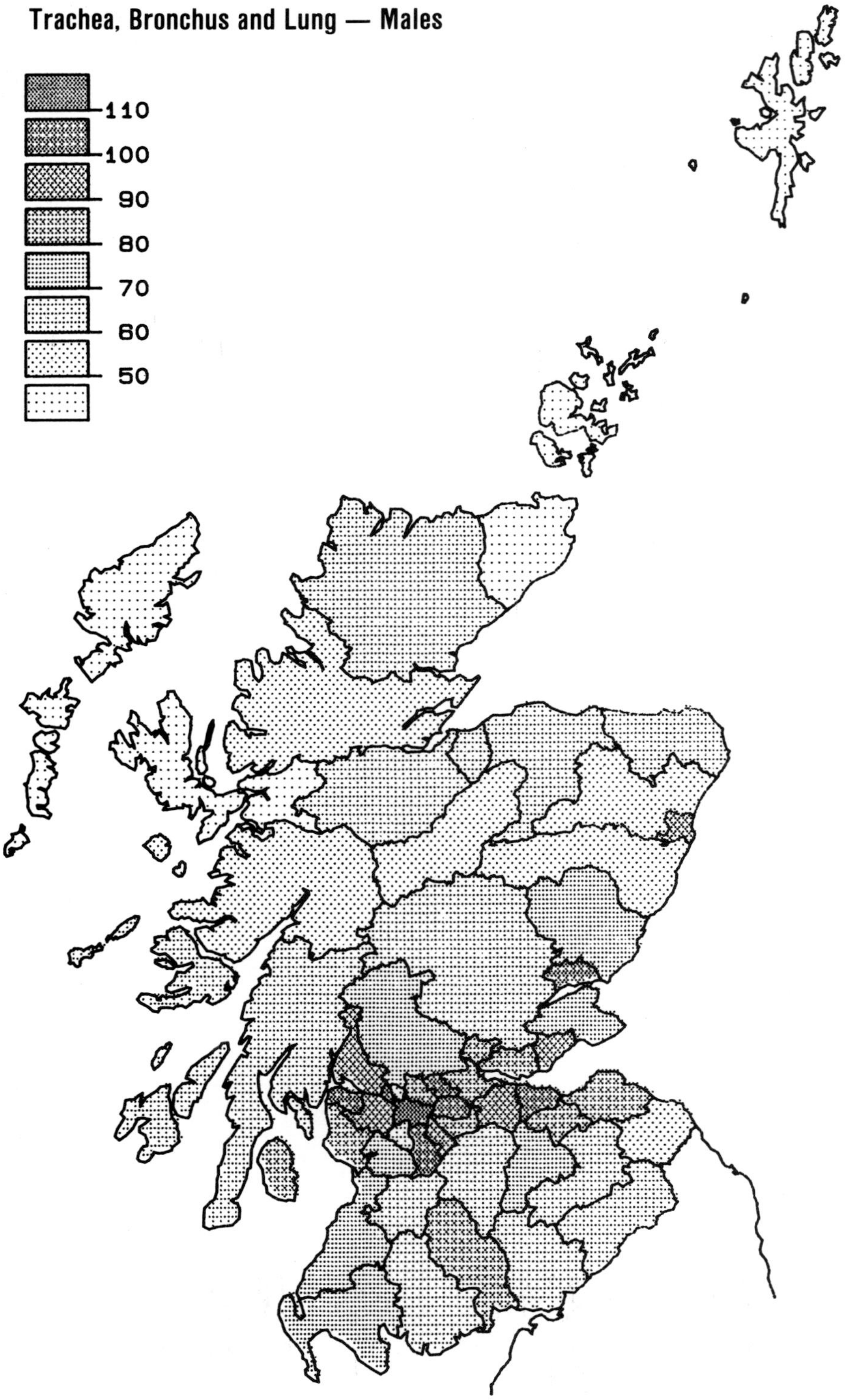

Legend (choropleth scale):
- 110
- 100
- 90
- 80
- 70
- 60
- 50

		CASES	C.R.	S.R.	S.E.	RK
1	CAITHNESS	48	57.7	42.8--	6.29	54
2	SUTHERLAND	38	101.3	64.9-	11.01	40
3	ROSS-CROMARTY	83	64.2	52.3--	5.87	51
4	SKYE-LOCHALSH	18	63.6	40.7--	10.27	55
5	LOCHABER	43	72.7	56.6--	8.71	46
6	INVERNESS	138	83.5	63.9--	5.55	42
7	BADENOCH	21	77.8	55.0--	12.56	47
8	NAIRN	29	96.7	66.2	13.24	35
9	MORAY	207	84.3	67.0--	4.76	39
10	BANFF-BUCHAN	230	99.4	69.3--	4.68	33
11	GORDON	125	75.5	52.8--	4.91	50
12	ABERDEEN	862	151.2	96.5+	3.40	9
13	KINCARDINE	94	84.2	53.4--	5.75	49
14	ANGUS	297	112.8	73.8--	4.41	27
15	DUNDEE	777	142.0	100.0+	3.65	8
16	PERTH-KINROSS	384	111.0	69.7--	3.67	31
17	KIRKCALDY	578	133.8	94.2	3.99	12
18	N-E FIFE	234	126.2	78.7--	5.33	24
19	DUNFERMLINE	432	114.0	86.6	4.24	15
20	WEST LOTHIAN	390	101.9	97.0	4.94	11
21	EDINBURGH	2067	162.1	103.2++	2.32	7
22	MIDLOTHIAN	232	82.9	83.3	5.61	16
23	EAST LOTHIAN	298	126.9	84.0	4.98	16
24	TWEEDDALE	49	126.6	73.6	11.03	28
25	ETTRICK	106	114.7	67.3--	6.68	36
26	ROXBURGH	118	114.9	67.9--	6.44	38
27	BERWICKSHIRE	53	102.5	54.7--	7.79	48
28	CLACKMANNAN	147	104.0	83.7	6.94	17
29	STIRLING	249	106.6	77.0--	5.00	26
30	FALKIRK	502	117.7	81.7--	3.71	20
31	ARGYLL-BUTE	196	102.7	61.0--	4.64	44
32	DUMBARTON	280	121.1	97.6	5.90	10
33	GLASGOW	4579	167.7	130.6++	2.01	1
34	CLYDEBANK	233	146.5	106.0+	7.15	4
35	BEARSDEN	82	74.1	56.5--	6.51	45
36	STRATHKELVIN	225	91.2	81.2	5.51	21
37	CUMBERNAULD	145	80.9	109.1	9.25	3
38	MONKLANDS	390	122.0	103.9+	5.32	6
39	MOTHERWELL	483	107.5	83.1-	3.86	19
40	HAMILTON	343	109.9	83.3	5.18	13
41	EAST KILBRIDE	220	89.1	104.3	7.25	5
42	EASTWOOD	156	106.6	69.3--	5.69	32
43	CLYDESDALE	137	83.6	61.6--	5.34	43
44	RENFREW	693	112.2	83.2	3.58	14
45	INVERCLYDE	438	147.9	109.9++	5.35	2
46	CUNNINGHAME	385	100.9	80.6--	4.11	22
47	KILMARNOCK	259	106.7	77.1--	4.85	25
48	KYLE-CARRICK	363	113.7	73.5--	3.94	29
49	CUMNOCK-DOON	129	92.7	67.6--	6.01	37
50	WIGTOWN	94	106.7	73.2--	7.72	30
51	STEWARTRY	73	111.5	63.9--	7.77	41
52	NITHSDALE	187	114.2	80.3	5.98	23
53	ANNANDALE	112	106.3	66.6--	6.84	34
54	ORKNEY	36	67.7	40.2--	6.97	56
55	SHETLAND	39	62.3	46.1--	7.70	53
56	WESTERN ISLES	63	71.7	46.3--	6.14	52
	ALL SCOTLAND	19239	126.4	91.4	0.67	

SCOTTISH CITIES

Trachea, Bronchus and Lung — Females

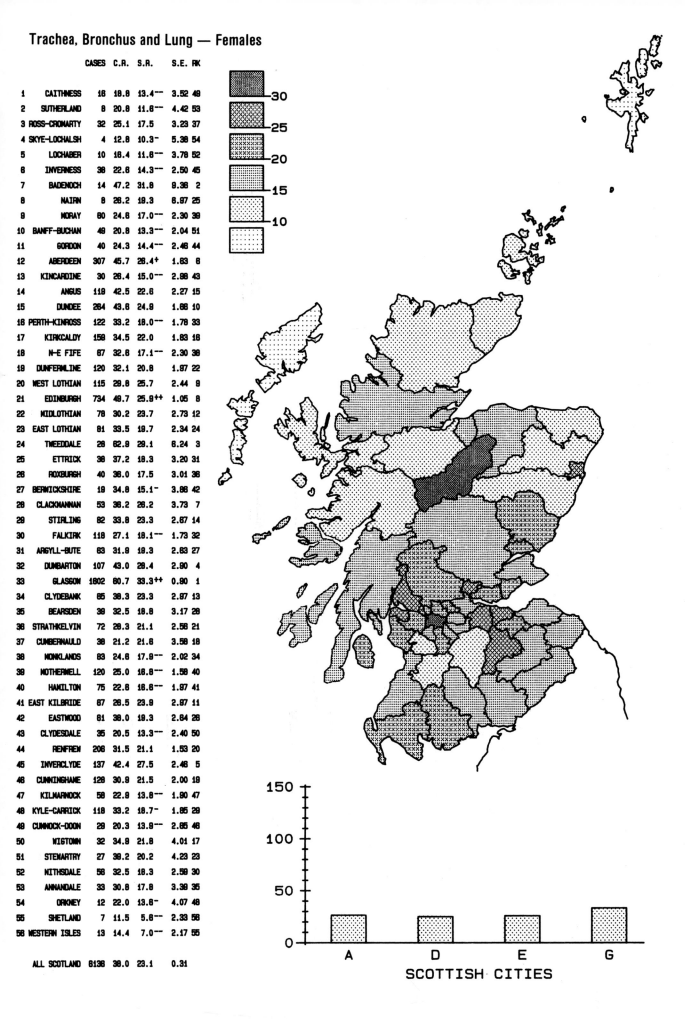

		CASES	C.R.	S.R.	S.E.	RK
1	CAITHNESS	16	18.8	13.4---	3.52	49
2	SUTHERLAND	8	20.8	11.6---	4.42	53
3	ROSS-CROMARTY	32	25.1	17.5	3.23	37
4	SKYE-LOCHALSH	4	12.8	10.3-	5.38	54
5	LOCHABER	10	18.4	11.6---	3.78	52
6	INVERNESS	38	22.8	14.3---	2.50	45
7	BADENOCH	14	47.2	31.8	9.36	2
8	NAIRN	8	26.2	19.3	6.97	25
9	MORAY	60	24.6	17.0---	2.30	39
10	BANFF-BUCHAN	49	20.8	13.3---	2.04	51
11	GORDON	40	24.3	14.4---	2.46	44
12	ABERDEEN	307	45.7	26.4+	1.63	6
13	KINCARDINE	30	26.4	15.0---	2.98	43
14	ANGUS	119	42.5	22.6	2.27	15
15	DUNDEE	264	43.6	24.8	1.66	10
16	PERTH-KINROSS	122	33.2	18.0---	1.78	33
17	KIRKCALDY	159	34.5	22.0	1.83	16
18	N-E FIFE	67	32.6	17.1---	2.30	38
19	DUNFERMLINE	120	32.1	20.8	1.97	22
20	WEST LOTHIAN	115	29.8	25.7	2.44	9
21	EDINBURGH	734	49.7	25.9++	1.05	8
22	MIDLOTHIAN	78	30.2	23.7	2.73	12
23	EAST LOTHIAN	81	33.5	19.7	2.34	24
24	TWEEDDALE	26	62.9	29.1	6.24	3
25	ETTRICK	38	37.2	18.3	3.20	31
26	ROXBURGH	40	36.0	17.5	3.01	36
27	BERWICKSHIRE	19	34.8	15.1-	3.86	42
28	CLACKMANNAN	53	36.2	26.2	3.73	7
29	STIRLING	82	33.8	23.3	2.67	14
30	FALKIRK	118	27.1	18.1---	1.73	32
31	ARGYLL-BUTE	83	31.9	19.3	2.63	27
32	DUMBARTON	107	43.0	26.4	2.90	4
33	GLASGOW	1602	60.7	33.3++	0.90	1
34	CLYDEBANK	65	38.3	23.3	2.97	13
35	BEARSDEN	39	32.5	18.8	3.17	28
36	STRATHKELVIN	72	28.3	21.1	2.56	21
37	CUMBERNAULD	38	21.2	21.6	3.56	18
38	MONKLANDS	83	24.6	17.9---	2.02	34
39	MOTHERWELL	120	25.0	18.8---	1.58	40
40	HAMILTON	75	22.8	16.6---	1.97	41
41	EAST KILBRIDE	67	26.5	23.9	2.97	11
42	EASTWOOD	81	38.0	19.3	2.64	26
43	CLYDESDALE	35	20.5	13.3---	2.40	50
44	RENFREW	206	31.5	21.1	1.53	20
45	INVERCLYDE	137	42.4	27.5	2.46	5
46	CUNNINGHAME	128	30.9	21.5	2.00	19
47	KILMARNOCK	58	22.9	13.8---	1.90	47
48	KYLE-CARRICK	118	33.2	18.7-	1.85	29
49	CUMNOCK-DOON	29	20.3	13.9---	2.85	46
50	WIGTOWN	32	34.9	21.8	4.01	17
51	STEWARTRY	27	38.2	20.2	4.23	23
52	NITHSDALE	56	32.5	18.3	2.59	30
53	ANNANDALE	33	30.6	17.8	3.38	35
54	ORKNEY	12	22.0	13.6-	4.07	48
55	SHETLAND	7	11.5	5.6---	2.33	56
56	WESTERN ISLES	13	14.4	7.0---	2.17	55
	ALL SCOTLAND	6136	38.0	23.1	0.31	

SCOTTISH CITIES

Trachea, Bronchus and Lung

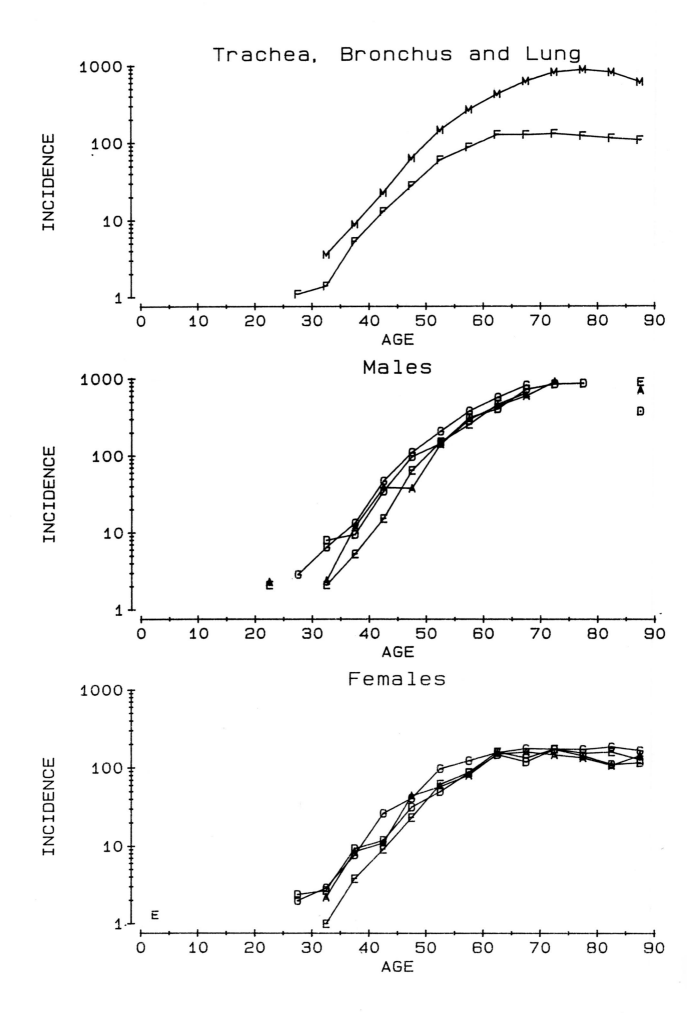

MALIGNANT MELANOMA OF THE SKIN
(ICD8 172)

Malignant melanoma of the skin is numerically not one of the major cancer problems of Scotland. However, the dramatic increase in incidence that is taking place in Scotland is causing concern, as are increasing incidence rates observed in many other countries.

Males (0.7%; 0.23%)

The Scottish male incidence rate (2.4) is lower than that in the Nordic countries but similar to that found in many regions of Britain; thus, Sweden (5.2), Denmark (4.6), Finland (3.9), Oxford (2.2), North-western (1.4) and Birmingham (1.3). By contrast, the rates in darker-skinned ethnic groups are much lower, e.g., Detroit blacks (0.6) and Japanese in Miyagi (0.4).

Within Scotland, the lowest rates are found in Tweeddale (0.0), the Shetland Islands (0.0), the Western Isles (0.4), Clackmannan (0.8) and Kincardine and Deeside (0.8); the highest rates are in Sutherland (7.7), Skye and Lochalsh (7.4), North-east Fife (5.4), Gordon (5.0) and Cumbernauld and Kilsyth (4.4). No clear geographical pattern is apparent, although it appears that more districts with higher rates are found in the north of the country.

Females (1.3%; 0.37%)

The Scottish female incidence rate (3.8) is lower than that of many regions of Europe, e.g., Denmark (6.2), Sweden (5.7) and Norway (9.1), although it is the same as in Finland (3.8). The rates in the English registries such as Oxford (3.7), Birmingham (2.5) and North-western (2.5) are slightly lower. The highest incidence rates are found in New Zealand non-Maoris (18.8), New South Wales (18.1), South Australia (17.6) and the white populations of Los Angeles (8.7) and Alameda County (8.2). In Detroit blacks and Miyagi, the rates were 0.6 and 0.4, respectively.

Within Scotland, the lowest incidence rates are reported from Nairn (0.0), Badenoch and Strathspey (0.0), the Shetland Islands (0.4), Berwickshire (1.2) and Cumbernauld and Kilsyth (1.2); the highest are in Skye and Lochalsh (8.5), Tweeddale (8.3), Nithsdale (7.4), Eastwood (6.9) and Inverness (6.8). As for males, no strong pattern is present, although, once again, more districts with high rates appear to be found in the North.

Scottish cities

Among females, the rate in Edinburgh (5.1) is higher than average (3.8), but those in Dundee (3.6),

Aberdeen (3.1) and Glasgow (2.5) are below. Among males, the highest rate recorded is in Dundee (3.7), with Edinburgh (3.2), Aberdeen (2.3) and Glasgow (1.5) recording lower incidence rates.

Spatial aggregation

There is no evidence of significant clustering of districts with similar risks of malignant melanoma in either males (D = 19.46, p = 0.658) or females (D = 18.30, p = 0.253).

Comment

Although a relatively rare cancer in comparison to cancers of the lung and breast, malignant melanoma of the skin has been the subject of much clinical and epidemiological research during the last decade. Epidemiological studies have documented a rapid rise in both the mortality and incidence of this form of cancer among white-skinned individuals in many countries (Jensen & Bolander, 1980), notably Norway, Sweden, Iceland, Finland and Denmark (Magnus, 1977), England and Wales and Australia (Holman et al., 1980), Canada (Elwood & Lee, 1974) and parts of the USA such as Connecticut (Houghton et al., 1980). Although these rising trends have been accompanied by a strong increase in clinical and pathological awareness, the increase cannot be attributed only to improvements in diagnostic practice (Truax et al., 1966; Lee & Carter, 1970; Cosman et al., 1976).

White populations living near the Equator have higher incidence rates than those living near the Poles (Scotto et al., 1976). European countries show complicated gradients. The general pattern is for high rates in northern countries and lower rates in southern countries (Jensen & Bolander, 1980). In Norway, however, there is a lower incidence in the north than in the south (Magnus, 1973). A similar gradient has been reported in Sweden (Eklund & Malec 1978), but not in Finland (Teppo et al., 1978).

Analysis of the time-trend data over lengthy periods has revealed that the change in melanoma rates can best be explained in terms of increasing rates over successive birth cohorts, i.e., a birth cohort effect (Holman et al., 1980; Magnus, 1981; Boyle et al., 1983b). The implications of this finding, allied to analyses by subsite, sex and occupational grouping, are that sunlight is the most important etiological factor. However, the association appears to be more complex than that which relates exposure to sunlight to non-melanoma skin cancers, for which total dose appears to be the relevant factor (Urbach, 1978).

Temporal changes in the incidence of malignant melanoma have been related to changes in the ozone layer, atmospheric pollution and sunspot activity (Wigle, 1978; Leach et al, 1979). The higher incidence in females suggests that occupation is not a major factor. The only reported association between malignant melanoma and occupation is an increased risk in those working in offices with fluorescent lighting (Beral et al., 1982), but this report remains to be confirmed. There is agreement, however, that lack of ability to tan, the presence of freckles (Pack et al., 1963; Klepp & Magnus, 1979) and lack of skin pigment (Lancaster, 1956; Gellin et al., 1969) are associated with an increased risk of this cancer. In Australia, risk has been related to the number of naevi (Green et al., 1985), and, in one form of this disease, risk was much lower in those migrating to Australia after the age of 20 when compared to those born in Australia (Holman & Armstrong, 1984).

In the east of Scotland and in the north, the skin of Scots is characteristically lacking in pigment and is frequently freckled. The other characteristic of the race, easily recognized in any crowd in the east of Scotland, is the frequent occurrence of red hair, the melanin of which has recently been shown to be mutagenic (Harsanyi et al., 1980).

There is need for a great deal of further information about incidence, in particular, the location of the tumours on the body surface and whether the different pathological types have the same or different geographical distribution. Since Scotland has the highest overall incidence in the UK, it would be appropriate if further detailed analysis of geographical incidence were undertaken.

Malignant Melanoma of Skin — Males

D=19.46, p=0.658

Malignant Melanoma of Skin — Females

9.0
7.8
5.8
4.9
3.2
2.0
0.7
0.0

3.8

D=18.30, p=0.253

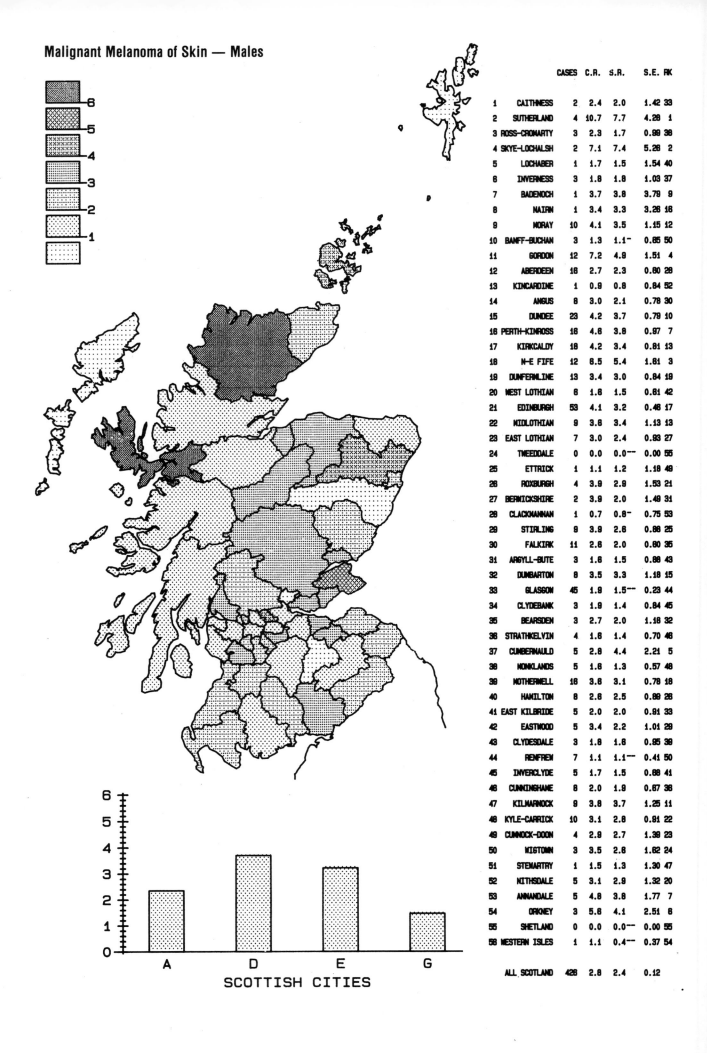

Malignant Melanoma of Skin — Males

		CASES	C.R.	S.R.	S.E.	RK
1	CAITHNESS	2	2.4	2.0	1.42	33
2	SUTHERLAND	4	10.7	7.7	4.28	1
3	ROSS-CROMARTY	3	2.3	1.7	0.99	38
4	SKYE-LOCHALSH	2	7.1	7.4	5.26	2
5	LOCHABER	1	1.7	1.5	1.54	40
6	INVERNESS	3	1.8	1.8	1.03	37
7	BADENOCH	1	3.7	3.8	3.79	9
8	NAIRN	1	3.4	3.3	3.26	16
9	MORAY	10	4.1	3.5	1.15	12
10	BANFF-BUCHAN	3	1.3	1.1⁻	0.65	50
11	GORDON	12	7.2	4.9	1.51	4
12	ABERDEEN	16	2.7	2.3	0.60	28
13	KINCARDINE	1	0.9	0.8	0.84	52
14	ANGUS	8	3.0	2.1	0.78	30
15	DUNDEE	23	4.2	3.7	0.79	10
16	PERTH-KINROSS	16	4.6	3.6	0.97	7
17	KIRKCALDY	18	4.2	3.4	0.81	13
18	N-E FIFE	12	6.5	5.4	1.61	3
19	DUNFERMLINE	13	3.4	3.0	0.84	19
20	WEST LOTHIAN	8	1.6	1.5	0.61	42
21	EDINBURGH	53	4.1	3.2	0.46	17
22	MIDLOTHIAN	9	3.6	3.4	1.13	13
23	EAST LOTHIAN	7	3.0	2.4	0.93	27
24	TWEEDDALE	0	0.0	0.0⁻⁻	0.00	55
25	ETTRICK	1	1.1	1.2	1.18	49
26	ROXBURGH	4	3.9	2.8	1.53	21
27	BERWICKSHIRE	2	3.9	2.0	1.49	31
28	CLACKMANNAN	1	0.7	0.8⁻	0.75	53
29	STIRLING	9	3.9	2.6	0.88	25
30	FALKIRK	11	2.6	2.0	0.60	35
31	ARGYLL-BUTE	3	1.6	1.5	0.88	43
32	DUMBARTON	8	3.5	3.3	1.18	15
33	GLASGOW	45	1.9	1.5⁻⁻	0.23	44
34	CLYDEBANK	3	1.9	1.4	0.84	45
35	BEARSDEN	3	2.7	2.0	1.18	32
36	STRATHKELVIN	4	1.6	1.4	0.70	46
37	CUMBERNAULD	5	2.8	4.4	2.21	5
38	MONKLANDS	5	1.6	1.3	0.57	48
39	MOTHERWELL	16	3.6	3.1	0.78	18
40	HAMILTON	8	2.6	2.5	0.89	26
41	EAST KILBRIDE	5	2.0	2.0	0.91	33
42	EASTWOOD	5	3.4	2.2	1.01	29
43	CLYDESDALE	3	1.6	1.6	0.95	39
44	RENFREW	7	1.1	1.1⁻⁻	0.41	50
45	INVERCLYDE	5	1.7	1.5	0.68	41
46	CUNNINGHAME	8	2.0	1.9	0.67	36
47	KILMARNOCK	9	3.8	3.7	1.25	11
48	KYLE-CARRICK	10	3.1	2.8	0.91	22
49	CUMNOCK-DOON	4	2.9	2.7	1.39	23
50	WIGTOWN	3	3.5	2.6	1.62	24
51	STEWARTRY	1	1.5	1.3	1.30	47
52	NITHSDALE	5	3.1	2.9	1.32	20
53	ANNANDALE	5	4.6	3.6	1.77	7
54	ORKNEY	3	5.8	4.1	2.51	8
55	SHETLAND	0	0.0	0.0⁻⁻	0.00	55
56	WESTERN ISLES	1	1.1	0.4⁻⁻	0.37	54
	ALL SCOTLAND	426	2.8	2.4	0.12	

SCOTTISH CITIES

Malignant Melanoma of Skin — Females

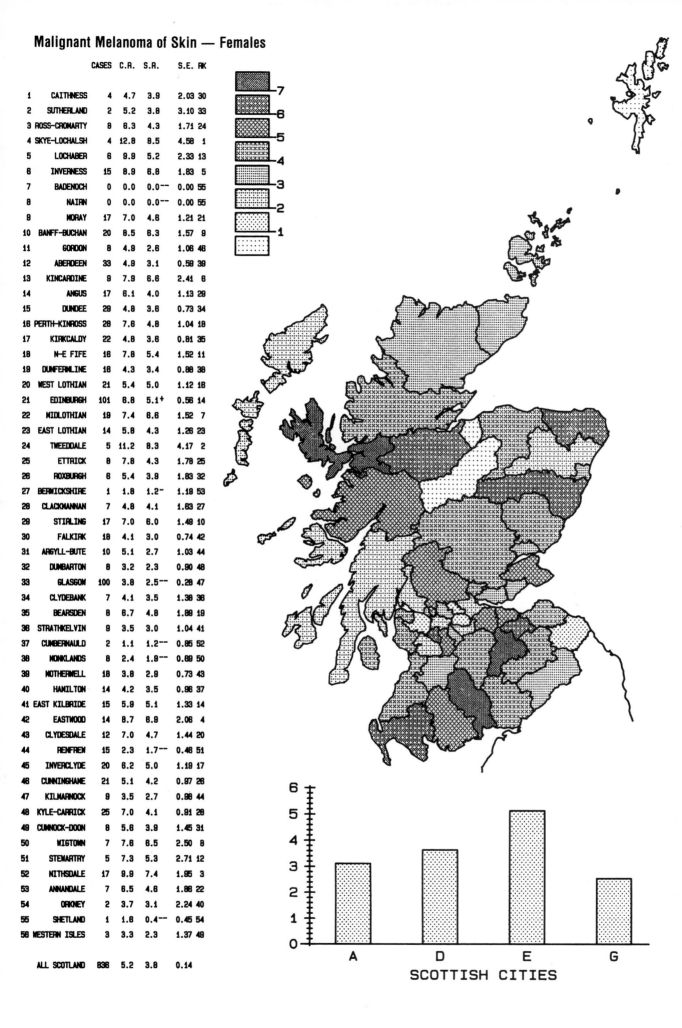

		CASES	C.R.	S.R.	S.E.	RK
1	CAITHNESS	4	4.7	3.9	2.03	30
2	SUTHERLAND	2	5.2	3.8	3.10	33
3	ROSS-CROMARTY	8	8.3	4.3	1.71	24
4	SKYE-LOCHALSH	4	12.8	8.5	4.58	1
5	LOCHABER	8	9.9	5.2	2.33	13
6	INVERNESS	15	8.9	6.8	1.83	5
7	BADENOCH	0	0.0	0.0--	0.00	55
8	NAIRN	0	0.0	0.0--	0.00	55
9	MORAY	17	7.0	4.6	1.21	21
10	BANFF-BUCHAN	20	8.5	6.3	1.57	9
11	GORDON	8	4.9	2.6	1.06	46
12	ABERDEEN	33	4.9	3.1	0.59	39
13	KINCARDINE	9	7.9	6.6	2.41	6
14	ANGUS	17	6.1	4.0	1.13	29
15	DUNDEE	29	4.8	3.6	0.73	34
16	PERTH-KINROSS	28	7.8	4.8	1.04	18
17	KIRKCALDY	22	4.8	3.6	0.81	35
18	N-E FIFE	18	7.8	5.4	1.52	11
19	DUNFERMLINE	18	4.3	3.4	0.88	38
20	WEST LOTHIAN	21	5.4	5.0	1.12	16
21	EDINBURGH	101	6.8	5.1+	0.56	14
22	MIDLOTHIAN	19	7.4	6.6	1.52	7
23	EAST LOTHIAN	14	5.8	4.3	1.26	23
24	TWEEDDALE	5	11.2	8.3	4.17	2
25	ETTRICK	8	7.8	4.3	1.78	25
26	ROXBURGH	8	5.4	3.9	1.63	32
27	BERWICKSHIRE	1	1.8	1.2-	1.19	53
28	CLACKMANNAN	7	4.8	4.1	1.63	27
29	STIRLING	17	7.0	6.0	1.49	10
30	FALKIRK	18	4.1	3.0	0.74	42
31	ARGYLL-BUTE	10	5.1	2.7	1.03	44
32	DUMBARTON	8	3.2	2.3	0.90	48
33	GLASGOW	100	3.8	2.5--	0.26	47
34	CLYDEBANK	7	4.1	3.5	1.38	36
35	BEARSDEN	8	6.7	4.8	1.89	19
36	STRATHKELVIN	9	3.5	3.0	1.04	41
37	CUMBERNAULD	2	1.1	1.2--	0.85	52
38	MONKLANDS	8	2.4	1.9--	0.69	50
39	MOTHERWELL	18	3.8	2.9	0.73	43
40	HAMILTON	14	4.2	3.5	0.96	37
41	EAST KILBRIDE	15	5.9	5.1	1.33	14
42	EASTWOOD	14	8.7	6.9	2.06	4
43	CLYDESDALE	12	7.0	4.7	1.44	20
44	RENFREW	15	2.3	1.7--	0.46	51
45	INVERCLYDE	20	6.2	5.0	1.19	17
46	CUNNINGHAME	21	5.1	4.2	0.97	26
47	KILMARNOCK	9	3.5	2.7	0.98	44
48	KYLE-CARRICK	25	7.0	4.1	0.81	28
49	CUMNOCK-DOON	8	5.6	3.9	1.45	31
50	WIGTOWN	7	7.8	6.5	2.50	8
51	STEWARTRY	5	7.3	5.3	2.71	12
52	NITHSDALE	17	9.9	7.4	1.95	3
53	ANNANDALE	7	6.5	4.6	1.86	22
54	ORKNEY	2	3.7	3.1	2.24	40
55	SHETLAND	1	1.6	0.4--	0.45	54
56	WESTERN ISLES	3	3.3	2.3	1.37	49
	ALL SCOTLAND	836	5.2	3.8	0.14	

SCOTTISH CITIES

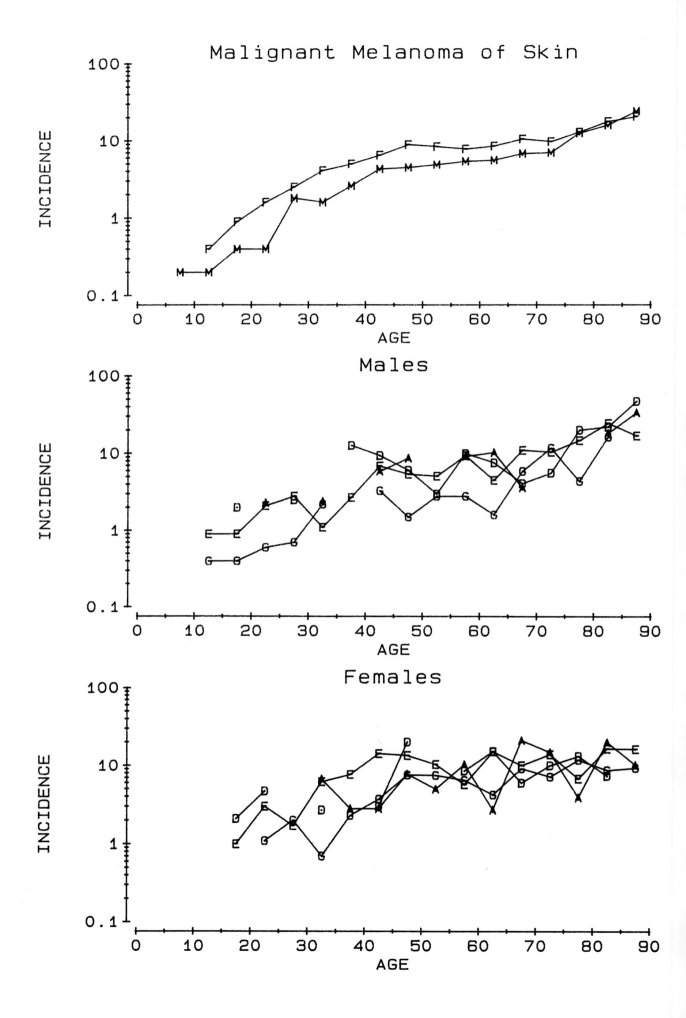

Malignant Melanoma of Skin

In Scotland, there is a four-fold variation between areas of the lowest (4.8) and highest (20.6) incidence rates. The lowest rates are found in Tweeddale (4.8), Strathkelvin (6.0), Stewartry (6.8) and Berwickshire (7.1); the highest are reported from Sutherland (20.6), the Shetland Islands (16.1), Ettrick and Lauderdale (15.1) and East Lothian (14.8).

There is an apparent aggregation of higher than average rates around the mouth of the River Forth, and extending north. These high rates commence on the southern bank in West Lothian (13.1) and Falkirk (12.9), then continue into Clackmannan (11.9), Kirkcaldy (12.3), Dunfermline (11.7) and Perth and Kinross (13.2), before extending further north into Lochaber (13.9), Badenoch and Strathspey (12.9), Nairn (12.8), Kincardine and Deeside (12.6) and Aberdeen City (14.6). There is a suggestion, also, of an aggregation of lower than average rates around the central area of southern Scotland, extending from Tweeddale (4.8) westwards through Clydesdale (11.2), Cumnock and Doon Valley (7.1) and southwards through Annandale and Eskdale (9.2), Nithsdale (8.8) to Stewartry (6.8). Further, there appears to be another grouping of low rates extending over large areas of the west of Scotland, centred on Glasgow and covering districts both north and south of the Firth of Clyde.

Scottish cities

Aberdeen (14.6) has the highest incidence rate of ovarian cancer of all Scottish cities. The rates in Edinburgh (10.7), Glasgow (10.2) and Dundee (10.0) are at, or slightly below, the all-Scottish rate (10.7).

Spatial aggregation

Although there appears to be pockets of high and low rates, this is not confirmed statistically ($D = 18.11$, $p = 0.200$) over the whole country.

Comment

The incidence of ovarian cancer has increased slightly but consistently in Scotland over the past 20 years. The geographical pattern of ovarian cancer suggests that aggregations of districts with higher than average incidence rates are centred on the Firth of Forth, and aggregations of lower than average rates are centred around the Firth of Clyde. So little is known about the possible etiology of this form of cancer that such spatial and temporal variations cannot easily be explained.

Women with ovarian cancer generally report having had fewer pregnancies and greater difficulty in becoming pregnant than other women, and death rates from this cancer are inversely correlated with the average number of children (Beral et al., 1978). Second and subsequent pregnancies appear to confer additional protection against ovarian cancer (Casagrande et al., 1979), an effect which cannot be attributed to any possible confounding effect of age at first birth.

Most evidence relating dietary factors to ovarian cancer is indirect. Ovarian and breast cancer incidence and mortality rates usually follow one another. The cancers share common reproductive risk factors; hence, by induction, if breast cancer risk is related to dietary fat intake, ovarian cancer risk should be similarly related. There is a little direct evidence associating dietary fat intake with an increased risk of ovarian cancer (National Academy of Sciences, 1982), but it is inconclusive.

Ovarian cancer, therefore, poses a problem: it is common, difficult to diagnose, causes death frequently, and incidence varies from east to west over central Scotland. With little or no knowledge of the etiology available, it would be important to uncover the reasons underlying the variation seen in Scotland.

Breast — Females

73
68
65
59
57.3
54
52
49
45

D=19.02, p=0.499

Ovary, Broad Ligament, Fallopian Tube

D=18.11, p=0.200

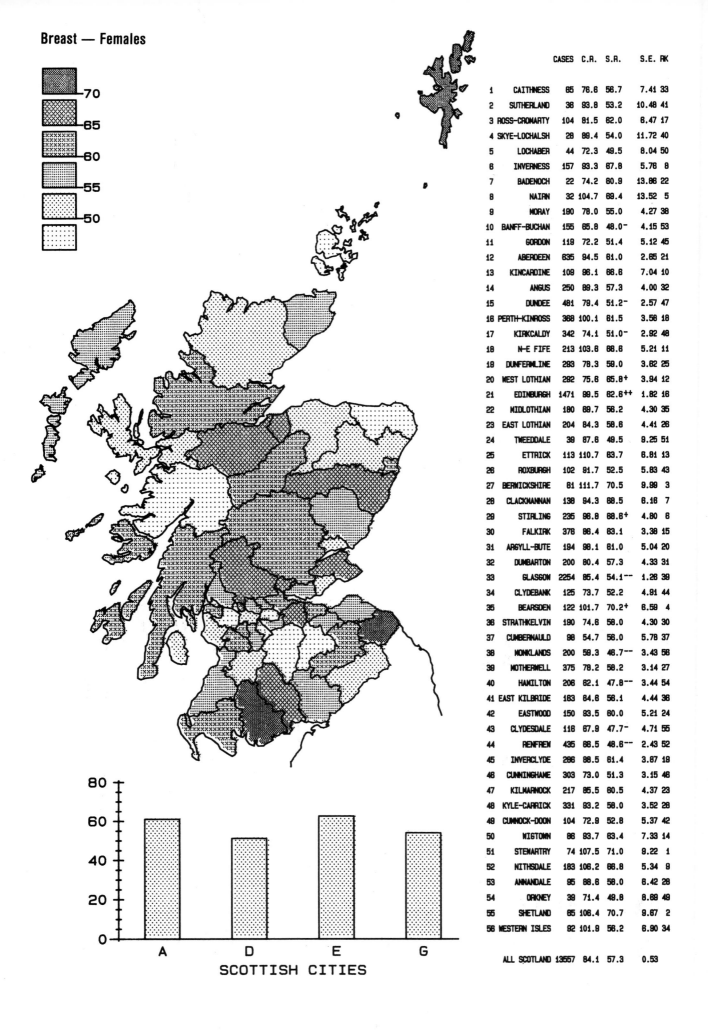

Breast — Females

		CASES	C.R.	S.R.	S.E.	RK
1	CAITHNESS	65	76.6	58.7	7.41	33
2	SUTHERLAND	38	93.8	53.2	10.48	41
3	ROSS-CROMARTY	104	81.5	62.0	6.47	17
4	SKYE-LOCHALSH	28	89.4	54.0	11.72	40
5	LOCHABER	44	72.3	49.5	8.04	50
6	INVERNESS	157	83.3	67.8	5.76	8
7	BADENOCH	22	74.2	60.9	13.86	22
8	NAIRN	32	104.7	69.4	13.52	5
9	MORAY	190	78.0	55.0	4.27	38
10	BANFF-BUCHAN	155	65.8	48.0⁻	4.15	53
11	GORDON	118	72.2	51.4	5.12	45
12	ABERDEEN	635	94.5	61.0	2.65	21
13	KINCARDINE	109	96.1	66.6	7.04	10
14	ANGUS	250	88.3	57.3	4.00	32
15	DUNDEE	481	79.4	51.2⁻	2.57	47
16	PERTH-KINROSS	368	100.1	61.5	3.58	18
17	KIRKCALDY	342	74.1	51.0⁻	2.92	48
18	N-E FIFE	213	103.6	66.6	5.21	11
19	DUNFERMLINE	293	78.3	59.0	3.62	25
20	WEST LOTHIAN	292	75.6	65.8⁺	3.94	12
21	EDINBURGH	1471	99.5	62.6⁺⁺	1.82	16
22	MIDLOTHIAN	180	69.7	56.2	4.30	35
23	EAST LOTHIAN	204	84.3	58.6	4.41	26
24	TWEEDDALE	39	87.6	49.5	9.25	51
25	ETTRICK	113	110.7	63.7	6.81	13
26	ROXBURGH	102	91.7	52.5	5.83	43
27	BERWICKSHIRE	61	111.7	70.5	9.99	3
28	CLACKMANNAN	138	94.3	66.5	6.16	7
29	STIRLING	235	96.8	66.6⁺	4.80	6
30	FALKIRK	376	88.4	63.1	3.38	15
31	ARGYLL-BUTE	194	98.1	61.0	5.04	20
32	DUMBARTON	200	80.4	57.3	4.33	31
33	GLASGOW	2254	85.4	54.1⁻⁻	1.26	39
34	CLYDEBANK	125	73.7	52.2	4.91	44
35	BEARSDEN	122	101.7	70.2⁺	6.59	4
36	STRATHKELVIN	190	74.6	58.0	4.30	30
37	CUMBERNAULD	98	54.7	58.0	5.78	37
38	MONKLANDS	200	59.3	48.7⁻⁻	3.43	56
39	MOTHERWELL	375	78.2	58.2	3.14	27
40	HAMILTON	206	82.1	47.8⁻⁻	3.44	54
41	EAST KILBRIDE	183	84.6	56.1	4.44	36
42	EASTWOOD	150	93.5	60.0	5.21	24
43	CLYDESDALE	118	67.9	47.7⁻	4.71	55
44	RENFREW	435	68.5	48.6⁻⁻	2.43	52
45	INVERCLYDE	266	88.5	61.4	3.87	19
46	CUNNINGHAME	303	73.0	51.3	3.15	46
47	KILMARNOCK	217	85.5	60.5	4.37	23
48	KYLE-CARRICK	331	93.2	58.0	3.52	28
49	CUMNOCK-DOON	104	72.9	52.8	5.37	42
50	WIGTOWN	86	93.7	63.4	7.33	14
51	STEWARTRY	74	107.5	71.0	9.22	1
52	NITHSDALE	183	106.2	66.8	5.34	9
53	ANNANDALE	95	88.6	58.0	6.42	29
54	ORKNEY	39	71.4	49.8	8.69	49
55	SHETLAND	65	106.4	70.7	9.67	2
56	WESTERN ISLES	82	101.9	58.2	6.90	34
	ALL SCOTLAND	13557	84.1	57.3	0.53	

SCOTTISH CITIES

Ovary, Broad Ligament, Fallopian Tube

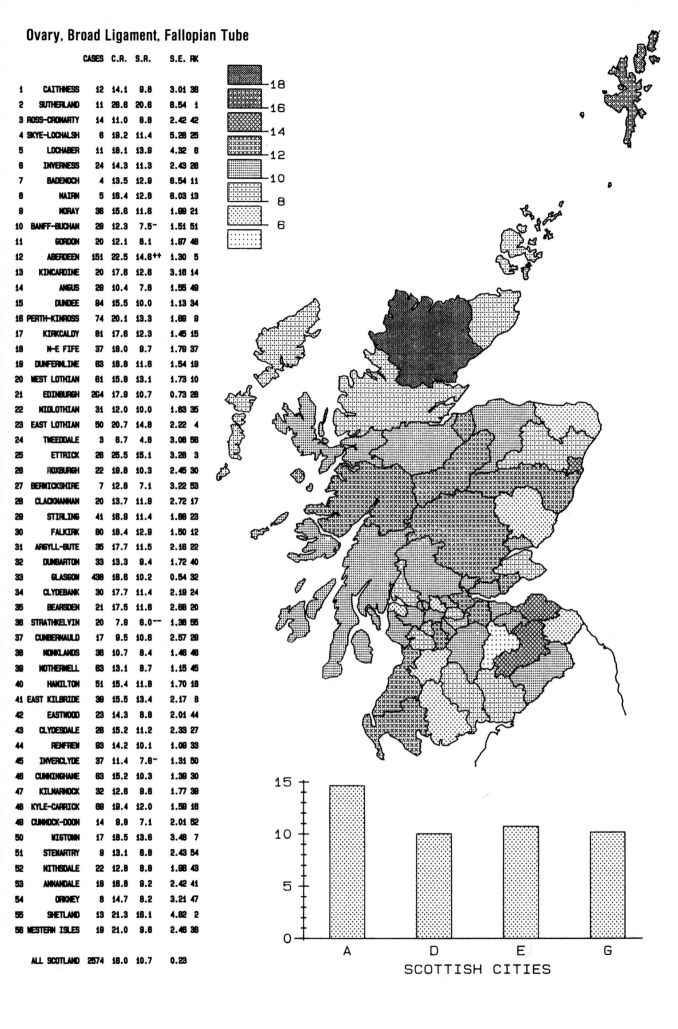

		CASES	C.R.	S.R.	S.E.	RK
1	CAITHNESS	12	14.1	9.8	3.01	36
2	SUTHERLAND	11	26.8	20.6	6.54	1
3	ROSS-CROMARTY	14	11.0	8.8	2.42	42
4	SKYE-LOCHALSH	6	19.2	11.4	5.26	25
5	LOCHABER	11	18.1	13.8	4.32	6
6	INVERNESS	24	14.3	11.3	2.43	26
7	BADENOCH	4	13.5	12.8	6.54	11
8	NAIRN	5	16.4	12.8	6.03	13
9	MORAY	38	15.8	11.6	1.99	21
10	BANFF-BUCHAN	29	12.3	7.5⁻	1.51	51
11	GORDON	20	12.1	8.1	1.97	48
12	ABERDEEN	151	22.5	14.6++	1.30	5
13	KINCARDINE	20	17.6	12.6	3.16	14
14	ANGUS	29	10.4	7.8	1.55	49
15	DUNDEE	94	15.5	10.0	1.13	34
16	PERTH-KINROSS	74	20.1	13.3	1.69	9
17	KIRKCALDY	81	17.8	12.3	1.45	15
18	N-E FIFE	37	18.0	9.7	1.79	37
19	DUNFERMLINE	63	16.6	11.6	1.54	19
20	WEST LOTHIAN	61	15.8	13.1	1.73	10
21	EDINBURGH	264	17.9	10.7	0.73	28
22	MIDLOTHIAN	31	12.0	10.0	1.83	35
23	EAST LOTHIAN	50	20.7	14.8	2.22	4
24	TWEEDDALE	3	6.7	4.8	3.06	56
25	ETTRICK	26	25.5	15.1	3.28	3
26	ROXBURGH	22	19.8	10.3	2.45	30
27	BERWICKSHIRE	7	12.6	7.1	3.22	53
28	CLACKMANNAN	20	13.7	11.9	2.72	17
29	STIRLING	41	16.9	11.4	1.86	23
30	FALKIRK	80	18.4	12.9	1.50	12
31	ARGYLL-BUTE	35	17.7	11.5	2.16	22
32	DUMBARTON	33	13.3	9.4	1.72	40
33	GLASGOW	438	16.8	10.2	0.54	32
34	CLYDEBANK	30	17.7	11.4	2.19	24
35	BEARSDEN	21	17.5	11.6	2.68	20
36	STRATHKELVIN	20	7.8	6.0⁻⁻	1.38	55
37	CUMBERNAULD	17	9.5	10.6	2.57	29
38	MONKLANDS	36	10.7	8.4	1.46	46
39	MOTHERWELL	63	13.1	8.7	1.15	45
40	HAMILTON	51	15.4	11.6	1.70	18
41	EAST KILBRIDE	38	15.5	13.4	2.17	8
42	EASTWOOD	23	14.3	8.8	2.01	44
43	CLYDESDALE	26	15.2	11.2	2.33	27
44	RENFREW	83	14.2	10.1	1.09	33
45	INVERCLYDE	37	11.4	7.6⁻	1.31	50
46	CUNNINGHAME	63	15.2	10.3	1.39	30
47	KILMARNOCK	32	12.8	9.6	1.77	39
48	KYLE-CARRICK	68	18.4	12.0	1.59	16
49	CUMNOCK-DOON	14	9.9	7.1	2.01	52
50	WIGTOWN	17	18.5	13.6	3.48	7
51	STEWARTRY	9	13.1	6.8	2.43	54
52	NITHSDALE	22	12.8	8.8	1.96	43
53	ANNANDALE	18	16.8	9.2	2.42	41
54	ORKNEY	8	14.7	8.2	3.21	47
55	SHETLAND	13	21.3	16.1	4.82	2
56	WESTERN ISLES	19	21.0	9.6	2.46	38
	ALL SCOTLAND	2574	16.0	10.7	0.23	

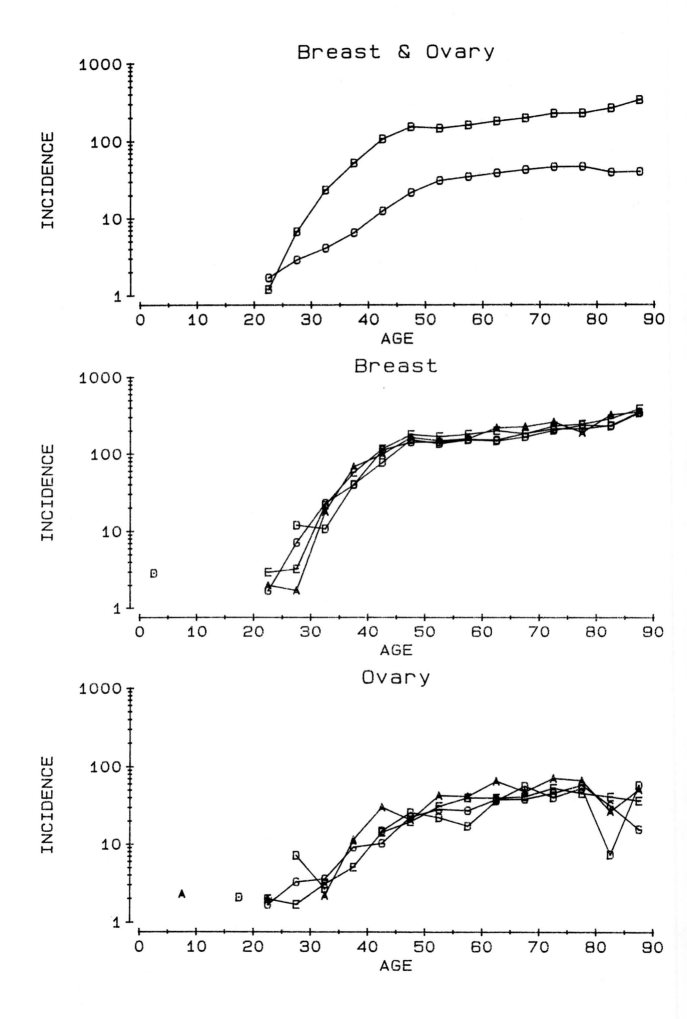

CANCER OF THE UTERINE CERVIX
(ICD8 180)

CANCER OF THE UTERINE CORPUS
(ICD8 182.0)

CANCER OF THE UTERINE CERVIX

The rubric cancer of the uterine cervix covers invasive cancer of the cervix, one of the commonest forms of cancer in women. It specifically excludes non-invasive cervical cancer, i.e., carcinoma *in situ* of the cervix.

Females (3.8%; 1.18%)

The all-Scottish incidence rate (11.3) is fairly typical of rates reported from around Britain; it is higher than the rates in South Thames (9.0) and Oxford (10.6) but lower than those in Birmingham (12.0) and Mersey (13.2). In Europe, rates are lower in Finland (8.5), but higher in Sweden (12.7), Norway (19.0) and Denmark (23.0). Incidence rates in Scotland are fairly similar to those in North American regions such as Alberta (9.1), Detroit (10.5) and Iowa (11.9), but lower than that in Ontario (19.6).

Within Scotland, the lowest rates are found in Skye and Lochalsh (3.6), Annandale and Eskdale (3.7), Bearsden and Milngavie (4.3) and Badenoch and Strathspey (4.9); the highest rates are reported in Perth and Kinross (30.4), Sutherland (21.8), Tweeddale (18.4), the Orkney Islands (17.6) and Caithness (17.2).

There are three patterns of possible significance in the geographical distribution of cervical cancer in Scotland. First, there is a collection of districts in the north-west with low rates: the Western Isles (7.7), Skye and Lochalsh (3.6), Lochaber (8.9), Ross and Cromarty (9.1), Inverness (10.2), Nairn (6.0), Badenoch and Strathspey (4.9). Second, there appears to be a set of areas with lower than average rates in and around Glasgow; and, third, there is an apparent aggregation of higher than average rates in Kincardine and Deeside (15.4), Perth and Kinross (30.4), Dundee City (12.0), Stirling (12.8), Clackmannan (14.2), Kirkcaldy (15.4) and Dunfermline (13.7).

Scottish cities

The lowest incidence of cervical cancer occurs in Aberdeen (9.5); those in Edinburgh (11.6), Glasgow (11.7) and Dundee (12.0) are higher. The age-specific rates in Aberdeen are consistently lower than those in the other three cities in women up to age 65.

Spatial aggregation

Although in small regions low or high rates appear together, there is only weak statistical evidence (D = 17.13, $p = 0.045$) that the pattern is non-random.

Comment

The general pattern of cervical cancer portrayed on the map is that of rural districts with both high and low incidence, levels in the north-west and west being slightly lower than average, and there being an aggregation of high rates in Fife and Perth. Cervical cancer probably has many contributing causes (Cramer, 1982), and their relative importance in different populations leads to complex geographical patterns of incidence.

Women who marry early are at an increased risk of cervical cancer (Lombard & Potter, 1950; Wynder *et al.*, 1954; Rotkin, 1966). The earlier the age at first coitus (Rotkin, 1967) and the greater the number of sexual partners (often reflected in statistics of multiple marriages, separations and divorces), the greater the risk.

It is commonly believed that the upbringing of young people in the more remote Highland and Island areas in Scotland has been somewhat stricter than elsewhere. If this were so, promiscuity could be expected to be less common, and the age at first coitus would tend to be higher. The age at which marriage occurs may be higher, also, in these areas, because of the relative scarcity of young men. Again, this would tend to retard the age of initial coitus. The proportion of women unmarried on attaining a specific age is known to be higher in the Western Isles than in Scotland as a whole. At the time of the 1981 census, 61.6% of women between 20 and 24 years of age in the Western Isles were unmarried, compared to the all-Scottish rate of 52.3%. Corresponding figures for women of 25-29 were 22.6% and 18.5%, and for 30-34 years of age, 14.3% and 8.9%, respectively (1981 census). There is a high proportion of Roman Catholics in the population of the west of Scotland, both in the industrialized areas and in certain of the rural areas, such as the Western Isles. Multiple marriages, separations and divorces tend, therefore, to occur at a lower rate in these areas. All of these factors might contribute to a lower level of cervical cancer.

Cervical cancer is commoner in lower socio-economic groups, which are found more usually in the west, so that the risk should be greater there. Women who smoke are at increased risk, as are users of oral contraceptive pills. Conversely, barrier contraceptives seem to reduce the risk, possibly by mechanical protection of the cervix.

The incidence of invasive cervical cancer can be influenced by a cytology screening scheme; screening has been carried out in Scotland for many years but the system does not operate evenly throughout the country. There is likely to be some bias toward the upper social classes, to the younger age groups and to areas with more highly developed medical services. One question worth considering is whether some of the higher incidence of cervical cancer recorded in several rural areas is related to inaccessibility of screening services. Yet, rates in other inaccessible rural areas are low. Finally, the somewhat lower incidence noted in Aberdeen may be related to the intensive screening programme that has been conducted in that city for many years.

Cervical cancer is believed by some to be caused by a virus (Durst *et al.*, 1983)

Features of the distribution of cervical cancer in Scotland merit further study (see also uterine corpus, below), particularly the elevated rate in Perth and Kinross (30.4) including the possibility of differences in classification.

CANCER OF THE UTERINE CORPUS

Cancer of the body of the uterus - the uterine corpus - generally arises from the lining or endometrium. The incidence of this form of cancer has been decreasing in Scotland for over a decade.

Females (2.6%; 0.77%)

The all-Scottish incidence rate (6.4) is lower than the rates in other parts of Britain, such as Mersey (8.1), Trent (9.0), Birmingham (9.4) and Oxford (10.7). Only North-western (5.4) in Britain reports a lower incidence. Rates are higher in most European countries, e.g., Finland (10.9), Norway (11.0), Denmark (13.0) and Sweden (13.0) and even higher in other parts of Europe such as Geneva (16.2), Bas-Rhin (15.9) and Varese (14.2), and in comparable regions of North America such as Ontario (14.2), Manitoba (19.2), Alberta (19.8), Iowa (21.0), Connecticut (21.3) and Detroit (24.6).

Within the Scottish districts, the lowest rates are reported from Dumbarton (3.0), Kilmarnock and Loudon (3.3), Clackmannan (3.4), Motherwell (3.7), Cunninghame (3.9) and Renfrew (4.0); the highest rates are found in Badenoch and Strathspey (17.6), Ross and Cromarty (12.6), Inverness (12.6), Nithsdale (12.3) and the Western Isles (11.6). The map clearly shows the juxtaposition of very high rates in Badenoch and Strathspey (17.6), Inverness (12.6) and Ross and Cromarty (12.6) alongside Lochaber (11.0) and the Western Isles (11.6). There is a smaller pocket of districts with high rates in the south-east of Scotland. However, the most remarkable feature of this map is the large aggregation of low rates centred around Glasgow in west-central Scotland.

Scottish cities

The highest rates are found in Aberdeen (8.2) and Edinburgh (7.9), the incidences in Dundee (5.4) and Glasgow (5.2) being lower than the average rate (6.4).

Spatial aggregation

The visual pattern is confirmed by strong statistical evidence of clustering of districts with similar rates of cancer of the uterine corpus (D = 15.80, p = 0.003).

Comment

The incidence of cancer of the uterine corpus, in international terms, is low in Scotland, and has been decreasing slowly since the mid-1960s. It is particularly low in the west of Scotland, in the area surrounding Glasgow. A similar pattern of low rates of cancer of the uterine cervix in Glasgow and the surrounding districts deserves further attention, as some of the risk factors for these two forms of cancer act in opposite directions. For example, child-bearing increases the risk of cervical cancer but decreases that of corpus cancer. The possibility of under-registration of gynaecological tumours in the west merits consideration; yet ovarian cancer occurs at the same rate as in Edinburgh.

Risk factors for cancer of the uterine corpus resemble those for breast cancer, although the evidence is based on fewer epidemiological studies. The proportion of nulliparous women who develop endometrial cancer is higher than that of women who bear children (Elwood et al., 1977). Patients with endometrial cancer tend to be of a higher social class, have fewer children, have an early menarche and a late menopause (Wynder et al., 1966), and tend to be overweight around the menopause (de Waard, 1973). Use of sequential-type oral contraceptives, also, may increase the risk (Silverberg et al., 1977), although this association remains unproven.

Links between cancer of the endometrium and diet are mainly indirect, being the results of correlations between levels of endometrial cancer and other cancers, such as those of the breast and colorectum that are also thought to have a dietary etiology (National Academy of Sciences, 1982). The more direct evidence necessary to establish causation with any degree of certainty is sparse. Armstrong and Doll (1975) found an association between endometrial cancer and per-caput consumption of total fat; and two case-control studies reported obesity as a risk factor (Wynder et al., 1966; Elwood et al., 1977).

The recent increases in the incidence of endometrial cancer in the USA and Canada, which have produced such extraordinarily high incidence rates, do not have a dietary etiology. The cause has been shown clearly to be the use of exogenous oestrogens at the time of menopause (Jick et al., 1979), a practice not nearly so widespread in Scotland as in North America.

Uterine Cervix

D=17.13, p=0.045

Uterine Corpus

13.0
11.9
10.3
8.0
6.4
5.1
4.3
3.4
2.0

1

D=15.80, p=0.003

Uterine Cervix

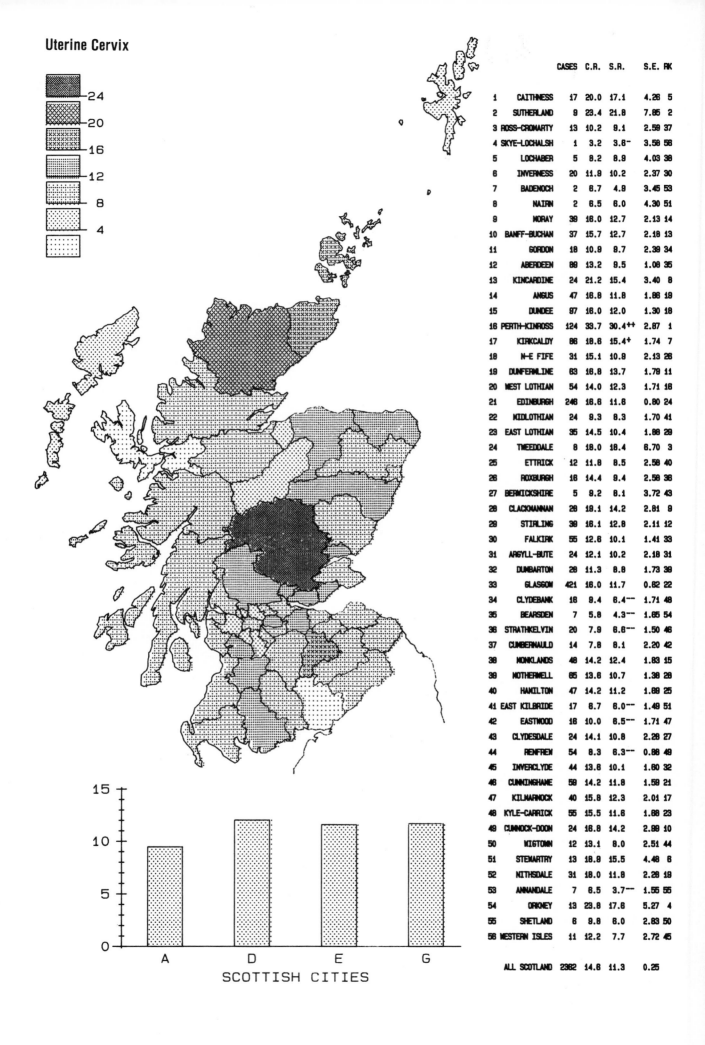

		CASES	C.R.	S.R.	S.E.	RK
1	CAITHNESS	17	20.0	17.1	4.26	5
2	SUTHERLAND	9	23.4	21.8	7.85	2
3	ROSS-CROMARTY	13	10.2	9.1	2.59	37
4	SKYE-LOCHALSH	1	3.2	3.6⁻	3.58	56
5	LOCHABER	5	8.2	8.9	4.03	38
6	INVERNESS	20	11.9	10.2	2.37	30
7	BADENOCH	2	6.7	4.9	3.45	53
8	NAIRN	2	6.5	6.0	4.30	51
9	MORAY	39	16.0	12.7	2.13	14
10	BANFF-BUCHAN	37	15.7	12.7	2.18	13
11	GORDON	18	10.9	9.7	2.39	34
12	ABERDEEN	89	13.2	9.5	1.06	35
13	KINCARDINE	24	21.2	15.4	3.40	8
14	ANGUS	47	16.8	11.8	1.88	19
15	DUNDEE	97	16.0	12.0	1.30	18
16	PERTH-KINROSS	124	33.7	30.4⁺⁺	2.87	1
17	KIRKCALDY	86	18.6	15.4⁺	1.74	7
18	N-E FIFE	31	15.1	10.8	2.13	26
19	DUNFERMLINE	63	16.8	13.7	1.79	11
20	WEST LOTHIAN	54	14.0	12.3	1.71	16
21	EDINBURGH	246	16.6	11.8	0.80	24
22	MIDLOTHIAN	24	9.3	8.3	1.70	41
23	EAST LOTHIAN	35	14.5	10.4	1.88	29
24	TWEEDDALE	8	16.0	16.4	6.70	3
25	ETTRICK	12	11.8	8.5	2.56	40
26	ROXBURGH	16	14.4	9.4	2.58	36
27	BERWICKSHIRE	5	9.2	8.1	3.72	43
28	CLACKMANNAN	26	19.1	14.2	2.81	9
29	STIRLING	38	16.1	12.8	2.11	12
30	FALKIRK	55	12.6	10.1	1.41	33
31	ARGYLL-BUTE	24	12.1	10.2	2.18	31
32	DUMBARTON	26	11.3	8.8	1.73	39
33	GLASGOW	421	16.0	11.7	0.62	22
34	CLYDEBANK	16	9.4	8.4⁻⁻	1.71	48
35	BEARSDEN	7	5.6	4.3⁻⁻	1.65	54
36	STRATHKELVIN	20	7.9	6.6⁻⁻	1.50	46
37	CUMBERNAULD	14	7.6	8.1	2.20	42
38	MONKLANDS	48	14.2	12.4	1.83	15
39	MOTHERWELL	65	13.6	10.7	1.38	28
40	HAMILTON	47	14.2	11.2	1.69	25
41	EAST KILBRIDE	17	8.7	6.0⁻⁻	1.49	51
42	EASTWOOD	16	10.0	6.5⁻⁻	1.71	47
43	CLYDESDALE	24	14.1	10.6	2.26	27
44	RENFREW	54	8.3	6.3⁻⁻	0.86	49
45	INVERCLYDE	44	13.6	10.1	1.60	32
46	CUNNINGHAME	59	14.2	11.8	1.59	21
47	KILMARNOCK	40	15.8	12.3	2.01	17
48	KYLE-CARRICK	55	15.5	11.6	1.68	23
49	CUMNOCK-DOON	24	16.8	14.2	2.99	10
50	WIGTOWN	12	13.1	8.0	2.51	44
51	STEWARTRY	13	18.9	15.5	4.48	6
52	NITHSDALE	31	16.0	11.8	2.26	18
53	ANNANDALE	7	6.5	3.7⁻⁻	1.55	55
54	ORKNEY	13	23.8	17.6	5.27	4
55	SHETLAND	6	9.8	6.0	2.63	50
56	WESTERN ISLES	11	12.2	7.7	2.72	45
	ALL SCOTLAND	2362	14.6	11.3	0.25	

Uterine Corpus

		CASES	C.R.	S.R.		S.E.	RK
1	CAITHNESS	11	13.0	9.5		2.93	10
2	SUTHERLAND	4	10.4	5.1		2.90	44
3	ROSS-CROMARTY	21	18.5	12.6 +		2.90	2
4	SKYE-LOCHALSH	4	12.8	7.3		4.05	21
5	LOCHABER	11	18.1	11.1		3.61	7
6	INVERNESS	31	18.4	12.6 +		2.45	3
7	BADENOCH	8	27.0	17.6		7.11	1
8	NAIRN	4	13.1	6.6		3.85	27
9	MORAY	26	10.7	6.6		1.40	27
10	BANFF-BUCHAN	32	13.6	8.7		1.64	15
11	GORDON	18	9.7	6.2		1.70	29
12	ABERDEEN	90	13.4	8.2 +		0.91	17
13	KINCARDINE	11	9.7	5.8		1.90	34
14	ANGUS	24	8.6	5.2		1.15	41
15	DUNDEE	59	9.7	5.4		0.76	38
16	PERTH-KINROSS	42	11.4	7.4		1.22	20
17	KIRKCALDY	54	11.7	7.1		1.02	23
18	N-E FIFE	33	16.0	9.4		1.81	11
19	DUNFERMLINE	45	12.0	7.8		1.20	19
20	WEST LOTHIAN	53	13.7	11.5 ++		1.62	6
21	EDINBURGH	194	13.1	7.9 +		0.82	18
22	MIDLOTHIAN	19	7.4	5.5		1.28	37
23	EAST LOTHIAN	42	17.4	10.5 +		1.76	8
24	TWEEDDALE	7	15.7	8.5		3.60	16
25	ETTRICK	15	14.7	8.8		2.43	14
26	ROXBURGH	18	16.2	9.0		2.49	13
27	BERWICKSHIRE	5	9.2	5.2		2.47	41
28	CLACKMANNAN	8	5.5	3.4 -		1.25	54
29	STIRLING	27	11.1	7.0		1.41	24
30	FALKIRK	43	8.9	6.2		0.99	30
31	ARGYLL-BUTE	18	9.1	4.9		1.25	46
32	DUMBARTON	12	4.6	3.0 --		0.90	56
33	GLASGOW	239	9.1	5.2 --		0.37	43
34	CLYDEBANK	15	8.8	6.1		1.87	31
35	BEARSDEN	9	7.5	5.4		1.84	39
36	STRATHKELVIN	21	8.2	6.9		1.50	26
37	CUMBERNAULD	7	3.9	4.2		1.60	49
38	MONKLANDS	18	5.3	4.1 -		0.98	50
39	MOTHERWELL	26	5.4	3.7 --		0.75	53
40	HAMILTON	33	10.0	7.3		1.30	22
41	EAST KILBRIDE	18	8.3	5.4		1.36	39
42	EASTWOOD	18	10.0	5.6		1.49	36
43	CLYDESDALE	17	10.0	5.9		1.55	33
44	RENFREW	40	6.1	4.0 --		0.66	51
45	INVERCLYDE	24	7.4	4.6		0.98	48
46	CUNNINGHAME	25	6.0	3.9 --		0.82	52
47	KILMARNOCK	12	4.7	3.3 --		0.98	55
48	KYLE-CARRICK	32	9.0	5.7		1.05	35
49	CUMNOCK-DOON	13	9.1	6.0		1.73	32
50	WIGTOWN	12	13.1	6.9		2.17	25
51	STEWARTRY	10	14.5	9.1		3.14	12
52	NITHSDALE	34	19.7	12.3 ++		2.20	4
53	ANNANDALE	9	8.4	4.6		1.59	47
54	ORKNEY	12	22.0	9.9		3.39	9
55	SHETLAND	5	8.2	5.1		2.47	45
56	WESTERN ISLES	17	18.8	11.8		3.21	5
	ALL SCOTLAND	1649	10.2	6.4		0.17	

Legend: 14, 12, 10, 8, 6, 4

SCOTTISH CITIES

15

10

5

0

A   D   E   G

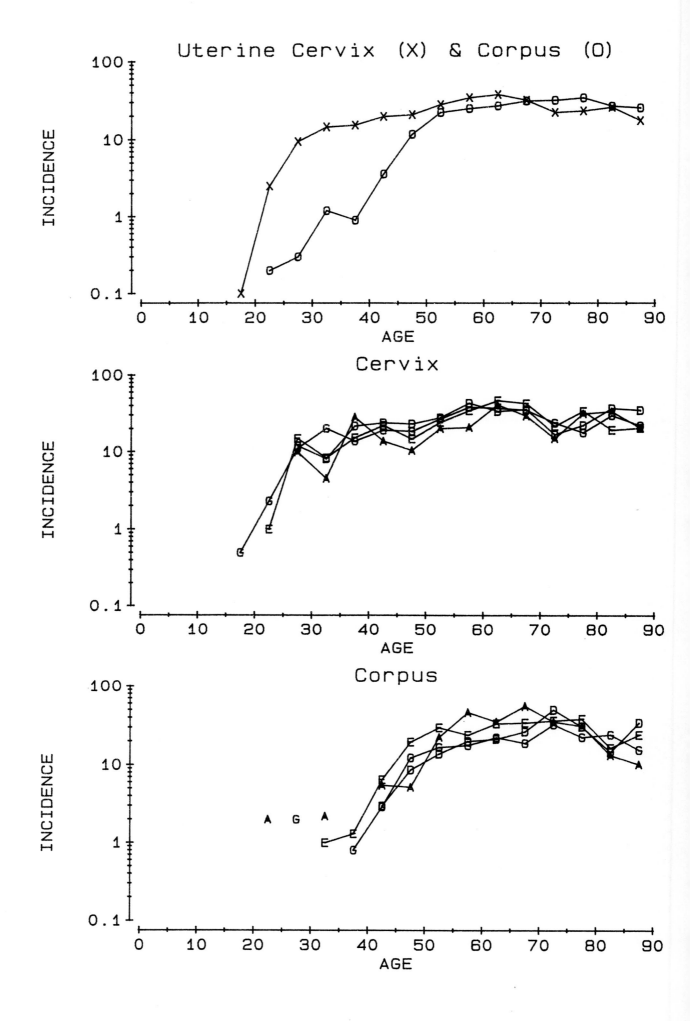

CANCER OF THE PROSTATE
(ICD8 185)

CANCER OF THE TESTIS
(ICD8 186)

CANCER OF THE PROSTATE

The incidence of cancer of the prostate gland has increased in Scotland throughout the period covered by cancer registration (1959 onwards). This numerically important cancer, which accounts for 7.6% of all male cancers in Scotland, generally affects old men, but is becoming increasingly common in younger males.

Males (7.6%; 2.27%)

In comparison with a number of countries, the incidence in Scotland (21.9) is moderately high. Prostatic cancer is particularly common in North America, e.g., Saskatchewan (46.2), Manitoba (43.2), San Francisco Bay Area (white, 47.4; black, 92.2), Los Angeles (white, 44.3, black, 79.1) and Connecticut (42.7). Closer to Scotland, rates in Norway (38.9) and Sweden (44.4) are higher, while rates in Finland (27.2) and Denmark (23.6) are nearer to the Scottish and English rates, e.g., Birmingham (18.6), South Thames (20.1) and Oxford (20.7). Rates in Japan and in Chinese populations are very low (around 5).

The range of incidence in Scotland is wide, from 12.9 to 38.4. High rates are found in Badenoch and Strathspey (38.4), Tweeddale (36.1), Midlothian (35.6), Stewartry (30.4) and Moray (29.6); and low rates are recorded in Caithness (12.9), Argyll and Bute (14.7), the Shetland Islands (14.9), Clydebank (15.0) and Hamilton (15.6). The incidence of prostatic cancer is higher in the north-east of Scotland and in some areas stretching south and west. A group of lower rates occurs in and around Glasgow, stretching south in both easterly and westerly directions.

Scottish cities

In the cities, the rate in Aberdeen (26.4) is highest, followed by those in Edinburgh (25.4) and Dundee (23.0); while that in Glasgow (17.8) is below the all-Scottish incidence level.

Spatial aggregation

The visual pattern is supported by statistical evidence of clustering (D = 15.77, p = 0.002) of districts with similar levels of prostatic cancer.

Comment

Even though it is one of the commonest cancers in males (in Sweden, the commonest), this form of malignancy has not been widely investigated (Zaridze et al., 1984). No convincing explanation has yet been suggested for the two-fold differences in risk between black and white populations in the USA. While sexual practices may influence risk by modulating the levels of male sex and other hormones, studies of these factors are difficult to conduct, and the answers obtained may not always be reliable. It has been suggested that marriage is a protective factor.

Microscopic examination of prostate glands removed at surgery for other prostatic disease, or at death, very commonly reveals very small areas of completely unsuspected cancer. These so-called latent cancers are so frequent that it is highly unlikely that more than a small proportion would grow sufficiently to cause symptoms. Therefore, current belief is that there may be several causal agents — one or more to start the cancer process and others which favour its progression from the latent form to clinically overt disease.

The rapid increase in risk observed in Japanese and Chinese who migrate to the USA is consistent with an effect of change in diet. Moreover, risk has recently increased in Japan, where diet is altering. The incidence of cancer of the prostate correlates with that of other cancers believed to be diet-associated (Berg, 1975). Therefore it is interesting to note similar patterns in Scotland of the risk for both prostatic and colorectal cancer, the high rates of each of these cancers in the north-east being particularly striking.

CANCER OF THE TESTIS

The incidence of germ-cell cancer of the testis has been reported to be increasing in many areas, particularly in men under the age of 45. Testicular cancer, although relatively uncommon, is the commonest malignancy among men between the ages of 20 and 34. There has been an unequivocal improvement in prognosis in recent years, although the incidence is increasing in Scotland.

Males (0.9%; 0.30%)

The incidence of testicular cancer is generally higher in Europe than elsewhere, with the incidence in Scotland (3.7) surpassed by rates from Vaud (9.9), Neuchatel (9.1), Denmark (6.7), Hamburg (5.4), Geneva (5.1) and Norway (4.4) (Waterhouse et al., 1982). Isolated higher rates occur in North America, e.g., the Northwest Territories and the Yukon (5.5), Utah (4.7) and Seattle (4.3). Incidence rates in the five Scottish registries fall in the upper half of the rates recorded worldwide.

There is an 11-fold variation in incidence in Scotland, ranging from 1.3 to 14.4. The highest incidence occurs in Berwickshire (14.4), Eastwood (8.3), Tweeddale (8.0), Stewartry (7.6) and the Orkney Islands (7.6), and the lowest rates in Falkirk (1.3), the Shetland Islands (1.5), Cumbernauld and Kilsyth (2.0), Wigtown (2.2) and Monklands (2.4). There is no overall pattern in testicular cancer incidence rates; low and high rates are found equally in urban and rural areas. The geographical pattern shows two regions of higher incidence, although these are somewhat blurred. One appears to extend from East Lothian across the Borders as far as the eastern part of the Strathclyde and Dumfries and Galloway. A second band of high incidence rates occurs in the southern Grampian and Highland regions.

Scottish cities

Among the four cities, Aberdeen (4.8) and Edinburgh (4.5) have higher than the Scottish average rate (3.7); while the incidence rate in Dundee (2.6) and Glasgow (2.5) is lower.

Spatial aggregation

There is only a weak suggestion of spatial aggregation of districts with similar levels of testicular cancer ($D = 17.14$, $p = 0.046$).

Comment

Worldwide, the descriptive epidemiology of testicular cancer is characterized by an overall increase in incidence and a fall in the peak of the age-incidence curve (Clemmesen, 1969; Davies, 1981). In other words, the rise occurs in younger men. In Scotland, the incidence of testicular cancer has increased by approximately 50% over the last two decades (Boyle et al., 1985b); this rise is unlikely to be due to improvements in diagnosis or registration. The peak of the age-incidence of the disease has fallen, and it is predominantly a disease of young men; in the age group 20-34, testicular cancer is now the commonest malignancy. These two trends are specific to the histological sub-type teratoma. Neither the overall incidence nor the peak age of pure seminoma has altered significantly over the last decades. This suggests that if a change in etiology has taken place, the malignancy now manifests itself more frequently at an earlier age in the form of teratoma. Several pathological studies suggest that the germ cells within the testis are capable, following malignant change, of expressing a range of microscopic features depending on the age of the bearer (Pugh, 1976; Brawn, 1983). In younger men, germ cells retain totipotentiality, i.e., may develop into a varied mixture of cell types grouped together as malignant teratoma. With advancing age, germ cells appear to lose this totipotentiality and, upon malignant change, develop into the more uniform histological pattern of seminoma.

The presence of an undescended testis is known to increase the risk of this cancer; approximately 10% of testicular cancers occur in patients with cryptorchidism (Henderson et al., 1979). A genetic component to testicular cancer has been reported in identical twin brothers (Villani, 1967), non-twin brothers (Kademian & Caldwell, 1976) and fathers and sons (Raghavan et al., 1980). In contrast to the risks of cancers of the stomach and lung, those of testicular cancer are higher in upper social classes, the ratio of risk in the highest class being 2.5 times that in the lowest (Schottenfeld & Warshauer, 1982). More recently, it has been claimed that those who engage regularly in a sporting activity such as cycling or horse-riding (Coldman et al., 1982) are at increased risk for testicular seminoma. Undescended testis, social class and selected sports do not account for more than a fraction of testicular cancer, however, and, without further knowledge of cause, prospects for prevention remain poor. Today, the best hope for a reduction in levels of mortality lies in improvements in treatment, some of which have taken place (Einhorn, 1981) and have reduced mortality in Scotland even though incidence rates continue to rise (Boyle et al, 1985b).

Prostate

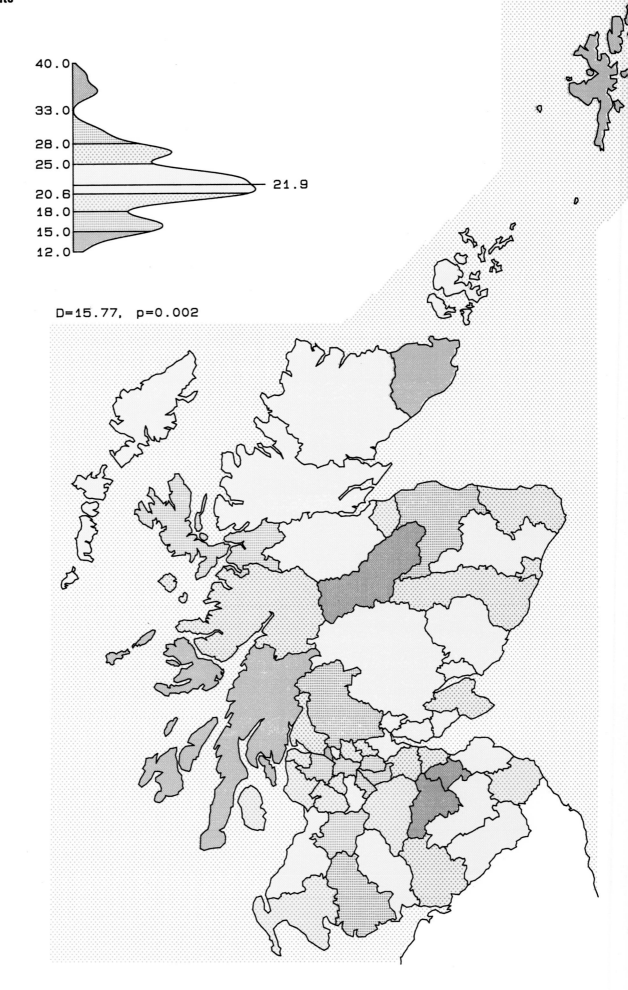

Testis

D=17.14, p=0.046

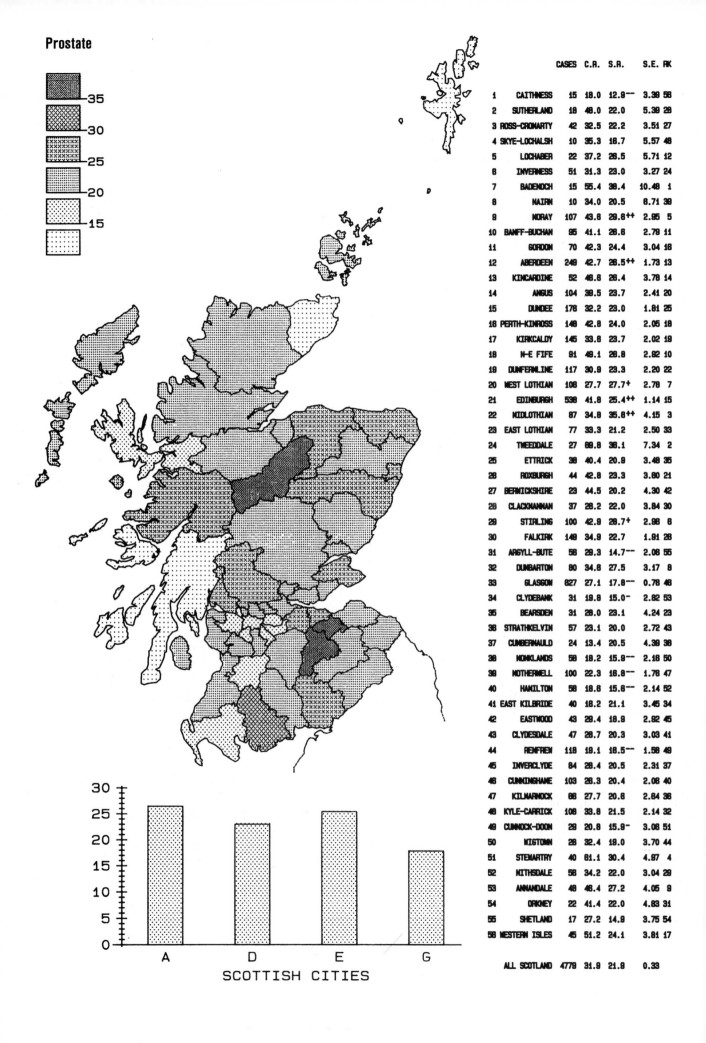

Prostate

		CASES	C.R.	S.R.	S.E.	RK
1	CAITHNESS	15	18.0	12.9 ̄ ̄	3.39	56
2	SUTHERLAND	18	48.0	22.0	5.39	29
3	ROSS-CROMARTY	42	32.5	22.2	3.51	27
4	SKYE-LOCHALSH	10	35.3	18.7	5.57	48
5	LOCHABER	22	37.2	26.5	5.71	12
6	INVERNESS	51	31.3	23.0	3.27	24
7	BADENOCH	15	55.4	38.4	10.48	1
8	NAIRN	10	34.0	20.5	6.71	39
9	MORAY	107	43.6	29.6 ++	2.95	5
10	BANFF-BUCHAN	95	41.1	26.6	2.79	11
11	GORDON	70	42.3	24.4	3.04	16
12	ABERDEEN	249	42.7	26.5 ++	1.73	13
13	KINCARDINE	52	46.6	26.4	3.78	14
14	ANGUS	104	39.5	23.7	2.41	20
15	DUNDEE	178	32.2	23.0	1.81	25
16	PERTH-KINROSS	148	42.8	24.0	2.05	18
17	KIRKCALDY	145	33.6	23.7	2.02	19
18	N-E FIFE	91	49.1	26.8	2.92	10
19	DUNFERMLINE	117	30.9	23.3	2.20	22
20	WEST LOTHIAN	106	27.7	27.7 +	2.78	7
21	EDINBURGH	536	41.6	25.4 ++	1.14	15
22	MIDLOTHIAN	87	34.8	35.6 ++	4.15	3
23	EAST LOTHIAN	77	33.3	21.2	2.50	33
24	TWEEDDALE	27	69.6	36.1	7.34	2
25	ETTRICK	36	40.4	20.9	3.48	35
26	ROXBURGH	44	42.8	23.3	3.60	21
27	BERWICKSHIRE	23	44.5	20.2	4.30	42
28	CLACKMANNAN	37	26.2	22.0	3.84	30
29	STIRLING	100	42.9	26.7 +	2.96	8
30	FALKIRK	149	34.9	22.7	1.91	26
31	ARGYLL-BUTE	56	29.3	14.7 ̄ ̄	2.06	55
32	DUMBARTON	80	34.6	27.5	3.17	6
33	GLASGOW	627	27.1	17.8 ̄ ̄	0.76	46
34	CLYDEBANK	31	19.8	15.0 ̄	2.82	53
35	BEARSDEN	31	26.0	23.1	4.24	23
36	STRATHKELVIN	57	23.1	20.0	2.72	43
37	CUMBERNAULD	24	13.4	20.5	4.39	38
38	MONKLANDS	56	18.2	15.9 ̄ ̄	2.16	50
39	MOTHERWELL	100	22.3	16.6 ̄ ̄	1.76	47
40	HAMILTON	56	18.6	15.6 ̄ ̄	2.14	52
41	EAST KILBRIDE	40	18.2	21.1	3.45	34
42	EASTWOOD	43	29.4	18.9	2.92	45
43	CLYDESDALE	47	26.7	20.3	3.03	41
44	RENFREW	118	19.1	16.5 ̄ ̄	1.58	49
45	INVERCLYDE	84	29.4	20.5	2.31	37
46	CUNNINGHAME	103	26.3	20.4	2.08	40
47	KILMARNOCK	68	27.7	20.8	2.64	36
48	KYLE-CARRICK	106	33.6	21.5	2.14	32
49	CUMNOCK-DOON	29	20.8	15.9 ̄	3.06	51
50	WIGTOWN	28	32.4	19.0	3.70	44
51	STEWARTRY	40	61.1	30.4	4.97	4
52	NITHSDALE	56	34.2	22.0	3.04	28
53	ANNANDALE	46	46.4	27.2	4.05	9
54	ORKNEY	22	41.4	22.0	4.83	31
55	SHETLAND	17	27.2	14.9	3.75	54
56	WESTERN ISLES	45	51.2	24.1	3.81	17
	ALL SCOTLAND	4779	31.9	21.9	0.33	

SCOTTISH CITIES

Testis

		CASES	C.R.	S.R.	S.E.	RK
1	CAITHNESS	3	3.8	2.8	1.68	41
2	SUTHERLAND	2	5.3	3.4	2.54	34
3	ROSS-CROMARTY	5	3.9	3.3	1.52	38
4	SKYE-LOCHALSH	1	3.5	3.9	3.88	29
5	LOCHABER	4	6.8	6.5	3.26	11
6	INVERNESS	5	3.1	2.7	1.21	42
7	BADENOCH	2	7.4	7.2	5.13	7
8	NAIRN	2	6.8	7.0	4.93	8
9	MORAY	8	3.3	3.2	1.15	39
10	BANFF-BUCHAN	13	5.6	5.3	1.49	13
11	GORDON	8	4.8	4.8	1.69	19
12	ABERDEEN	26	4.8	4.8	0.92	18
13	KINCARDINE	8	7.2	6.9	2.43	9
14	ANGUS	10	3.8	4.1	1.32	27
15	DUNDEE	15	2.7	2.6	0.68	46
16	PERTH-KINROSS	12	3.5	3.4	0.99	36
17	KIRKCALDY	21	4.9	4.8	1.08	15
18	N-E FIFE	8	4.3	4.4	1.57	23
19	DUNFERMLINE	20	5.3	5.2	1.17	14
20	WEST LOTHIAN	16	4.2	3.8	0.98	29
21	EDINBURGH	59	4.6	4.5	0.59	22
22	MIDLOTHIAN	10	4.0	3.7	1.19	31
23	EAST LOTHIAN	17	7.4	7.4+	1.81	6
24	TWEEDDALE	3	7.8	8.0	4.63	3
25	ETTRICK	4	4.2	4.1	2.06	25
26	ROXBURGH	8	5.8	4.8	1.98	21
27	BERWICKSHIRE	7	13.5	14.4+	5.47	1
28	CLACKMANNAN	7	5.0	4.8	1.82	17
29	STIRLING	11	4.7	4.7	1.42	20
30	FALKIRK	8	1.4	1.3—	0.55	56
31	ARGYLL-BUTE	5	2.6	2.7	1.19	43
32	DUMBARTON	8	3.5	3.4	1.20	37
33	GLASGOW	56	2.4	2.5—	0.35	47
34	CLYDEBANK	4	2.5	2.5	1.30	47
35	BEARSDEN	3	2.7	2.4	1.41	51
36	STRATHKELVIN	12	4.9	4.8	1.41	15
37	CUMBERNAULD	4	2.2	2.0	1.02	54
38	MONKLANDS	7	2.2	2.4	0.91	52
39	MOTHERWELL	12	2.7	2.8	0.78	45
40	HAMILTON	11	3.5	3.5	1.08	33
41	EAST KILBRIDE	10	4.1	3.8	1.22	29
42	EASTWOOD	12	8.2	8.3	2.42	2
43	CLYDESDALE	11	6.7	6.7	2.02	10
44	RENFREW	18	2.9	2.7	0.63	43
45	INVERCLYDE	8	2.7	2.5	0.90	47
46	CUNNINGHAME	13	3.3	3.0	0.85	40
47	KILMARNOCK	10	4.2	4.1	1.35	26
48	KYLE-CARRICK	9	2.8	2.5	0.88	50
49	CUMNOCK-DOON	5	3.6	3.4	1.54	35
50	WIGTOWN	2	2.3	2.3	1.59	53
51	STEWARTRY	5	7.8	7.8	3.41	5
52	NITHSDALE	10	6.1	5.8	1.84	12
53	ANNANDALE	4	3.9	3.6	1.82	32
54	ORKNEY	4	7.5	7.6	3.89	4
55	SHETLAND	1	1.6	1.5	1.45	55
56	WESTERN ISLES	4	4.6	4.3	2.24	24
	ALL SCOTLAND	569	3.8	3.7	0.16	

SCOTTISH CITIES

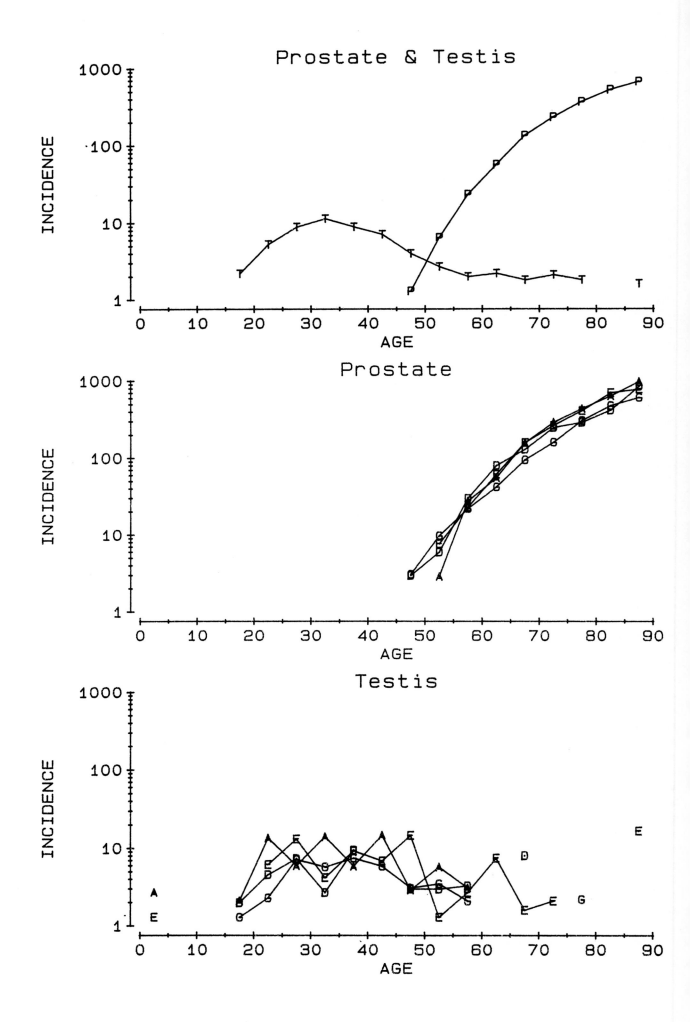

CANCER OF THE BLADDER
(ICD8 188)

The incidence of bladder cancer has been rising in people of each sex in Scotland, more steeply in males than in females.

Males (6.0%; 2.07%)

The Scottish male incidence rate for cancer of the bladder (18.0) is in the upper range of international rates. The highest incidence is reported in Switzerland (Geneva, 30.2), followed by Italy (Varese, 24.6) (Waterhouse *et al.*, 1982). Incidence in Denmark (22.0) is higher than in Scotland, but is lower in Norway (14.6), Sweden (13.7) and Finland (9.9).

The highest male rates in Scotland are recorded in Stirling (26.9), Kilmarnock and Loudon (26.5), the Orkney Islands (26.3), North-east Fife (25.3), Kyle and Carrick (24.3), and the lowest in Sutherland (3.2), Tweeddale (7.2), the Western Isles (9.6), Nairn (10.1) and Ettrick and Lauderdale (10.2). The areas of high incidence include some of the older manufacturing towns of Scotland, such as Kilmarnock and Stirling.

Females (2.7%; 0.65%)

The incidence rate (5.6) in Scottish females is a third of that in males and, again, is in the upper range of international incidence rates, the highest female rates being recorded in Canada (Northwest Territories and Yukon, 7.6). As in males, the rate in females equals that in Denmark (5.6), but is higher than in Sweden (4.2), Norway (4.0) and Finland (2.3).

The areas of high female incidence in Scotland are Ross and Cromarty (9.0), Aberdeen (8.6), Stewartry (8.1), Kilmarnock and Loudon (7.3) and Dunfermline (7.1); and areas of low incidence are Wigtown (0.9), Berwickshire (1.4), the Western Isles (1.9), Badenoch and Strathspey (2.3) and Sutherland (2.7). Again, areas such as Kilmarnock and Loudon and Dunfermline, where the older manufacturing towns are situated, have higher rates.

Scottish cities

The cities all show incidence rates higher than the average for both males and females. Aberdeen is among the areas of highest risk for both males (23.2) and females (8.6). For males in Dundee (19.9) the risk is relatively high, followed by those in Edinburgh (19.2) and Glasgow (18.3). In females, the highest risks after that in Aberdeen are as follows: in Edinburgh (6.8), Glasgow (6.7) and Dundee (6.4).

Spatial aggregation

Although weak, the evidence of clustering of bladder cancer is stronger in males (D = 17.30, $p = 0.060$) than females (D = 18.34, $p = 0.264$).

Comment

The incidence of bladder cancer in urban areas is generally higher than that in more rural settings. However, the areas with the highest incidence are not those with heavy industry, but old, established, medium-sized manufacturing towns, such as Kilmarnock, Dunfermline and Stirling. The causes are not obvious.

A number of occupations have been associated with bladder cancer: dyestuffs manufacture (Rehn, 1895; Case *et al.*, 1954), the rubber industry (Case & Hosker, 1954), textile work, leather work, metal work and painting (Lockwood, 1961; Wynder *et al.*, 1963). All of these occupations are carried on, or have been carried on, in Scotland at some time. The textile and rubber industries are now much less widespread than in the past. One of the difficulties of assessing their importance in causation is the long induction period for bladder cancer, i.e., the long period of time that elapses between first exposure and diagnosis of the disease. This has been estimated at 18 years (Case *et al.*, 1954) or as long as 45 years (Hoover & Cole, 1973). Occupational life histories are not easily obtained. Moreover, many of the occupations involving the use of dyestuffs are now closed down, and it is difficult to obtain histories of working practice or the health records of people employed in these industries.

While textile workers have long been thought to have an increased risk of bladder cancer, it is interesting that no exceptionally high rate, nor even aggregations of areas with increased rates, are present in those Border and Highland areas where weaving, spinning and textile production remain concentrated. However, carpets were manufactured in Kilmarnock and linen in Dunfermline.

The factor that most complicates investigation of the groups at risk is cigarette smoking, well established as a risk factor for bladder cancer (Lockwood, 1961; Cole *et al.*, 1971; Wynder & Goldsmith, 1977; Howe *et al.*, 1980). Cigarette smoking is ubiquitous in the urban populations in Scotland and is likely to be important in determining the level of bladder cancer in different areas.

Bladder — Males

D=17.30, p=0.060

Bladder — Females

D=18.34, p=0.264

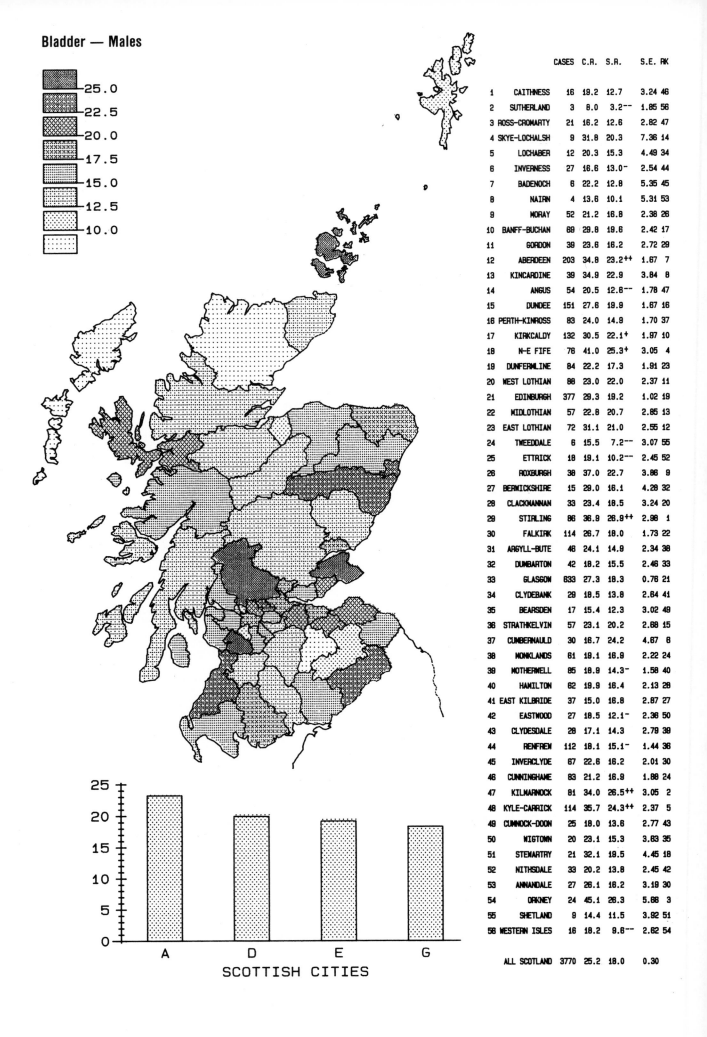

Bladder — Males

		CASES	C.R.	S.R.	S.E.	RK
1	CAITHNESS	16	19.2	12.7	3.24	48
2	SUTHERLAND	3	8.0	3.2--	1.85	56
3	ROSS-CROMARTY	21	16.2	12.6	2.82	47
4	SKYE-LOCHALSH	9	31.8	20.3	7.36	14
5	LOCHABER	12	20.3	15.3	4.49	34
6	INVERNESS	27	16.6	13.0-	2.54	44
7	BADENOCH	6	22.2	12.8	5.35	45
8	NAIRN	4	13.6	10.1	5.31	53
9	MORAY	52	21.2	16.8	2.38	26
10	BANFF-BUCHAN	69	29.8	19.6	2.42	17
11	GORDON	39	23.6	16.2	2.72	29
12	ABERDEEN	203	34.8	23.2++	1.67	7
13	KINCARDINE	39	34.9	22.9	3.84	8
14	ANGUS	54	20.5	12.6--	1.78	47
15	DUNDEE	151	27.6	19.9	1.67	16
16	PERTH-KINROSS	83	24.0	14.9	1.70	37
17	KIRKCALDY	132	30.5	22.1+	1.97	10
18	N-E FIFE	76	41.0	25.3+	3.05	4
19	DUNFERMLINE	84	22.2	17.3	1.91	23
20	WEST LOTHIAN	88	23.0	22.0	2.37	11
21	EDINBURGH	377	29.3	19.2	1.02	19
22	MIDLOTHIAN	57	22.8	20.7	2.85	13
23	EAST LOTHIAN	72	31.1	21.0	2.55	12
24	TWEEDDALE	6	15.5	7.2--	3.07	55
25	ETTRICK	18	19.1	10.2--	2.45	52
26	ROXBURGH	38	37.0	22.7	3.86	9
27	BERWICKSHIRE	15	29.0	16.1	4.28	32
28	CLACKMANNAN	33	23.4	18.5	3.24	20
29	STIRLING	86	36.9	26.9++	2.98	1
30	FALKIRK	114	26.7	18.0	1.73	22
31	ARGYLL-BUTE	48	24.1	14.9	2.34	38
32	DUMBARTON	42	18.2	15.5	2.46	33
33	GLASGOW	633	27.3	18.3	0.76	21
34	CLYDEBANK	29	18.5	13.8	2.64	41
35	BEARSDEN	17	15.4	12.3	3.02	49
36	STRATHKELVIN	57	23.1	20.2	2.68	15
37	CUMBERNAULD	30	16.7	24.2	4.67	6
38	MONKLANDS	61	19.1	16.9	2.22	24
39	MOTHERWELL	85	18.9	14.3-	1.58	40
40	HAMILTON	62	19.9	16.4	2.13	28
41	EAST KILBRIDE	37	15.0	16.8	2.87	27
42	EASTWOOD	27	18.5	12.1-	2.36	50
43	CLYDESDALE	28	17.1	14.3	2.79	39
44	RENFREW	112	18.1	15.1-	1.44	36
45	INVERCLYDE	67	22.6	16.2	2.01	30
46	CUNNINGHAME	83	21.2	16.9	1.88	24
47	KILMARNOCK	81	34.0	26.5++	3.05	2
48	KYLE-CARRICK	114	35.7	24.3++	2.37	5
49	CUMNOCK-DOON	25	18.0	13.6	2.77	43
50	WIGTOWN	20	23.1	15.3	3.63	35
51	STEWARTRY	21	32.1	19.5	4.45	18
52	NITHSDALE	33	20.2	13.8	2.45	42
53	ANNANDALE	27	26.1	16.2	3.19	30
54	ORKNEY	24	45.1	26.3	5.68	3
55	SHETLAND	9	14.4	11.5	3.92	51
56	WESTERN ISLES	16	18.2	9.8--	2.62	54
	ALL SCOTLAND	3770	25.2	18.0	0.30	

SCOTTISH CITIES

Bladder — Females

		CASES	C.R.	S.R.	S.E.	RK
1	CAITHNESS	4	4.7	3.1	1.62	45
2	SUTHERLAND	3	7.8	2.7	1.85	51
3	ROSS-CROMARTY	15	11.8	9.0	2.42	1
4	SKYE-LOCHALSH	3	9.6	2.4⁻	1.39	52
5	LOCHABER	4	6.6	4.6	2.34	29
6	INVERNESS	15	8.9	5.4	1.53	17
7	BADENOCH	2	8.7	2.3	1.69	53
8	NAIRN	2	6.5	5.0	3.53	21
9	MORAY	14	5.7	3.0⁻⁻	0.89	48
10	BANFF-BUCHAN	20	8.5	4.3	1.07	33
11	GORDON	13	7.9	5.3	1.61	19
12	ABERDEEN	112	16.7	8.6++	0.87	2
13	KINCARDINE	8	7.1	3.2	1.24	44
14	ANGUS	30	10.7	4.8	1.00	22
15	DUNDEE	73	12.0	6.3	0.81	11
16	PERTH-KINROSS	41	11.2	5.5	0.94	16
17	KIRKCALDY	48	10.4	6.2	0.94	13
18	N-E FIFE	17	8.3	4.6	1.22	30
19	DUNFERMLINE	45	12.0	7.1	1.13	5
20	WEST LOTHIAN	36	9.3	6.9	1.21	6
21	EDINBURGH	214	14.5	6.8+	0.52	7
22	MIDLOTHIAN	16	8.2	4.7	1.21	27
23	EAST LOTHIAN	20	8.3	4.1	0.95	35
24	TWEEDDALE	5	11.2	3.7	1.98	40
25	ETTRICK	7	6.9	2.7⁻	1.30	50
26	ROXBURGH	15	13.5	6.7	1.98	8
27	BERWICKSHIRE	3	5.5	1.4⁻⁻	0.84	55
28	CLACKMANNAN	11	7.5	4.5	1.40	31
29	STIRLING	22	8.1	4.8	1.07	25
30	FALKIRK	46	10.6	6.3	0.98	12
31	ARGYLL-BUTE	32	16.2	6.8	1.31	10
32	DUMBARTON	21	8.4	4.9	1.16	23
33	GLASGOW	375	14.2	6.7++	0.38	9
34	CLYDEBANK	15	8.8	4.9	1.30	24
35	BEARSDEN	9	7.5	3.9	1.40	37
36	STRATHKELVIN	19	7.5	5.3	1.24	20
37	CUMBERNAULD	11	6.1	6.1	1.89	14
38	MONKLANDS	25	7.4	4.3	0.89	34
39	MOTHERWELL	31	6.5	3.7⁻⁻	0.68	42
40	HAMILTON	18	5.4	3.5⁻	0.87	43
41	EAST KILBRIDE	16	6.3	5.4	1.38	18
42	EASTWOOD	14	8.7	3.7	1.06	41
43	CLYDESDALE	14	8.2	4.7	1.31	28
44	RENFREW	45	6.9	4.4	0.68	32
45	INVERCLYDE	21	6.5	3.9	0.80	37
46	CUNNINGHAME	30	7.2	4.0⁻	0.78	36
47	KILMARNOCK	32	12.6	7.3	1.38	4
48	KYLE-CARRICK	34	9.6	5.9	1.17	15
49	CUMNOCK-DOON	7	4.9	3.0⁻	1.19	47
50	WIGTOWN	2	2.2	0.9⁻⁻	0.84	56
51	STEWARTRY	10	14.5	8.1	2.98	3
52	NITHSDALE	15	8.7	3.8	1.04	39
53	ANNANDALE	8	8.4	4.7	1.73	26
54	ORKNEY	4	7.3	2.7	1.51	49
55	SHETLAND	3	4.8	3.1	2.00	46
56	WESTERN ISLES	3	3.3	1.9⁻⁻	1.26	54
	ALL SCOTLAND	1679	10.4	5.8	0.15	

SCOTTISH CITIES

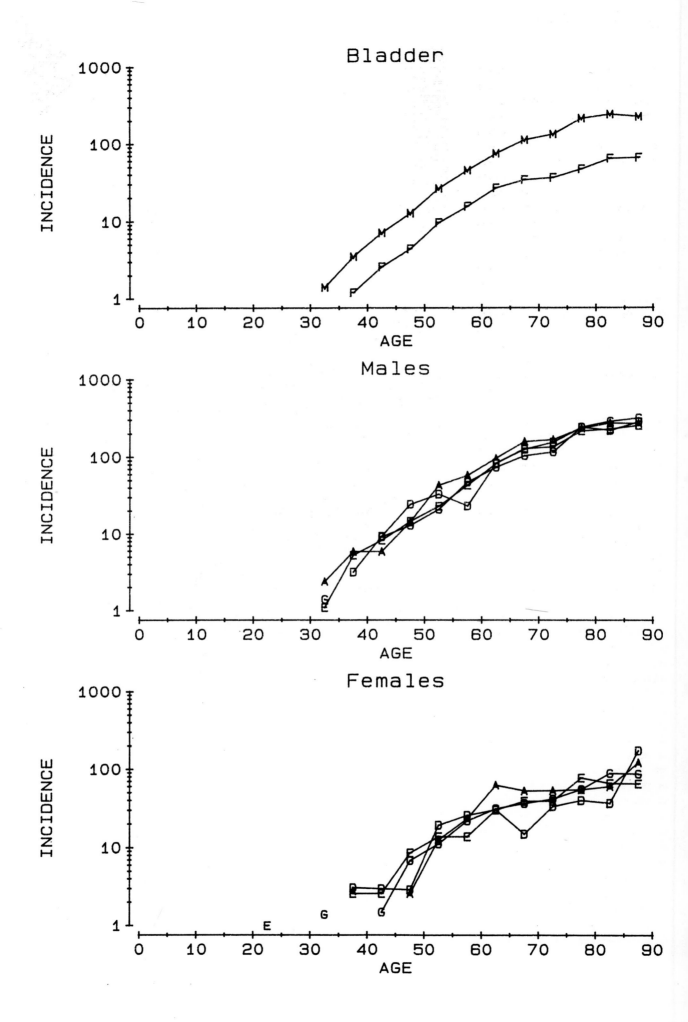

CANCER OF THE THYROID
(ICD8 193)

Cancer of the thyroid is unusual in that, like cancer of the gall-bladder, malignant melanoma and cancer of the breast, it is one of the few that are commoner in females than in males. The incidence is approximately three times higher in females than in males, and rates in people of both sexes have remained virtually unchanged since the mid-1960s.

Males (0.2%; 0.06%)

The average incidence of thyroid cancer in Scottish males (0.6) is remarkably similar to incidence rates in other regions of the UK such as Oxford (0.5), Trent (0.6) and Mersey (0.8). The incidence is slightly higher in Finland (1.6), Sweden (1.5), Norway (1.5) and Denmark (0.9), where the rates are typical of those reported from most other European and from North American centres. Rates of thyroid cancer in males are highest in Chinese population groups, such as those in Hawaii (7.8) and the San Francisco Bay Area (3.7).

Interpretation of the geographical pattern of thyroid cancer in males in Scotland is hampered by the fact that the disease is so rare: over the six-year period considered, no case was recorded in 19 districts, and only 112 cases from the remaining 35 districts. The highest incidence rate is recorded in Nairn (3.3), but the large standard error makes this impossible to interpret. Cases of male thyroid cancer tend to be found in districts such as Nairn (3.3), Caithness (2.3), Bearsden (2.3), Lochaber (1.7) and Ross and Cromarty (1.5), which abut the sea. However, no case was recorded in other coastal areas, such as Skye and Lochalsh and the Shetland Islands.

Females (0.7%; 0.18%)

The incidence of thyroid cancer in females (1.9) is three times higher than that among males (0.6), although it is fairly similar to the female incidence rates in other regions of Britain, such as Trent (1.3), Oxford (1.5) and Mersey (1.6). The incidence is similar in Denmark (1.4), but much greater in Sweden (3.8), Finland (3.9) and Norway (4.4) and in most other regions of Europe, e.g., Geneva (3.5), Varese (5.1) and Zaragoza (5.4). On an international scale, the highest female rates occur among various ethnic groups in Hawaii, such as native Hawaiians (17.6), Chinese (12.1) and Filipinos (16.6).

In Scotland, three areas — Lochaber, Stewartry and Caithness — report no cases in the six years studied. Other low rates are found in Roxburgh (0.2), Nithsdale (0.2) and Dumbarton (0.3). The highest

rates are recorded from Badenoch and Strathspey (9.2), Dunfermline (3.9), the Shetland Islands (3.6), Edinburgh City (3.3), Clydebank (3.0) and East Lothian (3.0). Two patterns emerge. First, there is a concentration of areas with high rates in East Lothian, Midlothian, Edinburgh City and West Lothian, extending across the River Forth to Dunfermline, Clackmannan, Perth and Kinross to Badenoch and Strathspey and Banff and Buchan in the north. Second, of equal interest is the collection of lower than average rates extending across the south of Scotland from east to west.

Scottish cities

For males, the highest rate is reported from Aberdeen (1.3), followed by those from Dundee (0.8), Edinburgh (0.5) and Glasgow (0.5). In females, the highest rate is found in Edinburgh (3.3), followed by those in Aberdeen (2.8), Dundee (1.9) and Glasgow (1.7). Overall, rates in the cities tend to be slightly higher than average.

Spatial aggregation

The map of thyroid cancer in males shows more evidence of clustering ($D = 17.39, p = 0.070$) than does the map for females ($D = 17.65, p = 0.105$), possibly due to the many zero rates (Appendix IIC).

Comment

The incidence of thyroid cancer in Scotland is three times higher in females than that in males, and this ratio has remained fairly unchanged over the last 20 years.

The histological type of thyroid cancer is of overwhelming importance to both survival and etiology. The relatively benign course of most, but not all, papillary tumours contrasts with the aggressive nature of the anaplastic form. Different cell types can be affected by completely opposite causes; excess iodine in the diet increases the risk of papillary carcinoma, while deficiencies of iodine in the diet increase the risk of follicular cancers.

Although thyroid cancer should not be treated as a homogeneous entity, the relative rarity of individual cell types in Scotland makes any individual analysis impossible.

Radiation is thought to be important in the etiology of thyroid cancer; a causal relationship has been established between thyroid cancer risk and exposure to ionizing radiation for enlarged thymus (Winship & Rosvoli, 1970), tinea capitis (Ron & Modan, 1980), tonsillitis (Colman *et al.*, 1978) and as a result of nuclear explosion (Okada *et al.*, 1975).

Endemic goitre, produced by gross dietary deficiency of iodine, has been associated with an excess risk of follicular and aplastic carcinomas (Correa *et al.*, 1969), while in iodine-rich areas there would appear to be an enhanced risk of papillary carcinoma (Williams *et al.*, 1977).

Several benign conditions of the thyroid gland have been associated with cancer risk, although the association has not yet been proved to be causal. For example, patients with Hashimoto's thyroiditis appear to have an increased rate of papillary carcinoma

of the thyroid (Hirabayashi & Lindsay, 1965). Such studies generally depend on the finding of tumours at autopsy; and there is evidence to suggest that the proportion of thyroid tumours discovered incidentally at autopsy among the general population could be similar to the occurrence rate among patients with Hashimoto's disease.

In summary, thyroid carcinoma comprises a heterogeneous group of tumours, each with a course and a cause that are strongly dependent on histological type; but the numbers in Scotland are too small to undertake analyses of only six years' of cases. There is a discernible pattern of high rates on and around the Forth estuary, while low rates occur all along the south of the country — differences that are worth further study.

Thyroid — Males

D=17.39, p=0.070

Thyroid — Females

D=17.65, p=0.105

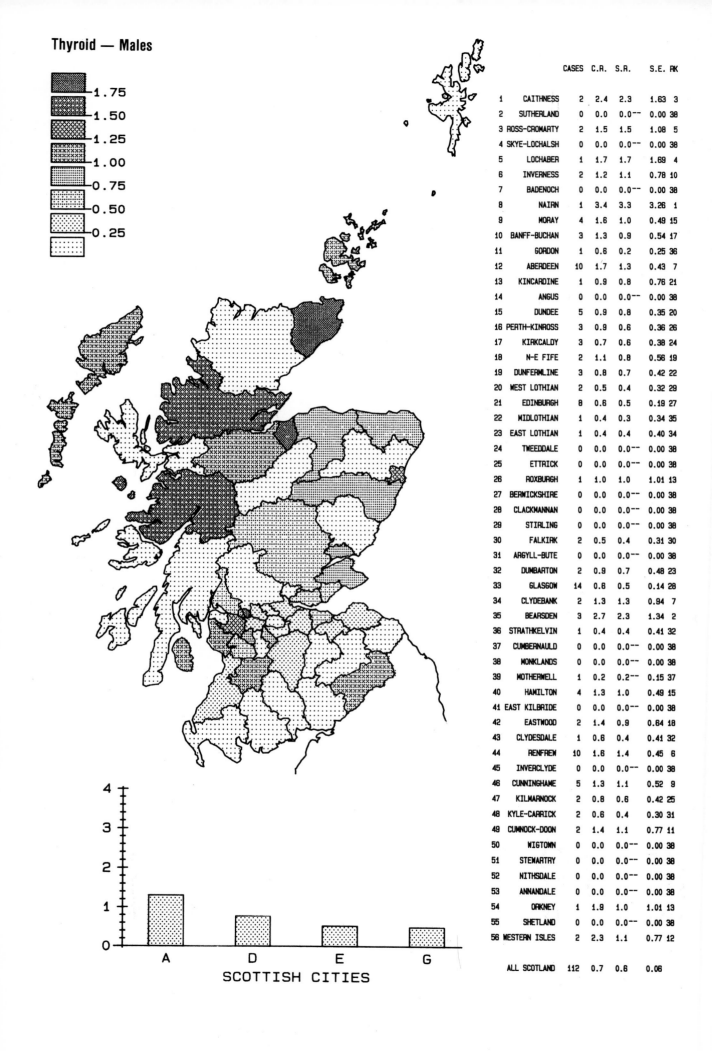

Thyroid — Males

		CASES	C.R.	S.R.	S.E.	RK
1	CAITHNESS	2	2.4	2.3	1.63	3
2	SUTHERLAND	0	0.0	0.0--	0.00	38
3	ROSS-CROMARTY	2	1.5	1.5	1.08	5
4	SKYE-LOCHALSH	0	0.0	0.0--	0.00	38
5	LOCHABER	1	1.7	1.7	1.69	4
6	INVERNESS	2	1.2	1.1	0.78	10
7	BADENOCH	0	0.0	0.0--	0.00	38
8	NAIRN	1	3.4	3.3	3.26	1
9	MORAY	4	1.6	1.0	0.49	15
10	BANFF-BUCHAN	3	1.3	0.9	0.54	17
11	GORDON	1	0.6	0.2	0.25	36
12	ABERDEEN	10	1.7	1.3	0.43	7
13	KINCARDINE	1	0.9	0.8	0.76	21
14	ANGUS	0	0.0	0.0--	0.00	38
15	DUNDEE	5	0.9	0.8	0.35	20
16	PERTH-KINROSS	3	0.9	0.6	0.36	26
17	KIRKCALDY	3	0.7	0.6	0.38	24
18	N-E FIFE	2	1.1	0.8	0.56	19
19	DUNFERMLINE	3	0.8	0.7	0.42	22
20	WEST LOTHIAN	2	0.5	0.4	0.32	29
21	EDINBURGH	8	0.6	0.5	0.19	27
22	MIDLOTHIAN	1	0.4	0.3	0.34	35
23	EAST LOTHIAN	1	0.4	0.4	0.40	34
24	TWEEDDALE	0	0.0	0.0--	0.00	38
25	ETTRICK	0	0.0	0.0--	0.00	38
26	ROXBURGH	1	1.0	1.0	1.01	13
27	BERWICKSHIRE	0	0.0	0.0--	0.00	38
28	CLACKMANNAN	0	0.0	0.0--	0.00	38
29	STIRLING	0	0.0	0.0--	0.00	38
30	FALKIRK	2	0.5	0.4	0.31	30
31	ARGYLL-BUTE	0	0.0	0.0--	0.00	38
32	DUMBARTON	2	0.9	0.7	0.48	23
33	GLASGOW	14	0.6	0.5	0.14	28
34	CLYDEBANK	2	1.3	1.3	0.94	7
35	BEARSDEN	3	2.7	2.3	1.34	2
36	STRATHKELVIN	1	0.4	0.4	0.41	32
37	CUMBERNAULD	0	0.0	0.0--	0.00	38
38	MONKLANDS	0	0.0	0.0--	0.00	38
39	MOTHERWELL	1	0.2	0.2--	0.15	37
40	HAMILTON	4	1.3	1.0	0.49	15
41	EAST KILBRIDE	0	0.0	0.0--	0.00	38
42	EASTWOOD	2	1.4	0.9	0.64	18
43	CLYDESDALE	1	0.6	0.4	0.41	32
44	RENFREW	10	1.6	1.4	0.45	6
45	INVERCLYDE	0	0.0	0.0--	0.00	38
46	CUNNINGHAME	5	1.3	1.1	0.52	9
47	KILMARNOCK	2	0.8	0.6	0.42	25
48	KYLE-CARRICK	2	0.6	0.4	0.30	31
49	CUMNOCK-DOON	2	1.4	1.1	0.77	11
50	WIGTOWN	0	0.0	0.0--	0.00	38
51	STEWARTRY	0	0.0	0.0--	0.00	38
52	NITHSDALE	0	0.0	0.0--	0.00	38
53	ANNANDALE	0	0.0	0.0--	0.00	38
54	ORKNEY	1	1.9	1.0	1.01	13
55	SHETLAND	0	0.0	0.0--	0.00	38
56	WESTERN ISLES	2	2.3	1.1	0.77	12
	ALL SCOTLAND	112	0.7	0.6	0.06	

Legend:
1.75
1.50
1.25
1.00
0.75
0.50
0.25

SCOTTISH CITIES: A D E G

Thyroid — Females

		CASES	C.R.	S.R.	S.E.	RK
1	CAITHNESS	0	0.0	0.0 --	0.00	54
2	SUTHERLAND	1	2.6	0.6 -	0.58	47
3	ROSS-CROMARTY	5	3.9	2.7	1.31	9
4	SKYE-LOCHALSH	1	3.2	2.5	2.51	15
5	LOCHABER	0	0.0	0.0 --	0.00	54
6	INVERNESS	2	1.2	0.6 -	0.49	45
7	BADENOCH	4	13.5	9.2	4.83	1
8	NAIRN	1	3.3	0.7	0.67	44
9	MORAY	7	2.8	2.2	0.89	18
10	BANFF-BUCHAN	4	1.7	1.4	0.70	33
11	GORDON	2	1.2	0.7 -	0.58	41
12	ABERDEEN	29	4.2	2.8	0.61	7
13	KINCARDINE	3	2.6	1.2	0.76	37
14	ANGUS	8	3.2	1.3	0.47	34
15	DUNDEE	18	2.6	1.9	0.53	23
16	PERTH-KINROSS	15	4.1	2.5	0.76	13
17	KIRKCALDY	10	2.2	1.4	0.49	31
18	N-E FIFE	5	2.4	1.4	0.75	32
19	DUNFERMLINE	19	5.1	3.9 +	0.93	2
20	WEST LOTHIAN	13	3.4	2.7	0.78	8
21	EDINBURGH	89	4.6	3.3 ++	0.44	4
22	MIDLOTHIAN	8	3.1	2.6	0.95	12
23	EAST LOTHIAN	11	4.5	3.0	1.00	5
24	TWEEDDALE	1	2.2	0.5 --	0.52	48
25	ETTRICK	4	3.9	2.1	1.30	21
26	ROXBURGH	1	0.9	0.2 --	0.17	53
27	BERWICKSHIRE	1	1.8	0.6 -	0.62	46
28	CLACKMANNAN	4	2.7	2.4	1.29	16
29	STIRLING	5	2.1	1.5	0.89	30
30	FALKIRK	11	2.5	1.7	0.55	26
31	ARGYLL-BUTE	8	3.0	1.7	0.63	27
32	DUMBARTON	1	0.4	0.3 --	0.29	51
33	GLASGOW	74	2.6	1.7	0.23	25
34	CLYDEBANK	8	3.5	3.0	1.26	5
35	BEARSDEN	4	3.3	2.3	1.26	17
36	STRATHKELVIN	8	2.4	1.9	0.78	24
37	CUMBERNAULD	5	2.8	2.2	0.99	19
38	MONKLANDS	4	1.2	0.7 --	0.36	43
39	MOTHERWELL	9	1.9	1.5	0.51	29
40	HAMILTON	8	2.4	2.2	0.78	20
41	EAST KILBRIDE	7	2.6	2.5	0.97	13
42	EASTWOOD	3	1.9	1.2	0.76	38
43	CLYDESDALE	1	0.6	0.4 --	0.42	50
44	RENFREW	12	1.8	1.5	0.44	28
45	INVERCLYDE	3	0.9	0.5 --	0.29	49
46	CUNNINGHAME	4	1.0	0.7 --	0.39	41
47	KILMARNOCK	5	2.0	1.2	0.61	36
48	KYLE-CARRICK	9	2.5	1.3	0.49	34
49	CUMNOCK-DOON	1	0.7	0.8	0.79	40
50	WIGTOWN	2	2.2	0.8	0.61	39
51	STEWARTRY	0	0.0	0.0 --	0.00	54
52	NITHSDALE	1	0.8	0.3 --	0.25	52
53	ANNANDALE	2	1.9	2.0	1.49	22
54	ORKNEY	3	5.5	2.8	1.53	11
55	SHETLAND	2	3.3	3.8	2.55	3
56	WESTERN ISLES	5	5.5	2.8	1.29	10
	ALL SCOTLAND	442	2.7	1.9	0.10	

SCOTTISH CITIES

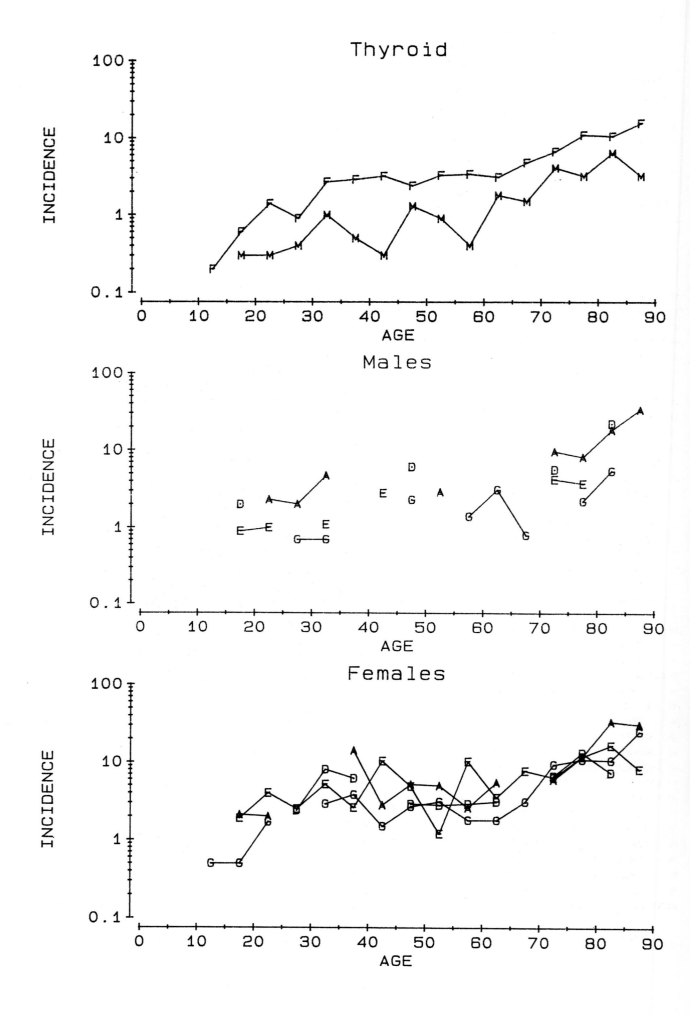

HODGKIN'S DISEASE
(ICD8 201)

Hodgkin's disease is a form of malignancy that affects the lymphatic tissue of the body, i.e., malignant lymphoma. Its incidence has changed little in Scotland since 1960.

The improved outlook for patients with Hodgkin's disease, which has been brought about by recent advances in radiotherapy and chemotherapy, is one of the great successes of modern medical science. As some believe Hodgkin's disease to be infectious, this comparatively rare form of cancer is of great interest, not only to clinicians but also to other specialist groups, such as immunologists, pathologists, bacteriologists and epidemiologists.

Males (0.8%; 0.25%)

The Scottish rate (2.9) is fairly typical of rates in other regions of Britain, e.g., Birmingham (2.7), Mersey (2.9), Oxford (3.1) and Trent (3.1). The incidence is remarkably similar in Norway (2.6), Denmark (2.7), Finland (2.7) and Sweden (3.3). Incidence rates are fairly similar also in regions of North America, e.g., Manitoba (2.8), Alberta (3.4), Ontario (3.5), Detroit (3.3), Iowa (3.4) and Connecticut (4.0). There is very little international variation in the male incidence of Hodgkin's disease (the highest rate recorded is 4.9 in urban Vaud, Switzerland).

Three districts in Scotland report no cases of Hodgkin's disease among males in the six years studied: Nairn, Badenoch and Strathspey and Wigtown. The next lowest rate is recorded from Inverness (0.6). The highest rates are in Berwickshire (13.0), Tweeddale (6.9), Cumnock and Doon Valley (5.5), Eastwood (5.5) and the Orkney Islands (5.5). There appears to be a concentration of low rates in the north and west, i.e., Nairn (0.0), Badenoch and Strathspey (0.0), Inverness (0.6), Moray (1.4), Skye and Lochalsh (1.4), Ross and Cromarty (1.5) and Lochaber (1.7). The high rates recorded in the neighbouring districts of Berwickshire (13.0) and East Lothian (4.9); in Banff and Buchan (5.0) and Gordon (4.4); and in Tweeddale (6.9), Clydesdale (3.7) and Cumnock and Doon Valley (5.5) are also interesting.

Females (0.6%; 0.17%)

The rate among Scottish females (1.9) is less than that in males (2.9), but is similar to that in females in other regions of Britain, e.g., Mersey (1.6), Birmingham (1.9), Oxford (1.9) and Trent (2.1). Comparable rates are recorded in neighbouring countries, e.g., Norway (1.6), Denmark (1.7), Finland

(1.7) and Sweden (1.8). Incidence rates of Hodgkin's disease are fairly similar in regions of North America, e.g., Manitoba (1.7), Alberta (2.2), Ontario (2.5), Detroit (2.3), Iowa (2.3) and Connecticut (2.8).

Seven districts in Scotland report no case of Hodgkin's disease in females over the six-year period studied: Nairn, Skye and Lochalsh, Badenoch and Strathspey, Stewartry, Sutherland, Orkney Islands and Berwickshire. The highest rates are recorded in Banff and Buchan (5.7), Tweeddale (3.8), Clydesdale (3.8), Stirling (3.5) and the Shetland Islands (3.5). The outstanding feature of the geographical distribution of Hodgkin's disease in females is the almost total absence of cases over a large number of districts of north-west Scotland: Nairn (0.0), Skye and Lochalsh (0.0), Badenoch and Strathspey (0.0), Sutherland (0.0), the Western Isles (0.5), Ross and Cromarty (0.6), Moray (0.8) and Inverness (1.5). There appear to be small pockets of districts with high rates, e.g., Tweeddale (3.8) and Clydesdale (3.8), and Banff and Buchan (5.7), Gordon (3.3) and Kincardine and Deeside (2.6).

Scottish cities

Incidence rates in males are remarkably similar in each of the Scottish cities: Edinburgh (3.3), Aberdeen (2.9), Dundee (2.8) and Glasgow (2.7). Among females, the highest recorded rate is found in Aberdeen (2.0), followed by Glasgow (1.7), Edinburgh (1.4) and Dundee (1.1).

Spatial aggregation

There is no statistical evidence of clustering of Hodgkin's disease among males (D = 17.96, p = 0.164) in the Scottish districts, and only weak evidence in females (D = 17.03, p = 0.038) (see, however, the end of Chapter 5).

Comment

Even though this cancer is not common, the absence of cases in the north-west, apparent in each sex, is the striking feature of Hodgkin's disease in Scotland. This is unlikely to be a problem of misclassification, either by cancer registry staff or by pathologists, since the incidence of non-Hodgkin's lymphoma, a group of rather similar forms of cancer, is also very low in these districts. The finding is unlikely to reflect under-registration, since very high levels of an equally uncommon cancer (lip cancer) have been reported from those districts. There is no possibility that a lack of health service resources could produce under-diagnosis on such a scale. This virtual absence of all forms of lymphoma in the north-west of Scotland appears to be a real phenomenon.

Hodgkin's disease in Scotland has a number of other interesting features. The high rates found in people of each sex in the contiguous districts of Clydesdale and Tweeddale (which are in different cancer registry regions) are quite striking, as is a similar situation in Banff and Buchan and Gordon. The high rates recorded among males in East Lothian and Berwickshire, where rates in females are low, may indicate an occupational influence on the risk of Hodgkin's disease in these coastal areas.

Considerable interest has been aroused by the hypothesis that Hodgkin's disease may be trans-missible (Grufferman, 1982). There have been reports of multiple occurrences of the disease within families (Grufferman, 1977) and reports of clusters of cases (George et al., 1965; Vianna & Polan, 1973). However, the evidence of space-time clustering of Hodgkin's disease is, at best, equivocal (Smith, 1982). It has been suggested that Hodgkin's disease is a late and rare sequel to infection with a common virus. Although this hypothesis receives some degree of support from the association between an increasing risk of Hodgkin's disease and increasing sibship (Gutensohn & Cole, 1980), it remains tenuous.

In summary, there is a suggestion of a geographical pattern of Hodgkin's disease in Scotland, but little epidemiological information available to explain the pattern observed.

Hodgkin's Disease — Males

D=17.96, p=0.164

Hodgkin's Disease — Females

D=17.03, p=0.038

Hodgkin's Disease — Males

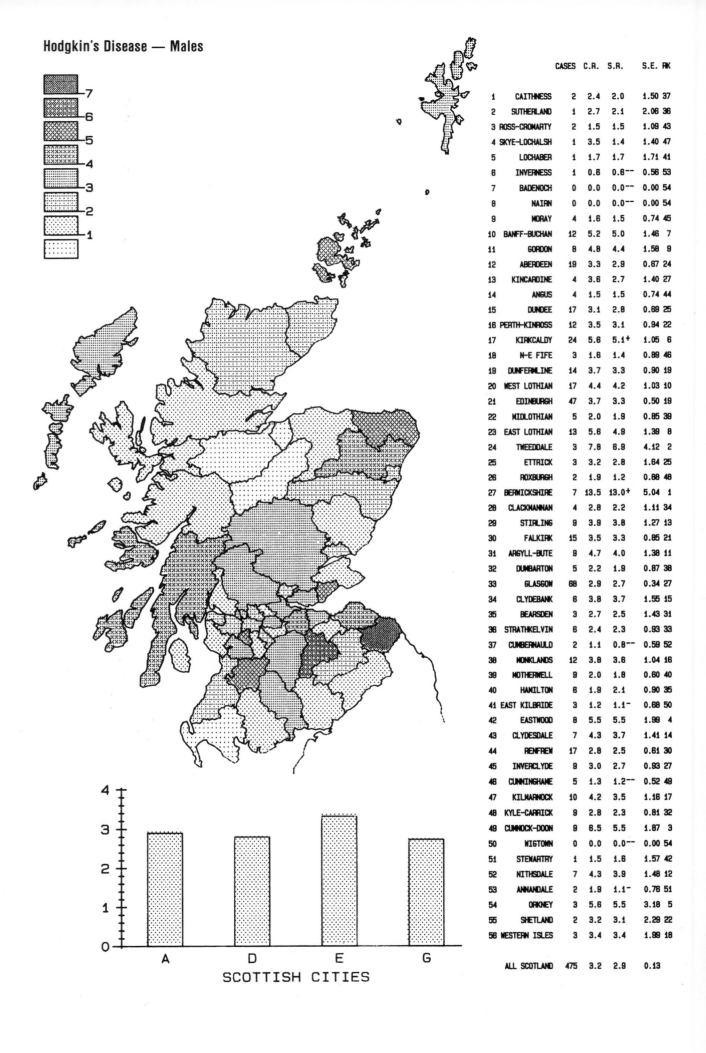

		CASES	C.R.	S.R.	S.E.	RK
1	CAITHNESS	2	2.4	2.0	1.50	37
2	SUTHERLAND	1	2.7	2.1	2.06	36
3	ROSS-CROMARTY	2	1.5	1.5	1.09	43
4	SKYE-LOCHALSH	1	3.5	1.4	1.40	47
5	LOCHABER	1	1.7	1.7	1.71	41
6	INVERNESS	1	0.6	0.6--	0.56	53
7	BADENOCH	0	0.0	0.0--	0.00	54
8	NAIRN	0	0.0	0.0--	0.00	54
9	MORAY	4	1.6	1.5	0.74	45
10	BANFF-BUCHAN	12	5.2	5.0	1.46	7
11	GORDON	8	4.8	4.4	1.58	9
12	ABERDEEN	19	3.3	2.9	0.67	24
13	KINCARDINE	4	3.6	2.7	1.40	27
14	ANGUS	4	1.5	1.5	0.74	44
15	DUNDEE	17	3.1	2.8	0.69	25
16	PERTH-KINROSS	12	3.5	3.1	0.94	22
17	KIRKCALDY	24	5.6	5.1+	1.05	6
18	N-E FIFE	3	1.6	1.4	0.89	46
19	DUNFERMLINE	14	3.7	3.3	0.90	19
20	WEST LOTHIAN	17	4.4	4.2	1.03	10
21	EDINBURGH	47	3.7	3.3	0.50	19
22	MIDLOTHIAN	5	2.0	1.9	0.85	39
23	EAST LOTHIAN	13	5.6	4.9	1.39	8
24	TWEEDDALE	3	7.8	6.9	4.12	2
25	ETTRICK	3	3.2	2.8	1.64	25
26	ROXBURGH	2	1.9	1.2	0.88	48
27	BERWICKSHIRE	7	13.5	13.0+	5.04	1
28	CLACKMANNAN	4	2.8	2.2	1.11	34
29	STIRLING	9	3.9	3.8	1.27	13
30	FALKIRK	15	3.5	3.3	0.85	21
31	ARGYLL-BUTE	9	4.7	4.0	1.38	11
32	DUMBARTON	5	2.2	1.9	0.87	38
33	GLASGOW	68	2.9	2.7	0.34	27
34	CLYDEBANK	6	3.8	3.7	1.55	15
35	BEARSDEN	3	2.7	2.5	1.43	31
36	STRATHKELVIN	6	2.4	2.3	0.93	33
37	CUMBERNAULD	2	1.1	0.8--	0.59	52
38	MONKLANDS	12	3.8	3.6	1.04	16
39	MOTHERWELL	9	2.0	1.8	0.60	40
40	HAMILTON	6	1.9	2.1	0.90	35
41	EAST KILBRIDE	3	1.2	1.1-	0.68	50
42	EASTWOOD	8	5.5	5.5	1.99	4
43	CLYDESDALE	7	4.3	3.7	1.41	14
44	RENFREW	17	2.8	2.5	0.61	30
45	INVERCLYDE	9	3.0	2.7	0.93	27
46	CUNNINGHAME	5	1.3	1.2--	0.52	49
47	KILMARNOCK	10	4.2	3.5	1.16	17
48	KYLE-CARRICK	9	2.8	2.3	0.81	32
49	CUMNOCK-DOON	9	6.5	5.5	1.87	3
50	WIGTOWN	0	0.0	0.0--	0.00	54
51	STEWARTRY	1	1.5	1.6	1.57	42
52	NITHSDALE	7	4.3	3.9	1.48	12
53	ANNANDALE	2	1.9	1.1-	0.76	51
54	ORKNEY	3	5.6	5.5	3.18	5
55	SHETLAND	2	3.2	3.1	2.29	22
56	WESTERN ISLES	3	3.4	3.4	1.99	18
	ALL SCOTLAND	475	3.2	2.9	0.13	

SCOTTISH CITIES

Hodgkin's Disease — Females

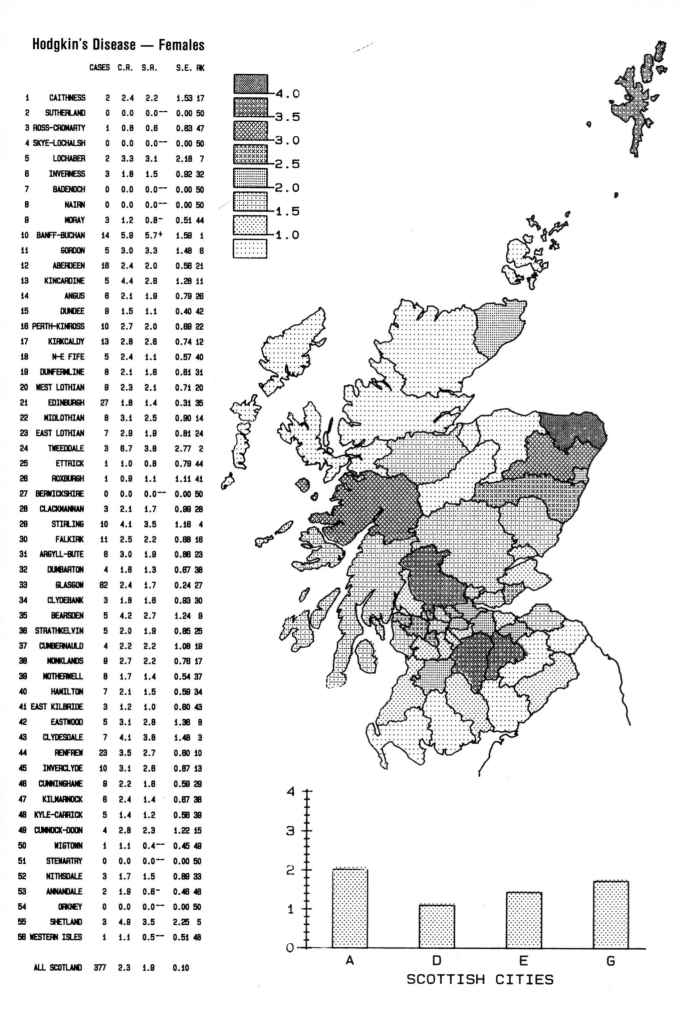

		CASES	C.R.	S.R.	S.E.	RK
1	CAITHNESS	2	2.4	2.2	1.53	17
2	SUTHERLAND	0	0.0	0.0--	0.00	50
3	ROSS-CROMARTY	1	0.8	0.6	0.63	47
4	SKYE-LOCHALSH	0	0.0	0.0--	0.00	50
5	LOCHABER	2	3.3	3.1	2.18	7
6	INVERNESS	3	1.8	1.5	0.92	32
7	BADENOCH	0	0.0	0.0--	0.00	50
8	NAIRN	0	0.0	0.0--	0.00	50
9	MORAY	3	1.2	0.8-	0.51	44
10	BANFF-BUCHAN	14	5.9	5.7+	1.59	1
11	GORDON	5	3.0	3.3	1.48	6
12	ABERDEEN	16	2.4	2.0	0.56	21
13	KINCARDINE	5	4.4	2.8	1.28	11
14	ANGUS	8	2.1	1.9	0.79	26
15	DUNDEE	9	1.5	1.1	0.40	42
16	PERTH-KINROSS	10	2.7	2.0	0.69	22
17	KIRKCALDY	13	2.8	2.6	0.74	12
18	N-E FIFE	5	2.4	1.1	0.57	40
19	DUNFERMLINE	8	2.1	1.6	0.61	31
20	WEST LOTHIAN	9	2.3	2.1	0.71	20
21	EDINBURGH	27	1.8	1.4	0.31	35
22	MIDLOTHIAN	8	3.1	2.5	0.90	14
23	EAST LOTHIAN	7	2.9	1.9	0.81	24
24	TWEEDDALE	3	6.7	3.8	2.77	2
25	ETTRICK	1	1.0	0.8	0.79	44
26	ROXBURGH	1	0.9	1.1	1.11	41
27	BERWICKSHIRE	0	0.0	0.0--	0.00	50
28	CLACKMANNAN	3	2.1	1.7	0.99	28
29	STIRLING	10	4.1	3.5	1.16	4
30	FALKIRK	11	2.5	2.2	0.68	16
31	ARGYLL-BUTE	6	3.0	1.9	0.86	23
32	DUMBARTON	4	1.6	1.3	0.67	38
33	GLASGOW	62	2.4	1.7	0.24	27
34	CLYDEBANK	3	1.8	1.6	0.83	30
35	BEARSDEN	5	4.2	2.7	1.24	9
36	STRATHKELVIN	5	2.0	1.9	0.85	25
37	CUMBERNAULD	4	2.2	2.2	1.08	19
38	MONKLANDS	9	2.7	2.2	0.76	17
39	MOTHERWELL	8	1.7	1.4	0.54	37
40	HAMILTON	7	2.1	1.5	0.59	34
41	EAST KILBRIDE	3	1.2	1.0	0.60	43
42	EASTWOOD	5	3.1	2.8	1.36	8
43	CLYDESDALE	7	4.1	3.8	1.48	3
44	RENFREW	23	3.5	2.7	0.60	10
45	INVERCLYDE	10	3.1	2.6	0.87	13
46	CUNNINGHAME	9	2.2	1.6	0.59	29
47	KILMARNOCK	6	2.4	1.4	0.67	36
48	KYLE-CARRICK	5	1.4	1.2	0.56	39
49	CUMNOCK-DOON	4	2.8	2.3	1.22	15
50	WIGTOWN	1	1.1	0.4--	0.45	49
51	STEWARTRY	0	0.0	0.0--	0.00	50
52	NITHSDALE	3	1.7	1.5	0.89	33
53	ANNANDALE	2	1.9	0.8-	0.48	46
54	ORKNEY	0	0.0	0.0--	0.00	50
55	SHETLAND	3	4.8	3.5	2.25	5
56	WESTERN ISLES	1	1.1	0.5--	0.51	48
	ALL SCOTLAND	377	2.3	1.9	0.10	

SCOTTISH CITIES

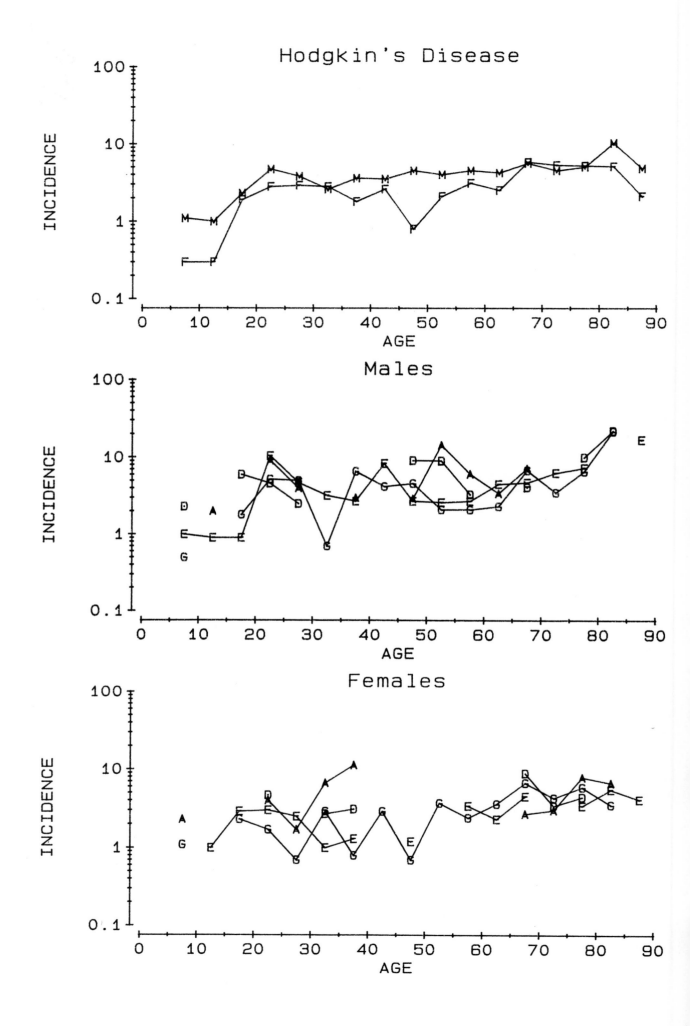

NON-HODGKIN'S LYMPHOMA
(ICD8 200, 202)

In addition to Hodgkin's disease, there are a number of other forms of malignant lymphoma. In view of the relatively small numbers of many of these forms of cancer and the rapid changes taking place in their pathological classifications, all cell types included in ICD rubrics 200 and 202 have been aggregated and are termed non-Hodgkin's lymphoma in the following text. The incidence of non-Hodgkin's lymphoma is increasing in people of each sex in Scotland

Males (1.9%; 0.68%)

The rate among males in Scotland (6.2) is consistently higher than rates in other regions of Britain, e.g., Oxford (5.6), Trent (4.9), Birmingham (4.8), Mersey (4.8), North-western (4.1) and South Thames (3.8). Among neighbouring countries, the Scottish rate is slightly lower than that in Sweden (6.3), but higher than rates in Norway (5.5), Denmark (5.1) and Finland (4.4). Rates for non-Hodgkin's lymphoma are consistently higher than the European rates among similar population groups of North America. Typical are rates reported from Ontario (7.1), Alberta (7.2), British Columbia (7.7), Los Angeles (7.9), Connecticut (8.0) and Detroit (8.7).

Within the districts of Scotland, the incidence of non-Hodgkin's lymphoma varies by a factor of ten. The lowest rates are found in Skye and Lochalsh (1.4), Inverclyde (2.9), Annandale and Eskdale (3.0) and the Western Isles (3.2), and the highest in Kincardine and Deeside (17.2), the Shetland Islands (13.9), Nairn (12.5), Berwickshire (11.9) and Tweeddale (10.2). The notable feature of the Scottish map is the aggregation of lower than average rates in the north and west of the country: Sutherland (3.7), Ross and Cromarty (3.6), Skye and Lochalsh (1.4), the Western Isles (3.2) and Lochaber (3.7). This phenomenon extends southward, and there is an aggregation of similar rates in Glasgow and the surrounding districts. There appears to be a band of elevated risk extending across the southern part of the country from Berwickshire (11.9) in the east to Kyle and Carrick in the West (7.4).

Females (1.9%; 0.50%)

The incidence rate in Scottish females (4.6) is considerably higher than that reported from regions of England and Wales: Trent (3.6), Oxford (3.6), Birmingham (3.5), North-western (3.2), Mersey (2.7) and South Thames (2.5). The Scottish rate is also higher than that in neighbouring countries, e.g., Sweden (4.3), Norway (3.9), Denmark (3.0) and Finland (2.7). Incidence rates are higher in North American population groups, e.g., British Columbia (5.2), Ontario (5.0), Alberta (4.8), Connecticut (6.7), Detroit (6.4) and Los Angeles (5.9).

Within Scotland, there is an eight-fold variation. The lowest rates are found in Cumnock and Doon Valley (1.1), Ettrick and Lauderdale (1.1), Monklands (1.8), Caithness (2.0) and Kilmarnock and Loudon (2.1), the highest in Kincardine and Deeside (8.3), Banff and Buchan (7.2), East Lothian (6.8), Aberdeen City (6.6) and West Lothian (6.3). The interesting feature of the geographic distribution of non-Hodgkin's lymphoma in Scottish females is the suggestion of high rates in the north-east and east, and low rates in the north-west and west.

Scottish cities

Males in Aberdeen (9.6) have the highest rate of non-Hodgkin's lymphoma of any Scottish city. The rate in Edinburgh (7.8) is slightly lower, but higher than the rates in either Dundee (5.5) or Glasgow (5.4). Among females, the highest rate is again found in Aberdeen (6.6), followed by those in Edinburgh (5.5), Glasgow (4.0) and Dundee (3.5).

Spatial aggregation

There appears to be no evidence of clustering of districts with similar levels of non-Hodgkin's lymphoma among females (D = 18.07, p = 0.191), but there is strong evidence of clustering among males (D = 16.28, p = 0.008).

Comment

So little is known about the etiology of the overwhelming majority of cases of non-Hodgkin's lymphoma that there is an urgent need for information about risk factors for this form of cancer. Examination of the geographical pattern in Scotland, for people of each sex, suggests some interesting fields of study. First, in people of each sex, rates are lower than average in large parts of the north-west and west of the country. Second, in people of each sex, high levels of this form of cancer are reported from Aberdeen and Kincardine and Deeside. Third, in people of each sex, higher than average rates are reported from the neighbouring districts of Clydesdale and Tweeddale. Finally, in people of each sex, incidence rates are significantly higher in Aberdeen and Edinburgh than in Glasgow and Dundee.

An increased risk of non-Hodgkin's lymphoma has been reported among certain groups with precisely defined, but nevertheless extremely rare, medical conditions such as primary immunodeficiency syndromes (Gatti & Good, 1971), acquired immuno-deficiency syndromes (AIDS) (Ziegler *et al.*, 1984), Sjogren's syndrome (Kassan *et al.*, 1978) and recipients of renal transplants (Kinlen *et al.*, 1979). The number of cases due to such causes comprises a very small proportion of all non-Hodgkin's lymphoma.

Other risk factors in the etiology of non-Hodgkin's lymphoma have been studied, but, in general, the findings have been either negative or equivocal. Among the latter are the reports of low-grade epidemicity (Smith, 1982), the role of ionizing radiation (Greene, 1982) [although patients irradiated thera-peutically for ankylosing spondylitis show an excess incidence of non-Hodgkin's lymphoma (Court-Brown & Doll, 1965)], occupational exposures, e.g., to anaesthetic agents (Vessey, 1978), the role of over-nutrition (Cunningham, 1976) and contaminated drinking water (Cantor *et al.*, 1978).

In view of the increasing incidence of non-Hodgkin's lymphoma in people of each sex in Scotland, the fact that rates are higher than in England and Wales and the lack of knowledge of the etiology of this condition, it appears worthwhile to utilize the spatial aggregation of non-Hodgkin's lymphoma in Scotland to search for etiological clues. A first step would be to record the characteristics of cases in areas of contrasting incidence rates (Chapter 8), concentrating especially on the features of cases in males.

Non-Hodgkin's Lymphoma — Males

□1

14.0
11.0
8.7
7.1
6.2
4.9
4.0
3.2
1.0

D=16.28, p=0.008

Non-Hodgkin's Lymphoma — Females

1

7.5
6.8
5.9
4.7 ——————— 4.6
3.5
2.7
1.5
1.0

D=18.07, p=0.191

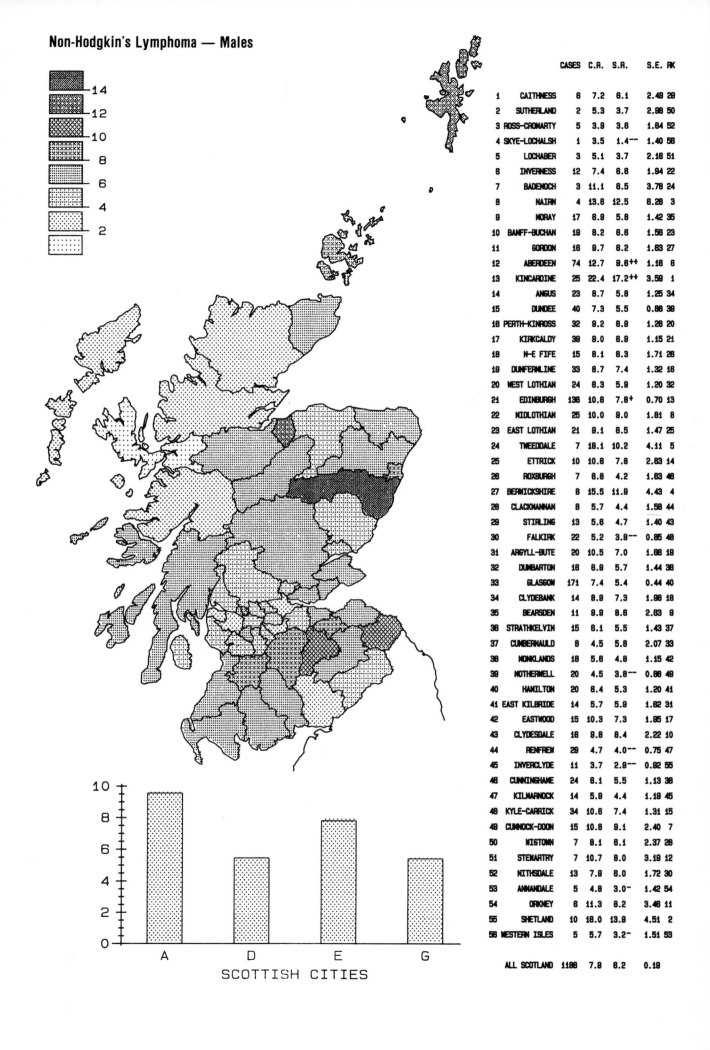

Non-Hodgkin's Lymphoma — Males

		CASES	C.R.	S.R.	S.E.	RK
1	CAITHNESS	8	7.2	8.1	2.49	29
2	SUTHERLAND	2	5.3	3.7	2.98	50
3	ROSS-CROMARTY	5	3.9	3.6	1.64	52
4	SKYE-LOCHALSH	1	3.5	1.4--	1.40	56
5	LOCHABER	3	5.1	3.7	2.16	51
6	INVERNESS	12	7.4	6.8	1.84	22
7	BADENOCH	3	11.1	8.5	3.78	24
8	NAIRN	4	13.6	12.5	6.28	3
9	MORAY	17	6.9	5.8	1.42	35
10	BANFF-BUCHAN	19	8.2	6.6	1.58	23
11	GORDON	16	8.7	6.2	1.63	27
12	ABERDEEN	74	12.7	9.6++	1.16	6
13	KINCARDINE	25	22.4	17.2++	3.59	1
14	ANGUS	23	8.7	5.8	1.25	34
15	DUNDEE	40	7.3	5.5	0.88	39
16	PERTH-KINROSS	32	9.2	6.9	1.26	20
17	KIRKCALDY	39	8.0	6.9	1.15	21
18	N-E FIFE	15	8.1	6.3	1.71	26
19	DUNFERMLINE	33	8.7	7.4	1.32	16
20	WEST LOTHIAN	24	6.3	5.9	1.20	32
21	EDINBURGH	136	10.6	7.8+	0.70	13
22	MIDLOTHIAN	25	10.0	8.0	1.81	8
23	EAST LOTHIAN	21	8.1	6.5	1.47	25
24	TWEEDDALE	7	18.1	10.2	4.11	5
25	ETTRICK	10	10.6	7.8	2.63	14
26	ROXBURGH	7	6.8	4.2	1.63	46
27	BERWICKSHIRE	8	15.5	11.9	4.43	4
28	CLACKMANNAN	8	5.7	4.4	1.58	44
29	STIRLING	13	5.6	4.7	1.40	43
30	FALKIRK	22	5.2	3.9--	0.85	48
31	ARGYLL-BUTE	20	10.5	7.0	1.68	19
32	DUMBARTON	18	6.9	5.7	1.44	36
33	GLASGOW	171	7.4	5.4	0.44	40
34	CLYDEBANK	14	8.9	7.3	1.98	18
35	BEARSDEN	11	9.9	8.6	2.63	9
36	STRATHKELVIN	15	6.1	5.5	1.43	37
37	CUMBERNAULD	8	4.5	5.8	2.07	33
38	MONKLANDS	18	5.6	4.9	1.15	42
39	MOTHERWELL	20	4.5	3.9--	0.88	49
40	HAMILTON	20	6.4	5.3	1.20	41
41	EAST KILBRIDE	14	5.7	5.9	1.62	31
42	EASTWOOD	15	10.3	7.3	1.95	17
43	CLYDESDALE	16	9.6	8.4	2.22	10
44	RENFREW	29	4.7	4.0--	0.75	47
45	INVERCLYDE	11	3.7	2.9--	0.92	55
46	CUNNINGHAME	24	6.1	5.5	1.13	38
47	KILMARNOCK	14	5.9	4.4	1.19	45
48	KYLE-CARRICK	34	10.6	7.4	1.31	15
49	CUMNOCK-DOON	15	10.8	9.1	2.40	7
50	WIGTOWN	7	8.1	6.1	2.37	28
51	STEWARTRY	7	10.7	8.0	3.19	12
52	NITHSDALE	13	7.8	6.0	1.72	30
53	ANNANDALE	5	4.8	3.0-	1.42	54
54	ORKNEY	6	11.3	8.2	3.48	11
55	SHETLAND	10	16.0	13.9	4.51	2
56	WESTERN ISLES	5	5.7	3.2-	1.51	53
	ALL SCOTLAND	1188	7.9	6.2	0.19	

SCOTTISH CITIES

Non-Hodgkin's Lymphoma — Females

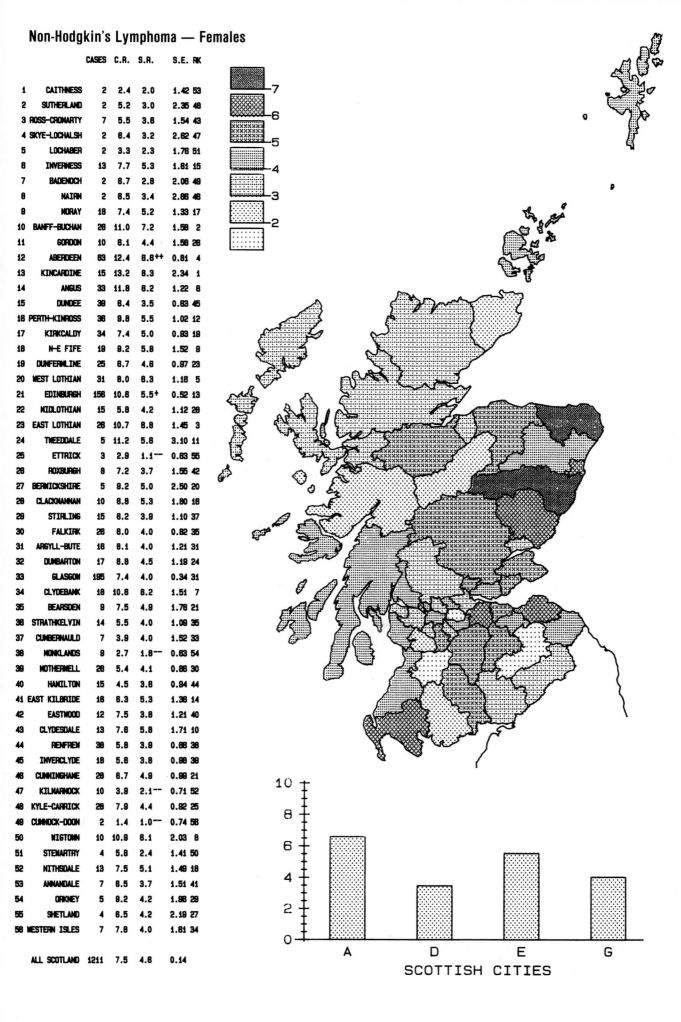

		CASES	C.R.	S.R.	S.E.	RK
1	CAITHNESS	2	2.4	2.0	1.42	53
2	SUTHERLAND	2	5.2	3.0	2.35	48
3	ROSS-CROMARTY	7	5.5	3.8	1.54	43
4	SKYE-LOCHALSH	2	6.4	3.2	2.62	47
5	LOCHABER	2	3.3	2.3	1.78	51
6	INVERNESS	13	7.7	5.3	1.61	15
7	BADENOCH	2	6.7	2.8	2.06	49
8	NAIRN	2	6.5	3.4	2.66	46
9	MORAY	18	7.4	5.2	1.33	17
10	BANFF-BUCHAN	26	11.0	7.2	1.58	2
11	GORDON	10	6.1	4.4	1.58	28
12	ABERDEEN	83	12.4	6.6++	0.81	4
13	KINCARDINE	15	13.2	8.3	2.34	1
14	ANGUS	33	11.8	6.2	1.22	6
15	DUNDEE	39	6.4	3.5	0.63	45
16	PERTH-KINROSS	36	8.8	5.5	1.02	12
17	KIRKCALDY	34	7.4	5.0	0.93	19
18	N-E FIFE	19	9.2	5.6	1.52	9
19	DUNFERMLINE	25	6.7	4.6	0.97	23
20	WEST LOTHIAN	31	8.0	6.3	1.16	5
21	EDINBURGH	156	10.6	5.5+	0.52	13
22	MIDLOTHIAN	15	5.8	4.2	1.12	26
23	EAST LOTHIAN	26	10.7	6.8	1.45	3
24	TWEEDDALE	5	11.2	5.6	3.10	11
25	ETTRICK	3	2.9	1.1--	0.63	55
26	ROXBURGH	8	7.2	3.7	1.55	42
27	BERWICKSHIRE	5	9.2	5.0	2.50	20
28	CLACKMANNAN	10	6.6	5.3	1.80	18
29	STIRLING	15	6.2	3.8	1.10	37
30	FALKIRK	26	6.0	4.0	0.82	35
31	ARGYLL-BUTE	16	6.1	4.0	1.21	31
32	DUMBARTON	17	6.8	4.5	1.19	24
33	GLASGOW	195	7.4	4.0	0.34	31
34	CLYDEBANK	18	10.6	6.2	1.51	7
35	BEARSDEN	9	7.5	4.9	1.76	21
36	STRATHKELVIN	14	5.5	4.0	1.09	35
37	CUMBERNAULD	7	3.9	4.0	1.52	33
38	MONKLANDS	9	2.7	1.8--	0.63	54
39	MOTHERWELL	26	5.4	4.1	0.86	30
40	HAMILTON	15	4.5	3.8	0.94	44
41	EAST KILBRIDE	16	6.3	5.3	1.38	14
42	EASTWOOD	12	7.5	3.8	1.21	40
43	CLYDESDALE	13	7.8	5.6	1.71	10
44	RENFREW	36	5.8	3.9	0.86	38
45	INVERCLYDE	18	5.6	3.8	0.98	39
46	CUNNINGHAME	26	6.7	4.9	0.99	21
47	KILMARNOCK	10	3.9	2.1--	0.71	52
48	KYLE-CARRICK	26	7.9	4.4	0.92	25
49	CUMNOCK-DOON	2	1.4	1.0--	0.74	56
50	WIGTOWN	10	10.9	6.1	2.03	8
51	STEWARTRY	4	5.8	2.4	1.41	50
52	NITHSDALE	13	7.5	5.1	1.49	16
53	ANNANDALE	7	8.5	3.7	1.51	41
54	ORKNEY	5	9.2	4.2	1.98	29
55	SHETLAND	4	6.5	4.2	2.19	27
56	WESTERN ISLES	7	7.8	4.0	1.61	34
	ALL SCOTLAND	1211	7.5	4.6	0.14	

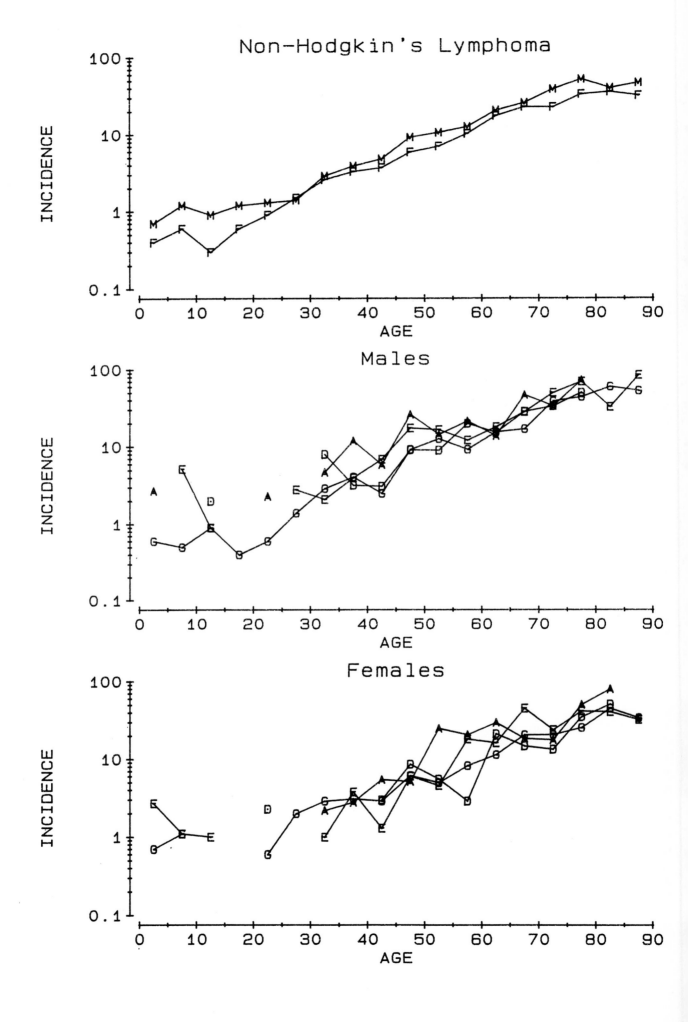

THE LEUKAEMIAS
(ICD8 204, 205, 206, 207)

Leukaemia is a generic name for a group of malignant diseases arising in the so-called white cells that circulate in the blood. These cells originate mainly in bone marrow. For this map, 'leukaemia' covers all the types of leukaemias in ICD8 rubrics 204-207. The major advantage of this definition is that, as the individual types of leukaemia are relatively rare, their aggregation as leukaemia provides more reliable rates for the local authority districts and, further, avoids the possibility of artefactual variation between some areas due to differences in the pathological classification of cell types. Rates for individual cell types are, however, presented later in this volume, with black-and-white maps for the two main types of lymphoid leukaemia (ICD8 204) and myeloid leukaemia (ICD8 205). The incidence of leukaemia in Scotland has been increasing in people of each sex, but still represents a very small proportion of all malignant diseases.

Males (2.0%; 0.64%)

The all-Scottish rate (7.1) is fairly typical of that found in other parts of the UK, e.g., Trent (7.1), South Thames (7.1), Mersey (6.9) and Birmingham (6.5). The North American rates are a little higher: the rate for Ontario (11.0) is typical for Canada, while in the USA, that for Detroit (whites) (10.5) is characteristic. The Scottish rate is close to that of Nagasaki City (7.7), while other Japanese rates are lower, e.g., Miyagi (4.5) and Osaka (4.4). The Nordic rates are slightly higher than the Scottish rate: Sweden (8.7), Denmark (8.4), Finland (8.0) and Norway (7.4).

Within Scotland, the highest rates are found in Tweeddale (13.6), the Western Isles (12.9) and Angus (12.6). The range in variation between highest and lowest is around six-fold, with the lowest rate being that of Skye and Lochalsh (2.5), excluding that of Badenoch and Strathspey (0.0).

Females (1.8%; 0.41%)

The Scottish female rate (4.7), also, is similar to that found in other parts of the UK, e.g., Trent (4.8), Mersey (4.6), South Thames (4.5) and Birmingham (4.1). The Scottish rate is slightly but consistently lower than those in North America, e.g., Ontario (6.5) and Saskatchewan (5.9); in the USA the rate for Alameda County (white) is 6.7, that for Connecticut, 6.6, that for Detroit (white), 6.1 and for Detroit (black), 6.0. The Scottish rate is similar to that for Nagasaki City (4.4) and slightly higher than that in Miyagi (3.9) and Osaka (3.3). The rates in Scandinavia

are slightly higher than the Scottish rate: Sweden (5.8), Denmark (5.7), Finland (5.1) and Norway (5.1).

The highest rates in Scotland are found in Berwickshire (11.4) and Skye and Lochalsh (10.1), and the lowest in Lochaber (0.8) and Nairn (0.9). The range in incidence is seven-fold — similar to that for males.

Scottish cities

The rates found in the Scottish cities are similar for people of each sex and typical of the average Scottish rate. Among males the rates are as follows: Aberdeen (7.3), Dundee (6.9), Edinburgh (7.3) and Glasgow (7.5), while among females they are Aberdeen (4.4), Dundee (4.4), Edinburgh (4.5) and Glasgow (4.5).

Spatial aggregation

There is no evidence of any clustering of districts with similar risks of leukaemia in either males (D = 20.05, p = 0.838) or females (D = 19.03, p = 0.502).

Comment

The most notable features of these maps are the complete lack of any discernible geographical pattern and the similarity of the rates in each of the four cities. While Aberdeen is known to have a high background level of naturally occurring radiation, this is not reflected in an increased leukaemia risk.

There is no evidence of geographical aggregation of areas with high or low rates, nor is there any evidence of systematic urban-rural or coastal-inland differences in incidence. Further, there is little correlation between the geographical pattern in males and females, e.g., the rate in Skye and Lochalsh is one of the lowest for males (2.5) yet one of the highest for females (10.1). It is probably safe to conclude that there is no geographical influence on the occurrence of leukaemia in Scotland, at least, not one that can be discerned from these maps, although individual cell types are not represented separately.

In 1980, an increase in the incidence of acute myeloid leukaemia was noted in Scotland (Kemp *et al.*, 1980). More recently, during an investigation of possible effects in Scotland of pollution from the nuclear reprocessing plant at Sellafield in Cumberland on the north-west coast of England (Independent Advisory Group, 1984), geographical differences in the occurrence of myeloid leukaemia in western Scottish coastal areas were reported for the period

1968-1974 (Heasman *et al.*, 1984); these differences disappeared between 1975 and 1980, the period covered by this atlas.

A most interesting finding is the large range of variation between rates in the districts of Scotland: six- to seven-fold in people of each sex. It has been generally accepted that the international range of incidence for this form of cancer is two-fold (and 1.5-fold within a country). This is believed to demonstrate a lack of variation in the risk factors for these forms of cancer [see Heath (1982) for a review]. Although the variation in the Scottish incidence could be due to chance, it might be used to investigate the role of etiological factors in leukaemia.

Leukaemia — Males

14.0
12.3
9.6
7.7
7.1
6.1
4.6
3.1
2.0

1

D=20.05, p=0.838

Leukaemia — Females

D=19.03, p=0.502

Leukaemia — Males

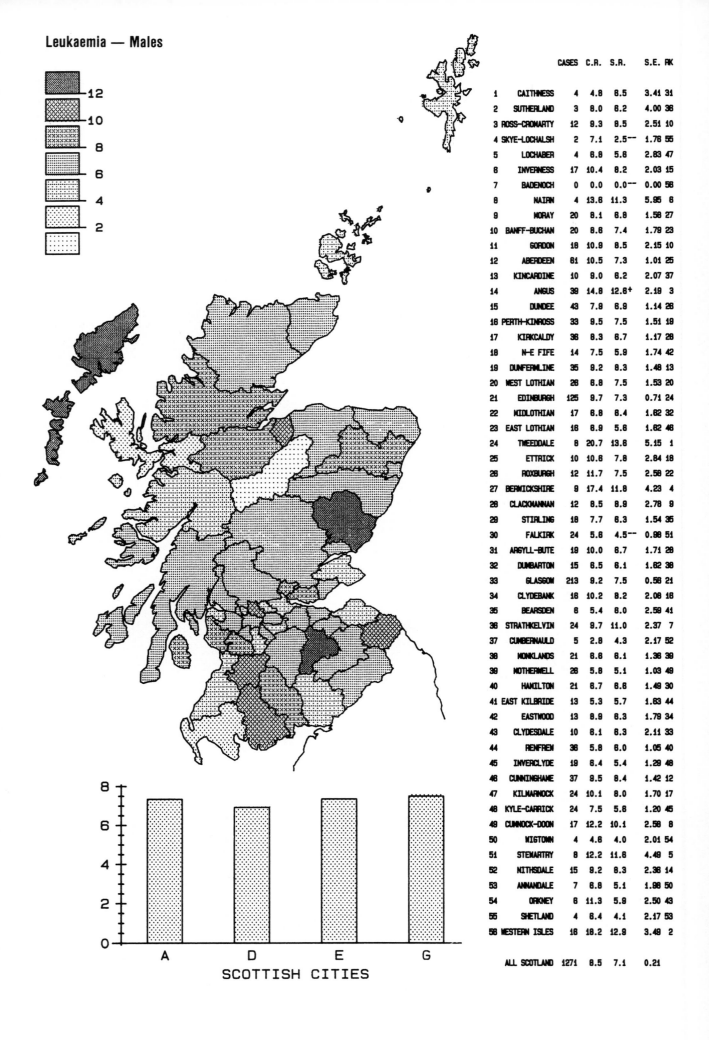

		CASES	C.R.	S.R.	S.E.	RK
1	CAITHNESS	4	4.8	6.5	3.41	31
2	SUTHERLAND	3	6.0	6.2	4.00	36
3	ROSS-CROMARTY	12	9.3	6.5	2.51	10
4	SKYE-LOCHALSH	2	7.1	2.5--	1.76	55
5	LOCHABER	4	6.8	5.6	2.83	47
6	INVERNESS	17	10.4	8.2	2.03	15
7	BADENOCH	0	0.0	0.0--	0.00	56
8	NAIRN	4	13.6	11.3	5.95	6
9	MORAY	20	8.1	6.8	1.58	27
10	BANFF-BUCHAN	20	8.6	7.4	1.79	23
11	GORDON	18	10.9	8.5	2.15	10
12	ABERDEEN	61	10.5	7.3	1.01	25
13	KINCARDINE	10	9.0	6.2	2.07	37
14	ANGUS	39	14.8	12.6+	2.19	3
15	DUNDEE	43	7.9	6.9	1.14	26
16	PERTH-KINROSS	33	9.5	7.5	1.51	19
17	KIRKCALDY	36	8.3	6.7	1.17	28
18	N-E FIFE	14	7.5	5.9	1.74	42
19	DUNFERMLINE	35	9.2	8.3	1.48	13
20	WEST LOTHIAN	26	8.8	7.5	1.53	20
21	EDINBURGH	125	9.7	7.3	0.71	24
22	MIDLOTHIAN	17	6.8	6.4	1.62	32
23	EAST LOTHIAN	16	8.9	5.6	1.62	46
24	TWEEDDALE	8	20.7	13.6	5.15	1
25	ETTRICK	10	10.6	7.8	2.84	18
26	ROXBURGH	12	11.7	7.5	2.56	22
27	BERWICKSHIRE	9	17.4	11.8	4.23	4
28	CLACKMANNAN	12	8.5	8.9	2.78	9
29	STIRLING	18	7.7	6.3	1.54	35
30	FALKIRK	24	5.6	4.5--	0.98	51
31	ARGYLL-BUTE	18	10.0	6.7	1.71	26
32	DUMBARTON	15	6.5	6.1	1.62	38
33	GLASGOW	213	9.2	7.5	0.58	21
34	CLYDEBANK	16	10.2	8.2	2.06	16
35	BEARSDEN	8	5.4	6.0	2.59	41
36	STRATHKELVIN	24	9.7	11.0	2.37	7
37	CUMBERNAULD	5	2.8	4.3	2.17	52
38	MONKLANDS	21	6.6	6.1	1.38	39
39	MOTHERWELL	26	5.8	5.1	1.03	49
40	HAMILTON	21	6.7	6.6	1.49	30
41	EAST KILBRIDE	13	5.3	5.7	1.63	44
42	EASTWOOD	13	8.9	6.3	1.79	34
43	CLYDESDALE	10	6.1	6.3	2.11	33
44	RENFREW	36	5.6	6.0	1.05	40
45	INVERCLYDE	19	6.4	5.4	1.29	48
46	CUNNINGHAME	37	9.5	6.4	1.42	12
47	KILMARNOCK	24	10.1	8.0	1.70	17
48	KYLE-CARRICK	24	7.5	5.6	1.20	45
49	CUMNOCK-DOON	17	12.2	10.1	2.58	8
50	WIGTOWN	4	4.6	4.0	2.01	54
51	STEWARTRY	8	12.2	11.6	4.49	5
52	NITHSDALE	15	9.2	8.3	2.36	14
53	ANNANDALE	7	8.8	5.1	1.96	50
54	ORKNEY	8	11.3	5.9	2.50	43
55	SHETLAND	4	6.4	4.1	2.17	53
56	WESTERN ISLES	16	18.2	12.9	3.49	2
	ALL SCOTLAND	1271	8.5	7.1	0.21	

Leukaemia — Females

		CASES	C.R.	S.R.	S.E.	RK
1	CAITHNESS	5	5.9	5.9	2.97	15
2	SUTHERLAND	2	5.2	2.8	2.28	52
3	ROSS-CROMARTY	12	8.4	7.1	2.16	8
4	SKYE-LOCHALSH	5	18.0	10.1	5.23	2
5	LOCHABER	1	1.8	0.8--	0.78	56
6	INVERNESS	8	4.8	2.7-	1.01	53
7	BADENOCH	3	10.1	6.5	4.17	11
8	NAIRN	1	3.3	0.9--	0.81	55
9	MORAY	24	8.9	6.3	1.88	3
10	BANFF-BUCHAN	17	7.2	4.9	1.45	26
11	GORDON	14	8.5	5.1	1.82	24
12	ABERDEEN	46	6.8	4.4	0.80	36
13	KINCARDINE	12	10.6	7.4	2.55	7
14	ANGUS	28	10.0	5.9	1.41	14
15	DUNDEE	43	7.1	4.4	0.80	37
16	PERTH-KINROSS	23	6.3	4.0	1.06	39
17	KIRKCALDY	32	8.9	4.9	1.04	29
18	N-E FIFE	14	6.8	3.3	0.99	50
19	DUNFERMLINE	15	4.0	3.9	1.11	41
20	WEST LOTHIAN	24	6.2	5.8	1.29	13
21	EDINBURGH	122	8.3	4.5	0.54	33
22	MIDLOTHIAN	18	6.2	5.1	1.29	26
23	EAST LOTHIAN	25	10.3	4.7	1.20	32
24	TWEEDDALE	3	6.7	6.6	5.26	9
25	ETTRICK	12	11.8	4.9	1.82	30
26	ROXBURGH	9	8.1	7.9	3.27	4
27	BERWICKSHIRE	12	22.0	11.4	3.67	1
28	CLACKMANNAN	11	7.5	5.1	1.80	25
29	STIRLING	14	5.8	3.9	1.20	45
30	FALKIRK	23	5.3	3.9	0.87	42
31	ARGYLL-BUTE	19	9.6	5.7	1.54	17
32	DUMBARTON	25	10.1	7.6	1.69	6
33	GLASGOW	174	6.6	4.5	0.41	33
34	CLYDEBANK	9	5.3	3.3	1.20	51
35	BEARSDEN	8	6.7	6.4	2.75	12
36	STRATHKELVIN	15	5.9	5.3	1.48	21
37	CUMBERNAULD	10	5.6	5.6	1.80	18
38	MONKLANDS	14	4.1	3.4	1.02	49
39	MOTHERWELL	23	4.8	3.6	0.80	46
40	HAMILTON	19	5.7	4.3	1.07	38
41	EAST KILBRIDE	10	4.0	3.5	1.12	48
42	EASTWOOD	16	10.0	5.8	1.73	16
43	CLYDESDALE	9	5.3	3.8	1.37	43
44	RENFREW	30	4.8	3.8	0.69	47
45	INVERCLYDE	20	6.2	3.8	0.95	43
46	CUNNINGHAME	25	6.0	5.2	1.17	22
47	KILMARNOCK	21	8.3	4.9	1.26	31
48	KYLE-CARRICK	25	7.0	5.3	1.24	20
49	CUMNOCK-DOON	8	4.2	4.0	1.80	40
50	WIGTOWN	11	12.0	6.5	2.27	10
51	STEWARTRY	3	4.4	2.2	1.30	54
52	NITHSDALE	14	8.1	4.4	1.34	35
53	ANNANDALE	12	11.2	5.4	1.74	19
54	ORKNEY	4	7.3	5.0	2.87	27
55	SHETLAND	6	9.8	5.1	2.29	23
56	WESTERN ISLES	11	12.2	7.6	3.01	5
	ALL SCOTLAND	1115	6.9	4.7	0.17	

SCOTTISH CITIES

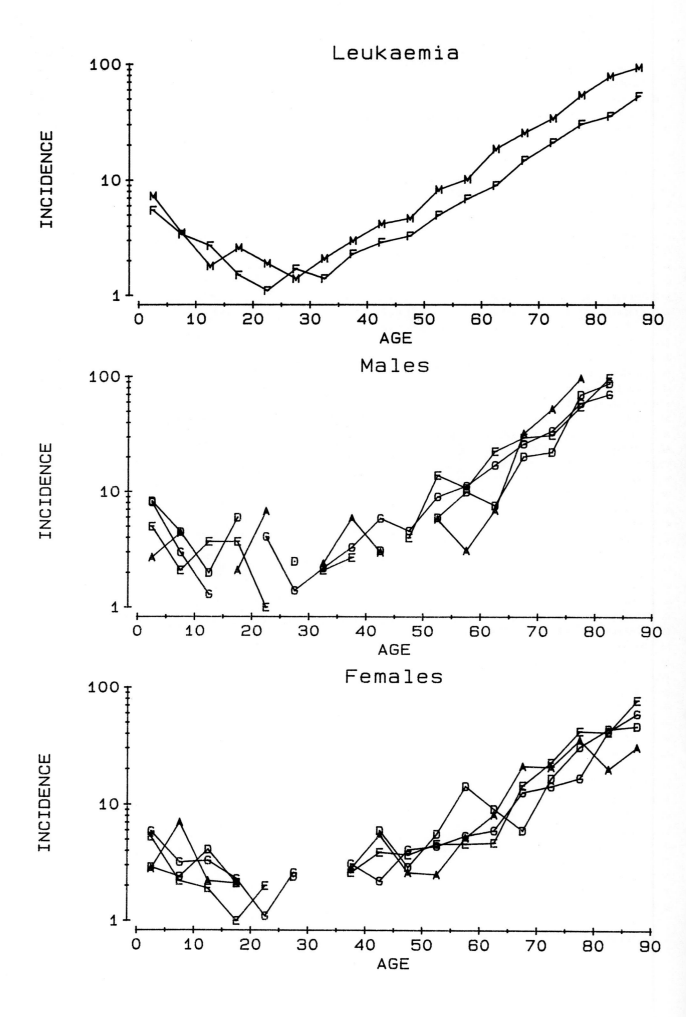

OTHER SITES

A series of maps relating to other cancer sites is now presented. The numbers of cases of these cancers are smaller than for the principal sites, and for this reason less importance can be attached to their geographical patterns of distribution. Nevertheless, they are shown here for completeness. The format used is that of the sixth and seventh pages of the previous presentations (*see* p. 55).

As the full title of several of these sites is frequently rather long, an abbreviated version has been used for the maps. When applicable, the abbreviated title, with the full title in parentheses, is given below as well as the relevant ICD8 rubric numbers.

Tongue	ICD8 141
Salivary gland	ICD8 142
Gum, floor of mouth, etc. (Gum, floor of mouth, and other and unspecified parts of mouth)	ICD8 143, 144, 145
Oropharynx, hypopharynx, etc. (Oropharynx, hypopharynx and pharynx, unspecified)	ICD8 146, 148, 149
Nasopharynx	ICD8 147
Small intestine (Small intestine, including duodenum)	ICD8 152
Colon (Large intestine, except rectum)	ICD8 153
Rectum (Rectum and rectosigmoid junction)	ICD8 154
Gallbladder, etc. (Gallbladder and bile ducts)	ICD8 156
Nasal cavities, etc (Nose, nasal cavities, middle ear and accessory sinuses)	ICD8 160
Bone	ICD8 170
Connective and other soft tissue	ICD8 171
Other female genital organs (Other and unspecified female genital organs)	ICD8 184
Kidney	ICD8 189.0, 189.1
Eye	ICD8 190
Brain and nervous system (Brain and other parts of nervous system)	ICD8 191, 192
Lymphatic leukaemia	ICD8 204
Myeloid leukaemia	ICD8 205

Tongue — Males

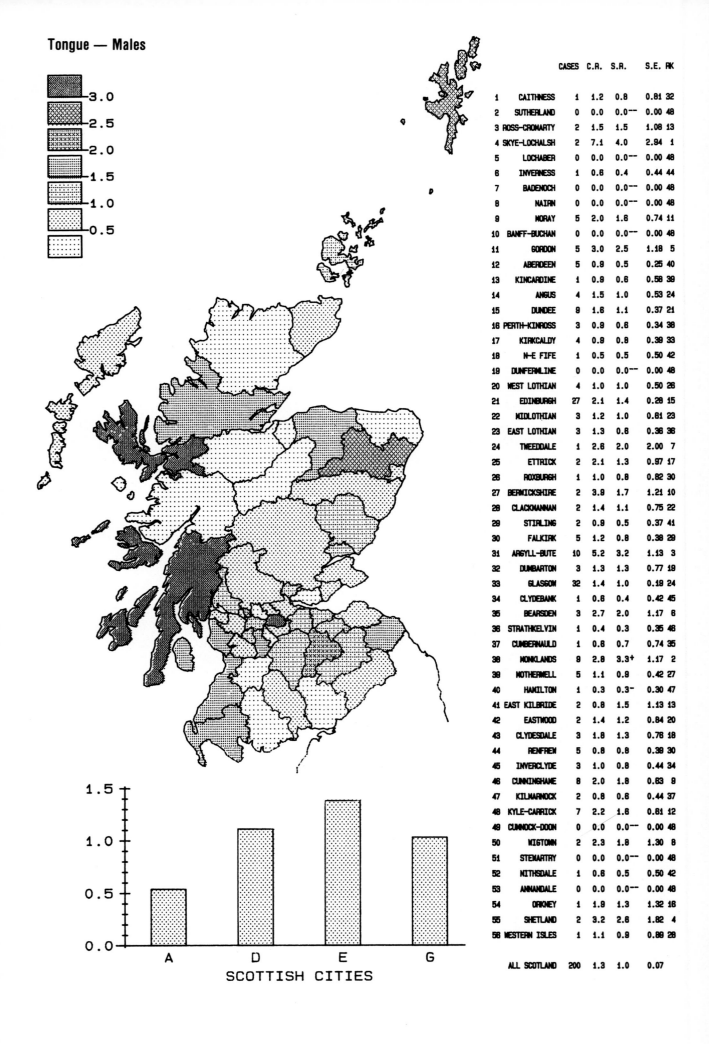

		CASES	C.R.	S.R.	S.E.	RK
1	CAITHNESS	1	1.2	0.8	0.81	32
2	SUTHERLAND	0	0.0	0.0--	0.00	48
3	ROSS-CROMARTY	2	1.5	1.5	1.08	13
4	SKYE-LOCHALSH	2	7.1	4.0	2.94	1
5	LOCHABER	0	0.0	0.0--	0.00	48
6	INVERNESS	1	0.6	0.4	0.44	44
7	BADENOCH	0	0.0	0.0--	0.00	48
8	NAIRN	0	0.0	0.0--	0.00	48
9	MORAY	5	2.0	1.6	0.74	11
10	BANFF-BUCHAN	0	0.0	0.0--	0.00	48
11	GORDON	5	3.0	2.5	1.18	5
12	ABERDEEN	5	0.9	0.5	0.25	40
13	KINCARDINE	1	0.9	0.6	0.58	39
14	ANGUS	4	1.5	1.0	0.53	24
15	DUNDEE	9	1.6	1.1	0.37	21
16	PERTH-KINROSS	3	0.9	0.6	0.34	38
17	KIRKCALDY	4	0.9	0.8	0.39	33
18	N-E FIFE	1	0.5	0.5	0.50	42
19	DUNFERMLINE	0	0.0	0.0--	0.00	48
20	WEST LOTHIAN	4	1.0	1.0	0.50	26
21	EDINBURGH	27	2.1	1.4	0.26	15
22	MIDLOTHIAN	3	1.2	1.0	0.61	23
23	EAST LOTHIAN	3	1.3	0.8	0.36	36
24	TWEEDDALE	1	2.6	2.0	2.00	7
25	ETTRICK	2	2.1	1.3	0.97	17
26	ROXBURGH	1	1.0	0.8	0.82	30
27	BERWICKSHIRE	2	3.6	1.7	1.21	10
28	CLACKMANNAN	2	1.4	1.1	0.75	22
29	STIRLING	2	0.9	0.5	0.37	41
30	FALKIRK	5	1.2	0.8	0.38	29
31	ARGYLL-BUTE	10	5.2	3.2	1.13	3
32	DUMBARTON	3	1.3	1.3	0.77	19
33	GLASGOW	32	1.4	1.0	0.19	24
34	CLYDEBANK	1	0.6	0.4	0.42	45
35	BEARSDEN	3	2.7	2.0	1.17	6
36	STRATHKELVIN	1	0.4	0.3	0.35	46
37	CUMBERNAULD	1	0.6	0.7	0.74	35
38	MONKLANDS	9	2.6	3.3+	1.17	2
39	MOTHERWELL	5	1.1	0.9	0.42	27
40	HAMILTON	1	0.3	0.3-	0.30	47
41	EAST KILBRIDE	2	0.8	1.5	1.13	13
42	EASTWOOD	2	1.4	1.2	0.84	20
43	CLYDESDALE	3	1.6	1.3	0.76	18
44	RENFREW	5	0.8	0.8	0.39	30
45	INVERCLYDE	3	1.0	0.8	0.44	34
46	CUNNINGHAME	8	2.0	1.6	0.63	9
47	KILMARNOCK	2	0.8	0.6	0.44	37
48	KYLE-CARRICK	7	2.2	1.6	0.61	12
49	CUMNOCK-DOON	0	0.0	0.0--	0.00	48
50	WIGTOWN	2	2.3	1.6	1.30	8
51	STEWARTRY	0	0.0	0.0--	0.00	48
52	NITHSDALE	1	0.6	0.5	0.50	42
53	ANNANDALE	0	0.0	0.0--	0.00	48
54	ORKNEY	1	1.9	1.3	1.32	16
55	SHETLAND	2	3.2	2.6	1.82	4
56	WESTERN ISLES	1	1.1	0.9	0.89	28
	ALL SCOTLAND	200	1.3	1.0	0.07	

SCOTTISH CITIES

Tongue — Females

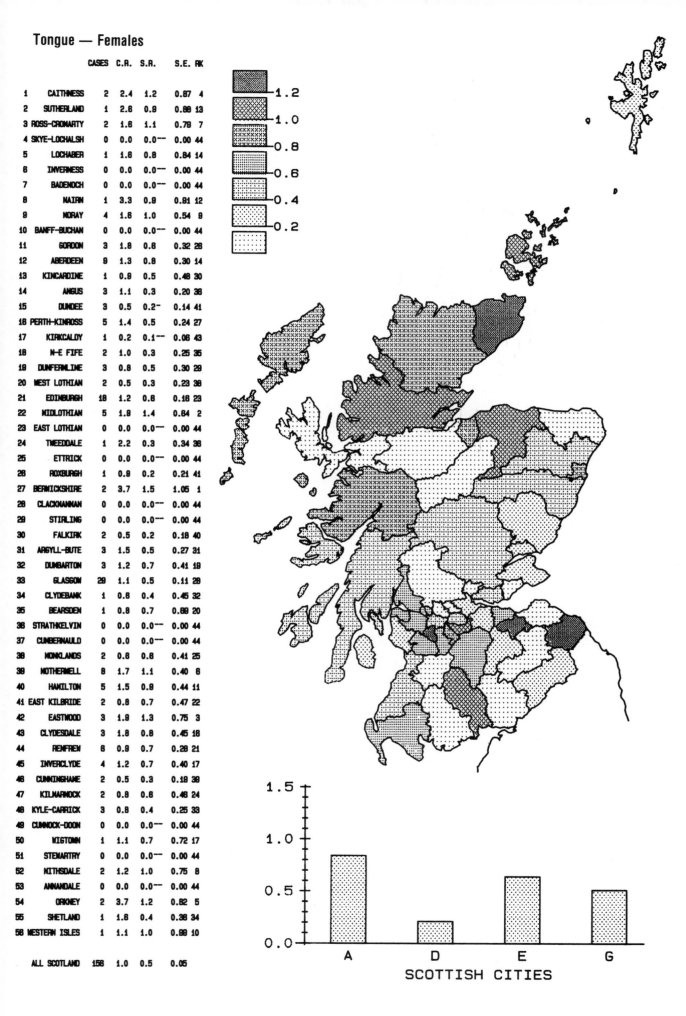

		CASES	C.R.	S.R.	S.E.	RK
1	CAITHNESS	2	2.4	1.2	0.87	4
2	SUTHERLAND	1	2.8	0.9	0.88	13
3	ROSS-CROMARTY	2	1.8	1.1	0.78	7
4	SKYE-LOCHALSH	0	0.0	0.0--	0.00	44
5	LOCHABER	1	1.8	0.8	0.84	14
6	INVERNESS	0	0.0	0.0--	0.00	44
7	BADENOCH	0	0.0	0.0--	0.00	44
8	NAIRN	1	3.3	0.9	0.81	12
9	MORAY	4	1.8	1.0	0.54	9
10	BANFF-BUCHAN	0	0.0	0.0--	0.00	44
11	GORDON	3	1.8	0.8	0.32	26
12	ABERDEEN	9	1.3	0.8	0.30	14
13	KINCARDINE	1	0.9	0.5	0.48	30
14	ANGUS	3	1.1	0.3	0.20	36
15	DUNDEE	3	0.5	0.2-	0.14	41
16	PERTH-KINROSS	5	1.4	0.5	0.24	27
17	KIRKCALDY	1	0.2	0.1--	0.06	43
18	N-E FIFE	2	1.0	0.3	0.25	35
19	DUNFERMLINE	3	0.8	0.5	0.30	29
20	WEST LOTHIAN	2	0.5	0.3	0.23	36
21	EDINBURGH	18	1.2	0.8	0.16	23
22	MIDLOTHIAN	5	1.9	1.4	0.64	2
23	EAST LOTHIAN	0	0.0	0.0--	0.00	44
24	TWEEDDALE	1	2.2	0.3	0.34	36
25	ETTRICK	0	0.0	0.0--	0.00	44
26	ROXBURGH	1	0.9	0.2	0.21	41
27	BERWICKSHIRE	2	3.7	1.5	1.05	1
28	CLACKMANNAN	0	0.0	0.0--	0.00	44
29	STIRLING	0	0.0	0.0--	0.00	44
30	FALKIRK	2	0.5	0.2	0.18	40
31	ARGYLL-BUTE	3	1.5	0.5	0.27	31
32	DUMBARTON	3	1.2	0.7	0.41	19
33	GLASGOW	29	1.1	0.5	0.11	28
34	CLYDEBANK	1	0.8	0.4	0.45	32
35	BEARSDEN	1	0.8	0.7	0.88	20
36	STRATHKELVIN	0	0.0	0.0--	0.00	44
37	CUMBERNAULD	0	0.0	0.0--	0.00	44
38	MONKLANDS	2	0.8	0.8	0.41	25
39	MOTHERWELL	8	1.7	1.1	0.40	6
40	HAMILTON	5	1.5	0.9	0.44	11
41	EAST KILBRIDE	2	0.8	0.7	0.47	22
42	EASTWOOD	3	1.9	1.3	0.75	3
43	CLYDESDALE	3	1.8	0.8	0.45	16
44	RENFREW	6	0.9	0.7	0.28	21
45	INVERCLYDE	4	1.2	0.7	0.40	17
46	CUNNINGHAME	2	0.5	0.3	0.19	39
47	KILMARNOCK	2	0.8	0.6	0.46	24
48	KYLE-CARRICK	3	0.8	0.4	0.25	33
49	CUMNOCK-DOON	0	0.0	0.0--	0.00	44
50	WIGTOWN	1	1.1	0.7	0.72	17
51	STEWARTRY	0	0.0	0.0--	0.00	44
52	NITHSDALE	2	1.2	1.0	0.75	8
53	ANNANDALE	0	0.0	0.0--	0.00	44
54	ORKNEY	2	3.7	1.2	0.82	5
55	SHETLAND	1	1.8	0.4	0.36	34
56	WESTERN ISLES	1	1.1	1.0	0.88	10
	ALL SCOTLAND	156	1.0	0.5	0.05	

SCOTTISH CITIES

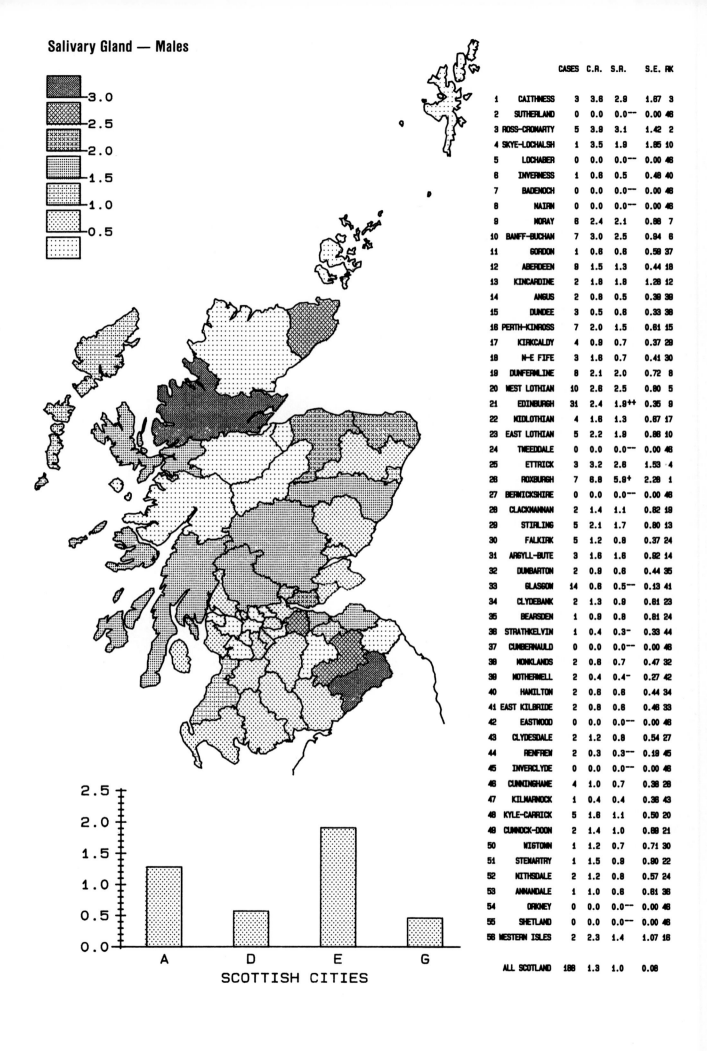

Salivary Gland — Males

		CASES	C.R.	S.R.	S.E.	RK
1	CAITHNESS	3	3.6	2.9	1.87	3
2	SUTHERLAND	0	0.0	0.0--	0.00	46
3	ROSS-CROMARTY	5	3.9	3.1	1.42	2
4	SKYE-LOCHALSH	1	3.5	1.9	1.85	10
5	LOCHABER	0	0.0	0.0--	0.00	46
6	INVERNESS	1	0.6	0.5	0.48	40
7	BADENOCH	0	0.0	0.0--	0.00	46
8	NAIRN	0	0.0	0.0--	0.00	46
9	MORAY	6	2.4	2.1	0.88	7
10	BANFF-BUCHAN	7	3.0	2.5	0.94	6
11	GORDON	1	0.6	0.6	0.59	37
12	ABERDEEN	9	1.5	1.3	0.44	18
13	KINCARDINE	2	1.8	1.8	1.26	12
14	ANGUS	2	0.6	0.5	0.39	39
15	DUNDEE	3	0.5	0.6	0.33	38
16	PERTH-KINROSS	7	2.0	1.5	0.61	15
17	KIRKCALDY	4	0.9	0.7	0.37	29
18	N-E FIFE	3	1.6	0.7	0.41	30
19	DUNFERMLINE	8	2.1	2.0	0.72	8
20	WEST LOTHIAN	10	2.6	2.5	0.80	5
21	EDINBURGH	31	2.4	1.9++	0.35	9
22	MIDLOTHIAN	4	1.6	1.3	0.67	17
23	EAST LOTHIAN	5	2.2	1.9	0.88	10
24	TWEEDDALE	0	0.0	0.0--	0.00	46
25	ETTRICK	3	3.2	2.6	1.53	4
26	ROXBURGH	7	6.6	5.9+	2.26	1
27	BERWICKSHIRE	0	0.0	0.0--	0.00	46
28	CLACKMANNAN	2	1.4	1.1	0.82	19
29	STIRLING	5	2.1	1.7	0.80	13
30	FALKIRK	5	1.2	0.8	0.37	24
31	ARGYLL-BUTE	3	1.8	1.6	0.92	14
32	DUMBARTON	2	0.9	0.6	0.44	35
33	GLASGOW	14	0.6	0.5--	0.13	41
34	CLYDEBANK	2	1.3	0.8	0.61	23
35	BEARSDEN	1	0.9	0.8	0.81	24
36	STRATHKELVIN	1	0.4	0.3-	0.33	44
37	CUMBERNAULD	0	0.0	0.0--	0.00	46
38	MONKLANDS	2	0.6	0.7	0.47	32
39	MOTHERWELL	2	0.4	0.4-	0.27	42
40	HAMILTON	2	0.6	0.6	0.44	34
41	EAST KILBRIDE	2	0.8	0.8	0.46	33
42	EASTWOOD	0	0.0	0.0--	0.00	46
43	CLYDESDALE	2	1.2	0.8	0.54	27
44	RENFREW	2	0.3	0.3--	0.19	45
45	INVERCLYDE	0	0.0	0.0--	0.00	46
46	CUNNINGHAME	4	1.0	0.7	0.38	26
47	KILMARNOCK	1	0.4	0.4	0.38	43
48	KYLE-CARRICK	5	1.6	1.1	0.50	20
49	CUMNOCK-DOON	2	1.4	1.0	0.69	21
50	WIGTOWN	1	1.2	0.7	0.71	30
51	STEWARTRY	1	1.5	0.9	0.90	22
52	NITHSDALE	2	1.2	0.8	0.57	24
53	ANNANDALE	1	1.0	0.6	0.61	36
54	ORKNEY	0	0.0	0.0--	0.00	46
55	SHETLAND	0	0.0	0.0--	0.00	46
56	WESTERN ISLES	2	2.3	1.4	1.07	16
	ALL SCOTLAND	188	1.3	1.0	0.08	

SCOTTISH CITIES

Salivary Gland — Females

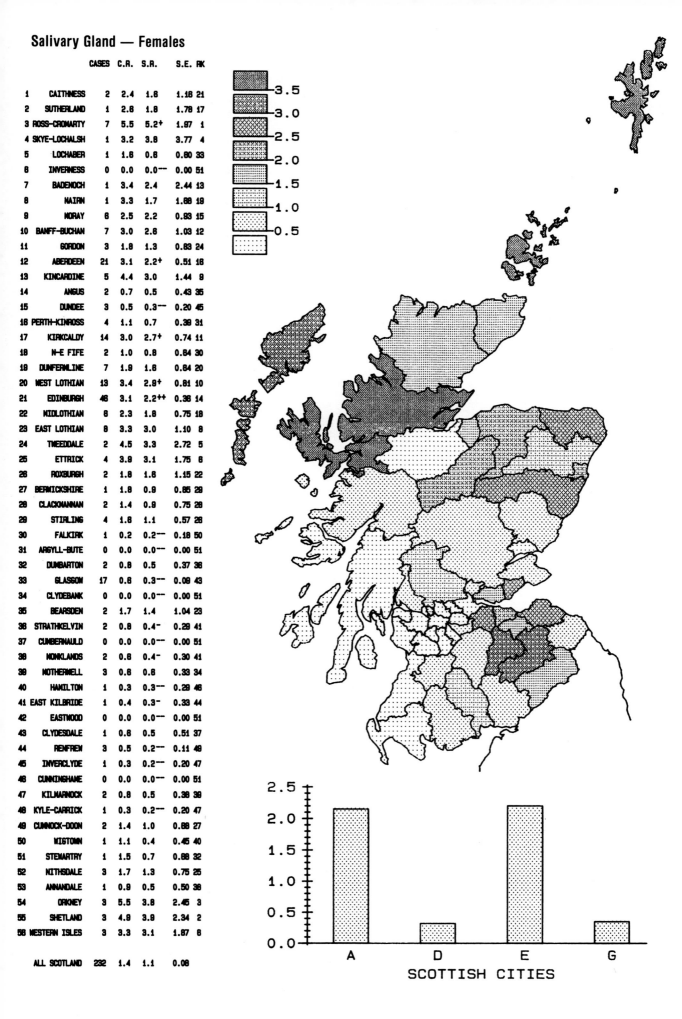

		CASES	C.R.	S.R.	S.E.	RK
1	CAITHNESS	2	2.4	1.8	1.18	21
2	SUTHERLAND	1	2.6	1.8	1.78	17
3	ROSS-CROMARTY	7	5.5	5.2+	1.97	1
4	SKYE-LOCHALSH	1	3.2	3.8	3.77	4
5	LOCHABER	1	1.6	0.6	0.60	33
6	INVERNESS	0	0.0	0.0--	0.00	51
7	BADENOCH	1	3.4	2.4	2.44	13
8	NAIRN	1	3.3	1.7	1.66	19
9	MORAY	6	2.5	2.2	0.93	15
10	BANFF-BUCHAN	7	3.0	2.6	1.03	12
11	GORDON	3	1.8	1.3	0.83	24
12	ABERDEEN	21	3.1	2.2+	0.51	16
13	KINCARDINE	5	4.4	3.0	1.44	9
14	ANGUS	2	0.7	0.5	0.43	35
15	DUNDEE	3	0.5	0.3--	0.20	45
16	PERTH-KINROSS	4	1.1	0.7	0.39	31
17	KIRKCALDY	14	3.0	2.7+	0.74	11
18	N-E FIFE	2	1.0	0.8	0.64	30
19	DUNFERMLINE	7	1.9	1.6	0.64	20
20	WEST LOTHIAN	13	3.4	2.9+	0.81	10
21	EDINBURGH	46	3.1	2.2++	0.36	14
22	MIDLOTHIAN	6	2.3	1.8	0.75	16
23	EAST LOTHIAN	6	3.3	3.0	1.10	8
24	TWEEDDALE	2	4.5	3.3	2.72	5
25	ETTRICK	4	3.9	3.1	1.75	6
26	ROXBURGH	2	1.8	1.6	1.15	22
27	BERWICKSHIRE	1	1.6	0.9	0.85	29
28	CLACKMANNAN	2	1.4	0.9	0.75	28
29	STIRLING	4	1.6	1.1	0.57	26
30	FALKIRK	1	0.2	0.2--	0.18	50
31	ARGYLL-BUTE	0	0.0	0.0--	0.00	51
32	DUMBARTON	2	0.8	0.5	0.37	38
33	GLASGOW	17	0.6	0.3--	0.09	43
34	CLYDEBANK	0	0.0	0.0--	0.00	51
35	BEARSDEN	2	1.7	1.4	1.04	23
36	STRATHKELVIN	2	0.6	0.4-	0.29	41
37	CUMBERNAULD	0	0.0	0.0--	0.00	51
38	MONKLANDS	2	0.6	0.4-	0.30	41
39	MOTHERWELL	3	0.6	0.6	0.33	34
40	HAMILTON	1	0.3	0.3--	0.29	46
41	EAST KILBRIDE	1	0.4	0.3-	0.33	44
42	EASTWOOD	0	0.0	0.0--	0.00	51
43	CLYDESDALE	1	0.6	0.5	0.51	37
44	RENFREW	3	0.5	0.2--	0.11	49
45	INVERCLYDE	1	0.3	0.2--	0.20	47
46	CUNNINGHAME	0	0.0	0.0--	0.00	51
47	KILMARNOCK	2	0.8	0.5	0.39	39
48	KYLE-CARRICK	1	0.3	0.2--	0.20	47
49	CUMNOCK-DOON	2	1.4	1.0	0.69	27
50	WIGTOWN	1	1.1	0.4	0.45	40
51	STEWARTRY	1	1.5	0.7	0.68	32
52	NITHSDALE	3	1.7	1.3	0.75	25
53	ANNANDALE	1	0.9	0.5	0.50	36
54	ORKNEY	3	5.5	3.8	2.45	3
55	SHETLAND	3	4.9	3.9	2.34	2
56	WESTERN ISLES	3	3.3	3.1	1.87	6
	ALL SCOTLAND	232	1.4	1.1	0.08	

SCOTTISH CITIES

Gum, Floor of Mouth, etc — Males

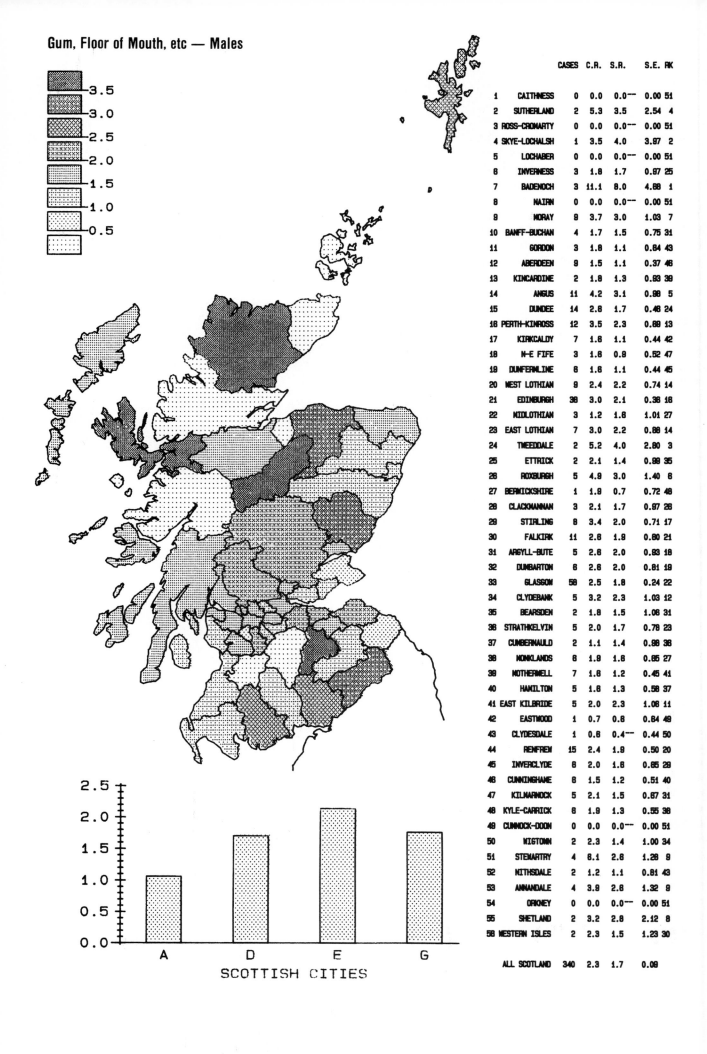

		CASES	C.R.	S.R.	S.E.	RK
1	CAITHNESS	0	0.0	0.0--	0.00	51
2	SUTHERLAND	2	5.3	3.5	2.54	4
3	ROSS-CROMARTY	0	0.0	0.0--	0.00	51
4	SKYE-LOCHALSH	1	3.5	4.0	3.97	2
5	LOCHABER	0	0.0	0.0--	0.00	51
6	INVERNESS	3	1.8	1.7	0.97	25
7	BADENOCH	3	11.1	8.0	4.88	1
8	NAIRN	0	0.0	0.0--	0.00	51
9	MORAY	9	3.7	3.0	1.03	7
10	BANFF-BUCHAN	4	1.7	1.5	0.75	31
11	GORDON	3	1.8	1.1	0.64	43
12	ABERDEEN	9	1.5	1.1	0.37	46
13	KINCARDINE	2	1.8	1.3	0.93	39
14	ANGUS	11	4.2	3.1	0.88	5
15	DUNDEE	14	2.6	1.7	0.46	24
16	PERTH-KINROSS	12	3.5	2.3	0.69	13
17	KIRKCALDY	7	1.6	1.1	0.44	42
18	N-E FIFE	3	1.8	0.9	0.52	47
19	DUNFERMLINE	6	1.6	1.1	0.44	45
20	WEST LOTHIAN	9	2.4	2.2	0.74	14
21	EDINBURGH	36	3.0	2.1	0.36	16
22	MIDLOTHIAN	3	1.2	1.6	1.01	27
23	EAST LOTHIAN	7	3.0	2.2	0.86	14
24	TWEEDDALE	2	5.2	4.0	2.80	3
25	ETTRICK	2	2.1	1.4	0.98	35
26	ROXBURGH	5	4.9	3.0	1.40	6
27	BERWICKSHIRE	1	1.8	0.7	0.72	48
28	CLACKMANNAN	3	2.1	1.7	0.97	26
29	STIRLING	8	3.4	2.0	0.71	17
30	FALKIRK	11	2.6	1.9	0.60	21
31	ARGYLL-BUTE	5	2.6	2.0	0.93	18
32	DUMBARTON	6	2.6	2.0	0.81	19
33	GLASGOW	58	2.5	1.6	0.24	22
34	CLYDEBANK	5	3.2	2.3	1.03	12
35	BEARSDEN	2	1.8	1.5	1.06	31
36	STRATHKELVIN	5	2.0	1.7	0.78	23
37	CUMBERNAULD	2	1.1	1.4	0.98	36
38	MONKLANDS	6	1.9	1.6	0.65	27
39	MOTHERWELL	7	1.6	1.2	0.45	41
40	HAMILTON	5	1.6	1.3	0.58	37
41	EAST KILBRIDE	5	2.0	2.3	1.06	11
42	EASTWOOD	1	0.7	0.8	0.84	49
43	CLYDESDALE	1	0.8	0.4--	0.44	50
44	RENFREW	15	2.4	1.9	0.50	20
45	INVERCLYDE	6	2.0	1.6	0.65	29
46	CUNNINGHAME	6	1.5	1.2	0.51	40
47	KILMARNOCK	5	2.1	1.5	0.67	31
48	KYLE-CARRICK	6	1.9	1.3	0.55	38
49	CUMNOCK-DOON	0	0.0	0.0--	0.00	51
50	WIGTOWN	2	2.3	1.4	1.00	34
51	STEWARTRY	4	8.1	2.8	1.28	9
52	NITHSDALE	2	1.2	1.1	0.81	43
53	ANNANDALE	4	3.9	2.8	1.32	9
54	ORKNEY	0	0.0	0.0--	0.00	51
55	SHETLAND	2	3.2	2.8	2.12	8
56	WESTERN ISLES	2	2.3	1.5	1.23	30
	ALL SCOTLAND	340	2.3	1.7	0.09	

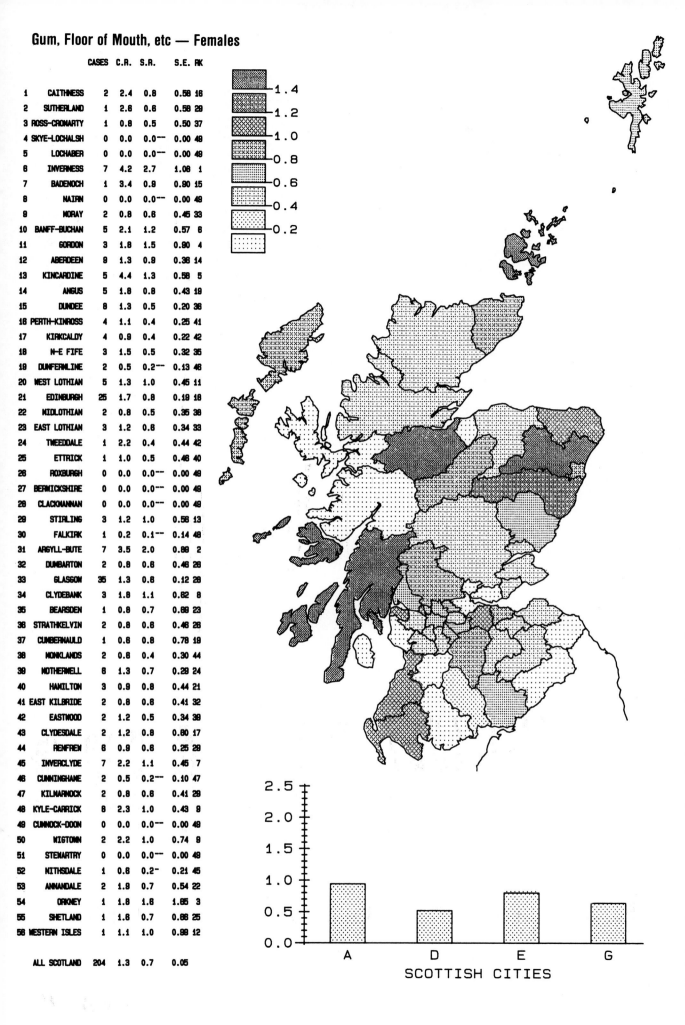

Gum, Floor of Mouth, etc — Females

		CASES	C.R.	S.R.	S.E.	RK
1	CAITHNESS	2	2.4	0.8	0.58	16
2	SUTHERLAND	1	2.6	0.6	0.58	29
3	ROSS-CROMARTY	1	0.8	0.5	0.50	37
4	SKYE-LOCHALSH	0	0.0	0.0--	0.00	49
5	LOCHABER	0	0.0	0.0--	0.00	49
6	INVERNESS	7	4.2	2.7	1.08	1
7	BADENOCH	1	3.4	0.9	0.90	15
8	NAIRN	0	0.0	0.0--	0.00	49
9	MORAY	2	0.8	0.6	0.45	33
10	BANFF-BUCHAN	5	2.1	1.2	0.57	6
11	GORDON	3	1.8	1.5	0.90	4
12	ABERDEEN	9	1.3	0.9	0.36	14
13	KINCARDINE	5	4.4	1.3	0.58	5
14	ANGUS	5	1.6	0.8	0.43	19
15	DUNDEE	8	1.3	0.5	0.20	36
16	PERTH-KINROSS	4	1.1	0.4	0.25	41
17	KIRKCALDY	4	0.9	0.4	0.22	42
18	N-E FIFE	3	1.5	0.5	0.32	35
19	DUNFERMLINE	2	0.5	0.2--	0.13	46
20	WEST LOTHIAN	5	1.3	1.0	0.45	11
21	EDINBURGH	25	1.7	0.8	0.19	18
22	MIDLOTHIAN	2	0.8	0.5	0.35	38
23	EAST LOTHIAN	3	1.2	0.6	0.34	33
24	TWEEDDALE	1	2.2	0.4	0.44	42
25	ETTRICK	1	1.0	0.5	0.46	40
26	ROXBURGH	0	0.0	0.0--	0.00	49
27	BERWICKSHIRE	0	0.0	0.0--	0.00	49
28	CLACKMANNAN	0	0.0	0.0--	0.00	49
29	STIRLING	3	1.2	1.0	0.58	13
30	FALKIRK	1	0.2	0.1--	0.14	48
31	ARGYLL-BUTE	7	3.5	2.0	0.89	2
32	DUMBARTON	2	0.8	0.6	0.46	26
33	GLASGOW	35	1.3	0.6	0.12	28
34	CLYDEBANK	3	1.8	1.1	0.62	8
35	BEARSDEN	1	0.8	0.7	0.69	23
36	STRATHKELVIN	2	0.8	0.6	0.46	26
37	CUMBERNAULD	1	0.8	0.8	0.78	19
38	MONKLANDS	2	0.8	0.4	0.30	44
39	MOTHERWELL	6	1.3	0.7	0.29	24
40	HAMILTON	3	0.9	0.6	0.44	21
41	EAST KILBRIDE	2	0.8	0.6	0.41	32
42	EASTWOOD	2	1.2	0.5	0.34	39
43	CLYDESDALE	2	1.2	0.8	0.60	17
44	RENFREW	6	0.9	0.6	0.25	28
45	INVERCLYDE	7	2.2	1.1	0.45	7
46	CUNNINGHAME	2	0.5	0.2--	0.10	47
47	KILMARNOCK	2	0.8	0.6	0.41	29
48	KYLE-CARRICK	6	2.3	1.0	0.43	9
49	CUMNOCK-DOON	0	0.0	0.0--	0.00	49
50	WIGTOWN	2	2.2	1.0	0.74	9
51	STEWARTRY	0	0.0	0.0--	0.00	49
52	NITHSDALE	1	0.6	0.2-	0.21	45
53	ANNANDALE	2	1.9	0.7	0.54	22
54	ORKNEY	1	1.6	1.6	1.65	3
55	SHETLAND	1	1.6	0.7	0.66	25
56	WESTERN ISLES	1	1.1	1.0	0.98	12
	ALL SCOTLAND	204	1.3	0.7	0.05	

Oropharynx, Hypopharynx, etc — Males

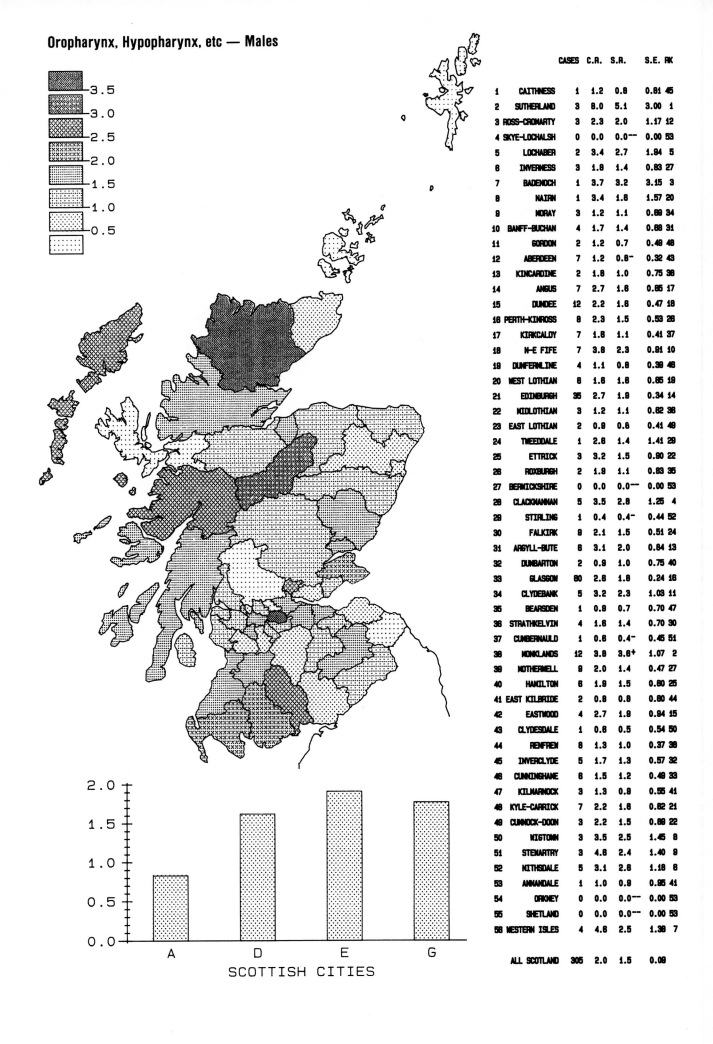

		CASES	C.R.	S.R.	S.E.	RK
1	CAITHNESS	1	1.2	0.8	0.81	45
2	SUTHERLAND	3	8.0	5.1	3.00	1
3	ROSS-CROMARTY	3	2.3	2.0	1.17	12
4	SKYE-LOCHALSH	0	0.0	0.0--	0.00	53
5	LOCHABER	2	3.4	2.7	1.94	5
6	INVERNESS	3	1.8	1.4	0.83	27
7	BADENOCH	1	3.7	3.2	3.15	3
8	NAIRN	1	3.4	1.6	1.57	20
9	MORAY	3	1.2	1.1	0.69	34
10	BANFF-BUCHAN	4	1.7	1.4	0.68	31
11	GORDON	2	1.2	0.7	0.49	48
12	ABERDEEN	7	1.2	0.8-	0.32	43
13	KINCARDINE	2	1.8	1.0	0.75	39
14	ANGUS	7	2.7	1.6	0.65	17
15	DUNDEE	12	2.2	1.6	0.47	18
16	PERTH-KINROSS	8	2.3	1.5	0.53	26
17	KIRKCALDY	7	1.8	1.1	0.41	37
18	N-E FIFE	7	3.8	2.3	0.91	10
19	DUNFERMLINE	4	1.1	0.8	0.39	46
20	WEST LOTHIAN	8	1.8	1.6	0.65	19
21	EDINBURGH	35	2.7	1.8	0.34	14
22	MIDLOTHIAN	3	1.2	1.1	0.82	36
23	EAST LOTHIAN	2	0.9	0.6	0.41	49
24	TWEEDDALE	1	2.6	1.4	1.41	29
25	ETTRICK	3	3.2	1.5	0.90	22
26	ROXBURGH	2	1.9	1.1	0.83	35
27	BERWICKSHIRE	0	0.0	0.0--	0.00	53
28	CLACKMANNAN	5	3.5	2.6	1.25	4
29	STIRLING	1	0.4	0.4-	0.44	52
30	FALKIRK	9	2.1	1.5	0.51	24
31	ARGYLL-BUTE	6	3.1	2.0	0.84	13
32	DUMBARTON	2	0.9	1.0	0.75	40
33	GLASGOW	60	2.6	1.8	0.24	16
34	CLYDEBANK	5	3.2	2.3	1.03	11
35	BEARSDEN	1	0.9	0.7	0.70	47
36	STRATHKELVIN	4	1.6	1.4	0.70	30
37	CUMBERNAULD	1	0.6	0.4-	0.45	51
38	MONKLANDS	12	3.8	3.8+	1.07	2
39	MOTHERWELL	9	2.0	1.4	0.47	27
40	HAMILTON	6	1.9	1.5	0.80	25
41	EAST KILBRIDE	2	0.8	0.8	0.60	44
42	EASTWOOD	4	2.7	1.9	0.94	15
43	CLYDESDALE	1	0.6	0.5	0.54	50
44	RENFREW	8	1.3	1.0	0.37	38
45	INVERCLYDE	5	1.7	1.3	0.57	32
46	CUNNINGHAME	6	1.5	1.2	0.49	33
47	KILMARNOCK	3	1.3	0.9	0.55	41
48	KYLE-CARRICK	7	2.2	1.6	0.62	21
49	CUMNOCK-DOON	3	2.2	1.5	0.89	22
50	WIGTOWN	3	3.5	2.5	1.45	8
51	STEWARTRY	3	4.6	2.4	1.40	9
52	NITHSDALE	5	3.1	2.6	1.16	6
53	ANNANDALE	1	1.0	0.9	0.95	41
54	ORKNEY	0	0.0	0.0--	0.00	53
55	SHETLAND	0	0.0	0.0--	0.00	53
56	WESTERN ISLES	4	4.6	2.5	1.39	7
	ALL SCOTLAND	305	2.0	1.5	0.09	

SCOTTISH CITIES

Oropharynx, Hypopharynx, etc — Females

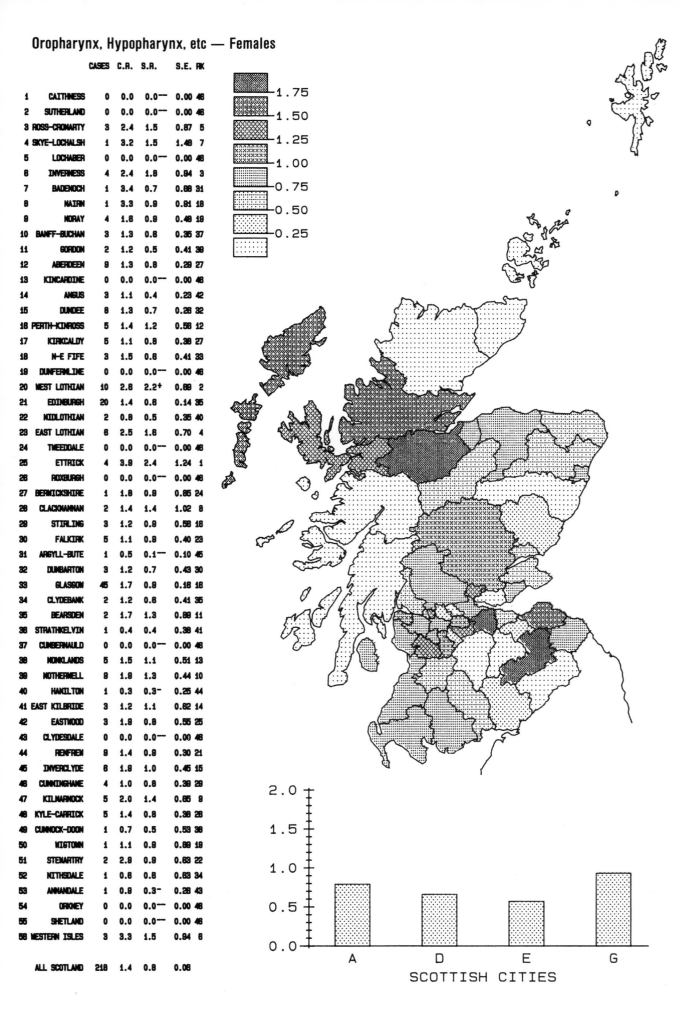

		CASES	C.R.	S.R.	S.E.	RK
1	CAITHNESS	0	0.0	0.0 --	0.00	46
2	SUTHERLAND	0	0.0	0.0 --	0.00	46
3	ROSS-CROMARTY	3	2.4	1.5	0.87	5
4	SKYE-LOCHALSH	1	3.2	1.5	1.48	7
5	LOCHABER	0	0.0	0.0 --	0.00	46
6	INVERNESS	4	2.4	1.8	0.94	3
7	BADENOCH	1	3.4	0.7	0.68	31
8	NAIRN	1	3.3	0.9	0.91	18
9	MORAY	4	1.6	0.9	0.46	19
10	BANFF-BUCHAN	3	1.3	0.6	0.35	37
11	GORDON	2	1.2	0.5	0.41	39
12	ABERDEEN	9	1.3	0.8	0.29	27
13	KINCARDINE	0	0.0	0.0 --	0.00	46
14	ANGUS	3	1.1	0.4	0.23	42
15	DUNDEE	8	1.3	0.7	0.26	32
16	PERTH-KINROSS	5	1.4	1.2	0.56	12
17	KIRKCALDY	5	1.1	0.8	0.36	27
18	N-E FIFE	3	1.5	0.6	0.41	33
19	DUNFERMLINE	0	0.0	0.0 --	0.00	46
20	WEST LOTHIAN	10	2.6	2.2 +	0.69	2
21	EDINBURGH	20	1.4	0.8	0.14	35
22	MIDLOTHIAN	2	0.8	0.5	0.35	40
23	EAST LOTHIAN	6	2.5	1.6	0.70	4
24	TWEEDDALE	0	0.0	0.0 --	0.00	46
25	ETTRICK	4	3.9	2.4	1.24	1
26	ROXBURGH	0	0.0	0.0 --	0.00	46
27	BERWICKSHIRE	1	1.6	0.9	0.85	24
28	CLACKMANNAN	2	1.4	1.4	1.02	8
29	STIRLING	3	1.2	0.9	0.56	18
30	FALKIRK	5	1.1	0.9	0.40	23
31	ARGYLL-BUTE	1	0.5	0.1 --	0.10	45
32	DUMBARTON	3	1.2	0.7	0.43	30
33	GLASGOW	45	1.7	0.9	0.16	16
34	CLYDEBANK	2	1.2	0.8	0.41	36
35	BEARSDEN	2	1.7	1.3	0.89	11
36	STRATHKELVIN	1	0.4	0.4	0.36	41
37	CUMBERNAULD	0	0.0	0.0 --	0.00	46
38	MONKLANDS	5	1.5	1.1	0.51	13
39	MOTHERWELL	8	1.8	1.3	0.44	10
40	HAMILTON	1	0.3	0.3 -	0.25	44
41	EAST KILBRIDE	3	1.2	1.1	0.62	14
42	EASTWOOD	3	1.8	0.9	0.55	25
43	CLYDESDALE	0	0.0	0.0 --	0.00	46
44	RENFREW	8	1.4	0.9	0.30	21
45	INVERCLYDE	6	1.8	1.0	0.45	15
46	CUNNINGHAME	4	1.0	0.8	0.39	29
47	KILMARNOCK	5	2.0	1.4	0.65	9
48	KYLE-CARRICK	5	1.4	0.8	0.36	26
49	CUMNOCK-DOON	1	0.7	0.5	0.53	38
50	WIGTOWN	1	1.1	0.9	0.89	19
51	STEWARTRY	2	2.9	0.9	0.63	22
52	NITHSDALE	1	0.6	0.8	0.63	34
53	ANNANDALE	1	0.9	0.3 -	0.26	43
54	ORKNEY	0	0.0	0.0 --	0.00	46
55	SHETLAND	0	0.0	0.0 --	0.00	46
56	WESTERN ISLES	3	3.3	1.5	0.94	6
	ALL SCOTLAND	218	1.4	0.8	0.06	

SCOTTISH CITIES

Nasopharynx — Males

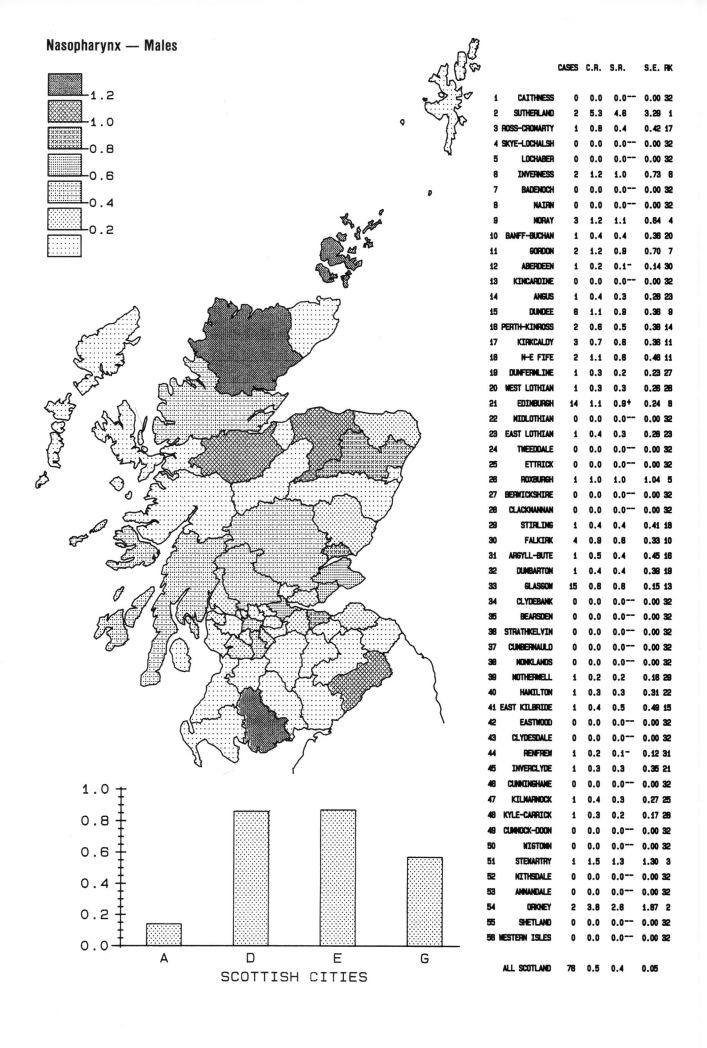

		CASES	C.R.	S.R.	S.E.	RK
1	CAITHNESS	0	0.0	0.0--	0.00	32
2	SUTHERLAND	2	5.3	4.6	3.29	1
3	ROSS-CROMARTY	1	0.8	0.4	0.42	17
4	SKYE-LOCHALSH	0	0.0	0.0--	0.00	32
5	LOCHABER	0	0.0	0.0--	0.00	32
6	INVERNESS	2	1.2	1.0	0.73	6
7	BADENOCH	0	0.0	0.0--	0.00	32
8	NAIRN	0	0.0	0.0--	0.00	32
9	MORAY	3	1.2	1.1	0.64	4
10	BANFF-BUCHAN	1	0.4	0.4	0.36	20
11	GORDON	2	1.2	0.9	0.70	7
12	ABERDEEN	1	0.2	0.1-	0.14	30
13	KINCARDINE	0	0.0	0.0--	0.00	32
14	ANGUS	1	0.4	0.3	0.26	23
15	DUNDEE	6	1.1	0.9	0.36	9
16	PERTH-KINROSS	2	0.6	0.5	0.36	14
17	KIRKCALDY	3	0.7	0.6	0.36	11
18	N-E FIFE	2	1.1	0.6	0.46	11
19	DUNFERMLINE	1	0.3	0.2	0.23	27
20	WEST LOTHIAN	1	0.3	0.3	0.26	26
21	EDINBURGH	14	1.1	0.9+	0.24	8
22	MIDLOTHIAN	0	0.0	0.0--	0.00	32
23	EAST LOTHIAN	1	0.4	0.3	0.26	23
24	TWEEDDALE	0	0.0	0.0--	0.00	32
25	ETTRICK	0	0.0	0.0--	0.00	32
26	ROXBURGH	1	1.0	1.0	1.04	5
27	BERWICKSHIRE	0	0.0	0.0--	0.00	32
28	CLACKMANNAN	0	0.0	0.0--	0.00	32
29	STIRLING	1	0.4	0.4	0.41	18
30	FALKIRK	4	0.9	0.8	0.33	10
31	ARGYLL-BUTE	1	0.5	0.4	0.45	16
32	DUMBARTON	1	0.4	0.4	0.39	19
33	GLASGOW	15	0.6	0.6	0.15	13
34	CLYDEBANK	0	0.0	0.0--	0.00	32
35	BEARSDEN	0	0.0	0.0--	0.00	32
36	STRATHKELVIN	0	0.0	0.0--	0.00	32
37	CUMBERNAULD	0	0.0	0.0--	0.00	32
38	MONKLANDS	0	0.0	0.0--	0.00	32
39	MOTHERWELL	1	0.2	0.2	0.16	29
40	HAMILTON	1	0.3	0.3	0.31	22
41	EAST KILBRIDE	1	0.4	0.5	0.49	15
42	EASTWOOD	0	0.0	0.0--	0.00	32
43	CLYDESDALE	0	0.0	0.0--	0.00	32
44	RENFREW	1	0.2	0.1-	0.12	31
45	INVERCLYDE	1	0.3	0.3	0.35	21
46	CUNNINGHAME	0	0.0	0.0--	0.00	32
47	KILMARNOCK	1	0.4	0.3	0.27	25
48	KYLE-CARRICK	1	0.3	0.2	0.17	28
49	CUMNOCK-DOON	0	0.0	0.0--	0.00	32
50	WIGTOWN	0	0.0	0.0--	0.00	32
51	STEWARTRY	1	1.5	1.3	1.30	3
52	NITHSDALE	0	0.0	0.0--	0.00	32
53	ANNANDALE	0	0.0	0.0--	0.00	32
54	ORKNEY	2	3.8	2.6	1.87	2
55	SHETLAND	0	0.0	0.0--	0.00	32
56	WESTERN ISLES	0	0.0	0.0--	0.00	32
	ALL SCOTLAND	76	0.5	0.4	0.05	

SCOTTISH CITIES

Nasopharynx — Females

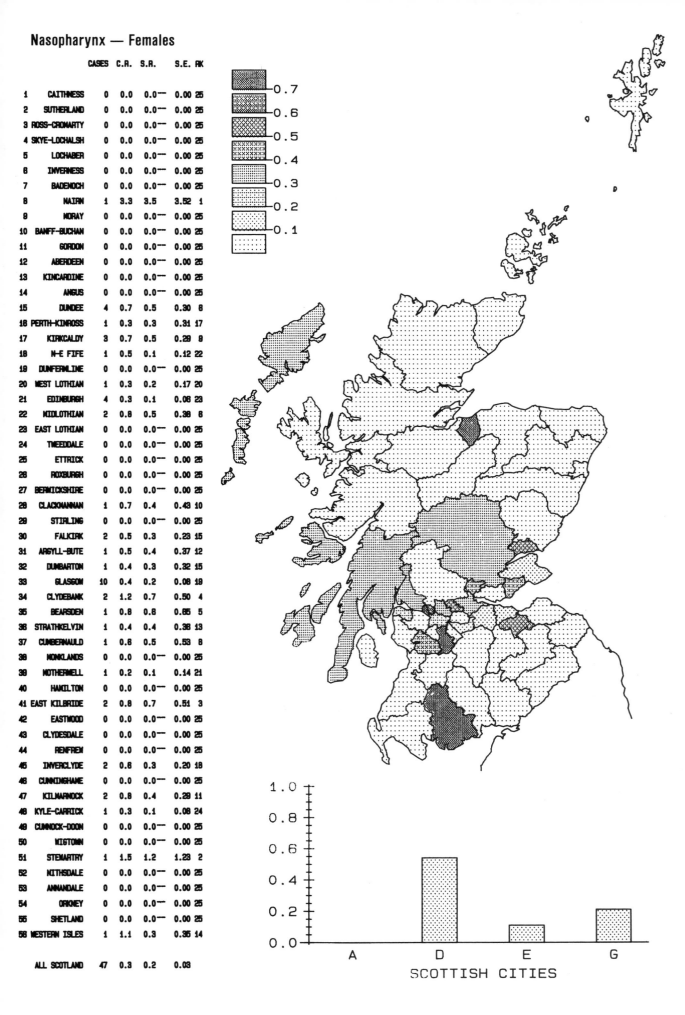

		CASES	C.R.	S.R.	S.E.	RK
1	CAITHNESS	0	0.0	0.0 —	0.00	25
2	SUTHERLAND	0	0.0	0.0 —	0.00	25
3	ROSS-CROMARTY	0	0.0	0.0 —	0.00	25
4	SKYE-LOCHALSH	0	0.0	0.0 —	0.00	25
5	LOCHABER	0	0.0	0.0 —	0.00	25
6	INVERNESS	0	0.0	0.0 —	0.00	25
7	BADENOCH	0	0.0	0.0 —	0.00	25
8	NAIRN	1	3.3	3.5	3.52	1
9	MORAY	0	0.0	0.0 —	0.00	25
10	BANFF-BUCHAN	0	0.0	0.0 —	0.00	25
11	GORDON	0	0.0	0.0 —	0.00	25
12	ABERDEEN	0	0.0	0.0 —	0.00	25
13	KINCARDINE	0	0.0	0.0 —	0.00	25
14	ANGUS	0	0.0	0.0 —	0.00	25
15	DUNDEE	4	0.7	0.5	0.30	6
16	PERTH-KINROSS	1	0.3	0.3	0.31	17
17	KIRKCALDY	3	0.7	0.5	0.29	6
18	N-E FIFE	1	0.5	0.1	0.12	22
19	DUNFERMLINE	0	0.0	0.0 —	0.00	25
20	WEST LOTHIAN	1	0.3	0.2	0.17	20
21	EDINBURGH	4	0.3	0.1	0.08	23
22	MIDLOTHIAN	2	0.8	0.5	0.38	6
23	EAST LOTHIAN	0	0.0	0.0 —	0.00	25
24	TWEEDDALE	0	0.0	0.0 —	0.00	25
25	ETTRICK	0	0.0	0.0 —	0.00	25
26	ROXBURGH	0	0.0	0.0 —	0.00	25
27	BERWICKSHIRE	0	0.0	0.0 —	0.00	25
28	CLACKMANNAN	1	0.7	0.4	0.43	10
29	STIRLING	0	0.0	0.0 —	0.00	25
30	FALKIRK	2	0.5	0.3	0.23	15
31	ARGYLL-BUTE	1	0.5	0.4	0.37	12
32	DUMBARTON	1	0.4	0.3	0.32	15
33	GLASGOW	10	0.4	0.2	0.08	19
34	CLYDEBANK	2	1.2	0.7	0.50	4
35	BEARSDEN	1	0.8	0.8	0.65	5
36	STRATHKELVIN	1	0.4	0.4	0.38	13
37	CUMBERNAULD	1	0.8	0.5	0.53	8
38	MONKLANDS	0	0.0	0.0 —	0.00	25
39	MOTHERWELL	1	0.2	0.1	0.14	21
40	HAMILTON	0	0.0	0.0 —	0.00	25
41	EAST KILBRIDE	2	0.8	0.7	0.51	3
42	EASTWOOD	0	0.0	0.0 —	0.00	25
43	CLYDESDALE	0	0.0	0.0 —	0.00	25
44	RENFREW	0	0.0	0.0 —	0.00	25
45	INVERCLYDE	2	0.8	0.3	0.20	18
46	CUNNINGHAME	0	0.0	0.0 —	0.00	25
47	KILMARNOCK	2	0.8	0.4	0.29	11
48	KYLE-CARRICK	1	0.3	0.1	0.08	24
49	CUMNOCK-DOON	0	0.0	0.0 —	0.00	25
50	WIGTOWN	0	0.0	0.0 —	0.00	25
51	STEWARTRY	1	1.5	1.2	1.23	2
52	NITHSDALE	0	0.0	0.0 —	0.00	25
53	ANNANDALE	0	0.0	0.0 —	0.00	25
54	ORKNEY	0	0.0	0.0 —	0.00	25
55	SHETLAND	0	0.0	0.0 —	0.00	25
56	WESTERN ISLES	1	1.1	0.3	0.35	14
	ALL SCOTLAND	47	0.3	0.2	0.03	

SCOTTISH CITIES

Small Intestine — Males

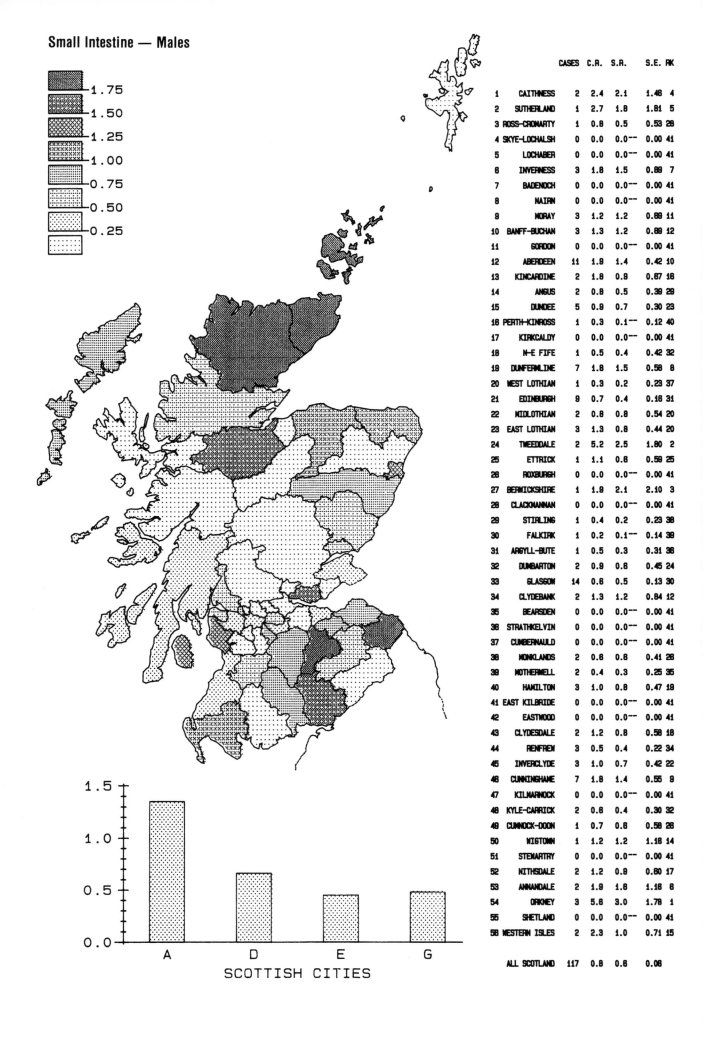

		CASES	C.R.	S.R.	S.E.	RK
1	CAITHNESS	2	2.4	2.1	1.46	4
2	SUTHERLAND	1	2.7	1.8	1.81	5
3	ROSS-CROMARTY	1	0.8	0.5	0.53	26
4	SKYE-LOCHALSH	0	0.0	0.0 --	0.00	41
5	LOCHABER	0	0.0	0.0 --	0.00	41
6	INVERNESS	3	1.8	1.5	0.89	7
7	BADENOCH	0	0.0	0.0 --	0.00	41
8	NAIRN	0	0.0	0.0 --	0.00	41
9	MORAY	3	1.2	1.2	0.89	11
10	BANFF-BUCHAN	3	1.3	1.2	0.89	12
11	GORDON	0	0.0	0.0 --	0.00	41
12	ABERDEEN	11	1.9	1.4	0.42	10
13	KINCARDINE	2	1.8	0.9	0.87	16
14	ANGUS	2	0.8	0.5	0.39	29
15	DUNDEE	5	0.9	0.7	0.30	23
16	PERTH-KINROSS	1	0.3	0.1 --	0.12	40
17	KIRKCALDY	0	0.0	0.0 --	0.00	41
18	N-E FIFE	1	0.5	0.4	0.42	32
19	DUNFERMLINE	7	1.8	1.5	0.58	8
20	WEST LOTHIAN	1	0.3	0.2	0.23	37
21	EDINBURGH	8	0.7	0.4	0.16	31
22	MIDLOTHIAN	2	0.8	0.8	0.54	20
23	EAST LOTHIAN	3	1.3	0.8	0.44	20
24	TWEEDDALE	2	5.2	2.5	1.80	2
25	ETTRICK	1	1.1	0.8	0.59	25
26	ROXBURGH	0	0.0	0.0 --	0.00	41
27	BERWICKSHIRE	1	1.9	2.1	2.10	3
28	CLACKMANNAN	0	0.0	0.0 --	0.00	41
29	STIRLING	1	0.4	0.2	0.23	38
30	FALKIRK	1	0.2	0.1 --	0.14	39
31	ARGYLL-BUTE	1	0.5	0.3	0.31	36
32	DUMBARTON	2	0.9	0.6	0.45	24
33	GLASGOW	14	0.6	0.5	0.13	30
34	CLYDEBANK	2	1.3	1.2	0.84	12
35	BEARSDEN	0	0.0	0.0 --	0.00	41
36	STRATHKELVIN	0	0.0	0.0 --	0.00	41
37	CUMBERNAULD	0	0.0	0.0 --	0.00	41
38	MONKLANDS	2	0.6	0.6	0.41	26
39	MOTHERWELL	2	0.4	0.3	0.25	35
40	HAMILTON	3	1.0	0.8	0.47	19
41	EAST KILBRIDE	0	0.0	0.0 --	0.00	41
42	EASTWOOD	0	0.0	0.0 --	0.00	41
43	CLYDESDALE	2	1.2	0.8	0.58	18
44	RENFREW	3	0.5	0.4	0.22	34
45	INVERCLYDE	3	1.0	0.7	0.42	22
46	CUNNINGHAME	7	1.8	1.4	0.55	9
47	KILMARNOCK	0	0.0	0.0 --	0.00	41
48	KYLE-CARRICK	2	0.8	0.4	0.30	32
49	CUMNOCK-DOON	1	0.7	0.6	0.58	26
50	WIGTOWN	1	1.2	1.2	1.16	14
51	STEWARTRY	0	0.0	0.0 --	0.00	41
52	NITHSDALE	2	1.2	0.9	0.60	17
53	ANNANDALE	2	1.9	1.6	1.16	6
54	ORKNEY	3	5.6	3.0	1.76	1
55	SHETLAND	0	0.0	0.0 --	0.00	41
56	WESTERN ISLES	2	2.3	1.0	0.71	15
	ALL SCOTLAND	117	0.8	0.6	0.06	

Small Intestine — Females

		CASES	C.R.	S.R.	S.E.	RK
1	CAITHNESS	0	0.0	0.0--	0.00	44
2	SUTHERLAND	0	0.0	0.0--	0.00	44
3	ROSS-CROMARTY	2	1.6	0.9	0.84	9
4	SKYE-LOCHALSH	0	0.0	0.0--	0.00	44
5	LOCHABER	0	0.0	0.0--	0.00	44
6	INVERNESS	2	1.2	0.8	0.67	13
7	BADENOCH	0	0.0	0.0--	0.00	44
8	NAIRN	0	0.0	0.0--	0.00	44
9	MORAY	2	0.8	0.3	0.21	36
10	BANFF-BUCHAN	1	0.4	0.4	0.38	31
11	GORDON	2	1.2	0.6	0.51	18
12	ABERDEEN	6	0.8	0.4	0.20	30
13	KINCARDINE	1	0.9	0.7	0.66	16
14	ANGUS	4	1.4	0.9	0.45	9
15	DUNDEE	7	1.2	0.5	0.22	24
16	PERTH-KINROSS	4	1.1	0.4	0.24	26
17	KIRKCALDY	2	0.4	0.3	0.20	37
18	N-E FIFE	1	0.5	0.2	0.18	40
19	DUNFERMLINE	4	1.1	0.5	0.26	23
20	WEST LOTHIAN	5	1.3	1.0	0.47	7
21	EDINBURGH	12	0.8	0.3	0.10	34
22	MIDLOTHIAN	2	0.8	0.5	0.38	21
23	EAST LOTHIAN	2	0.8	0.7	0.52	15
24	TWEEDDALE	2	4.5	1.0	0.79	6
25	ETTRICK	1	1.0	0.5	0.46	25
26	ROXBURGH	1	0.9	0.4	0.43	29
27	BERWICKSHIRE	2	3.7	2.3	1.85	2
28	CLACKMANNAN	0	0.0	0.0--	0.00	44
29	STIRLING	0	0.0	0.0--	0.00	44
30	FALKIRK	3	0.7	0.4	0.22	32
31	ARGYLL-BUTE	1	0.5	0.3	0.25	39
32	DUMBARTON	2	0.8	0.7	0.53	14
33	GLASGOW	24	0.8	0.4	0.11	27
34	CLYDEBANK	2	1.2	0.8	0.61	12
35	BEARSDEN	1	0.8	0.3	0.30	34
36	STRATHKELVIN	1	0.4	0.2	0.18	40
37	CUMBERNAULD	2	1.1	1.4	0.98	4
38	MONKLANDS	3	0.8	0.6	0.34	19
39	MOTHERWELL	6	1.3	0.9	0.37	9
40	HAMILTON	1	0.3	0.2	0.17	42
41	EAST KILBRIDE	1	0.4	0.3	0.35	33
42	EASTWOOD	3	1.9	1.2	0.74	5
43	CLYDESDALE	2	1.2	0.5	0.40	21
44	RENFREW	6	0.9	0.6	0.24	20
45	INVERCLYDE	0	0.0	0.0--	0.00	44
46	CUNNINGHAME	0	0.0	0.0--	0.00	44
47	KILMARNOCK	2	0.8	0.4	0.31	27
48	KYLE-CARRICK	1	0.3	0.1-	0.14	43
49	CUMNOCK-DOON	0	0.0	0.0--	0.00	44
50	WIGTOWN	0	0.0	0.0--	0.00	44
51	STEWARTRY	1	1.5	0.3	0.28	37
52	NITHSDALE	3	1.7	1.0	0.59	6
53	ANNANDALE	2	1.9	0.6	0.46	17
54	ORKNEY	3	5.5	2.9	1.82	1
55	SHETLAND	2	3.3	1.8	1.42	3
56	WESTERN ISLES	0	0.0	0.0--	0.00	44
	ALL SCOTLAND	137	0.8	0.4	0.04	

SCOTTISH CITIES

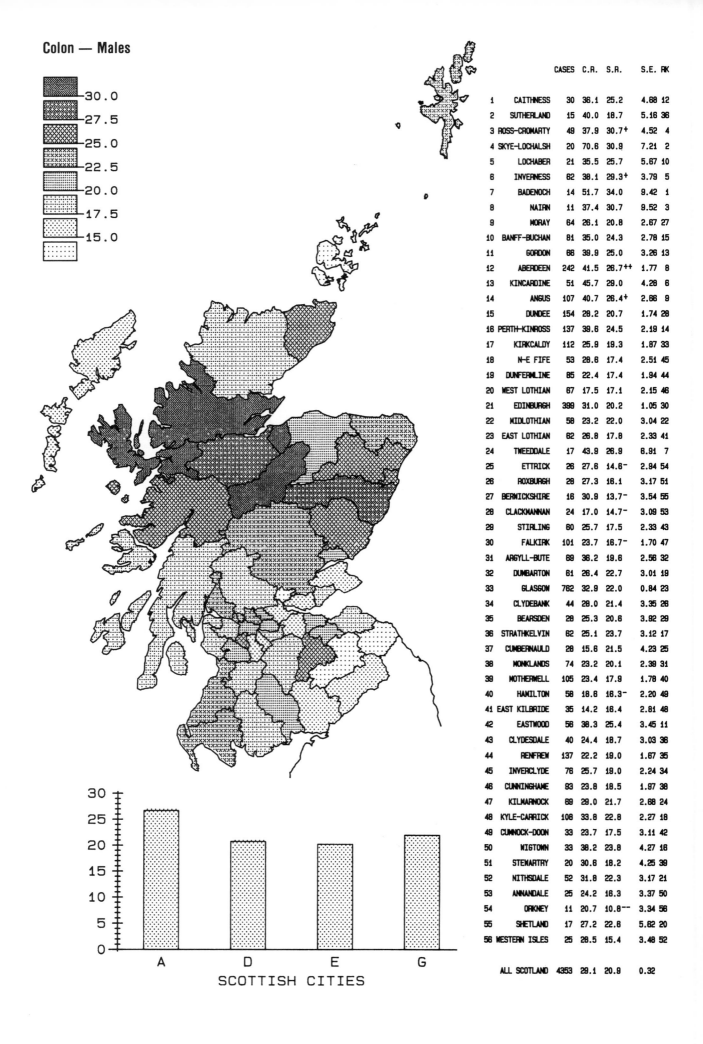

Colon — Males

		CASES	C.R.	S.R.	S.E.	RK
1	CAITHNESS	30	36.1	25.2	4.68	12
2	SUTHERLAND	15	40.0	18.7	5.16	36
3	ROSS-CROMARTY	49	37.9	30.7+	4.52	4
4	SKYE-LOCHALSH	20	70.6	30.9	7.21	2
5	LOCHABER	21	35.5	25.7	5.67	10
6	INVERNESS	62	38.1	29.3+	3.79	5
7	BADENOCH	14	51.7	34.0	9.42	1
8	NAIRN	11	37.4	30.7	9.52	3
9	MORAY	64	26.1	20.8	2.67	27
10	BANFF-BUCHAN	81	35.0	24.3	2.78	15
11	GORDON	66	39.9	25.0	3.26	13
12	ABERDEEN	242	41.5	26.7++	1.77	8
13	KINCARDINE	51	45.7	29.0	4.28	6
14	ANGUS	107	40.7	26.4+	2.66	9
15	DUNDEE	154	28.2	20.7	1.74	28
16	PERTH-KINROSS	137	39.6	24.5	2.19	14
17	KIRKCALDY	112	25.9	19.3	1.87	33
18	N-E FIFE	53	28.6	17.4	2.51	45
19	DUNFERMLINE	85	22.4	17.4	1.94	44
20	WEST LOTHIAN	67	17.5	17.1	2.15	46
21	EDINBURGH	399	31.0	20.2	1.05	30
22	MIDLOTHIAN	58	23.2	22.0	3.04	22
23	EAST LOTHIAN	62	26.8	17.8	2.33	41
24	TWEEDDALE	17	43.9	26.9	6.91	7
25	ETTRICK	26	27.6	14.6-	2.94	54
26	ROXBURGH	28	27.3	16.1	3.17	51
27	BERWICKSHIRE	18	30.9	13.7-	3.54	55
28	CLACKMANNAN	24	17.0	14.7-	3.09	53
29	STIRLING	60	25.7	17.5	2.33	43
30	FALKIRK	101	23.7	16.7-	1.70	47
31	ARGYLL-BUTE	69	36.2	19.6	2.56	32
32	DUMBARTON	61	26.4	22.7	3.01	19
33	GLASGOW	762	32.9	22.0	0.84	23
34	CLYDEBANK	44	28.0	21.4	3.35	26
35	BEARSDEN	28	25.3	20.6	3.92	29
36	STRATHKELVIN	62	25.1	23.7	3.12	17
37	CUMBERNAULD	28	15.6	21.5	4.23	25
38	MONKLANDS	74	23.2	20.1	2.39	31
39	MOTHERWELL	105	23.4	17.9	1.78	40
40	HAMILTON	58	18.6	16.3-	2.20	49
41	EAST KILBRIDE	35	14.2	16.4	2.81	48
42	EASTWOOD	56	38.3	25.4	3.45	11
43	CLYDESDALE	40	24.4	18.7	3.03	36
44	RENFREW	137	22.2	19.0	1.67	35
45	INVERCLYDE	76	25.7	19.0	2.24	34
46	CUNNINGHAME	83	23.8	18.5	1.97	38
47	KILMARNOCK	69	29.0	21.7	2.68	24
48	KYLE-CARRICK	108	33.8	22.8	2.27	18
49	CUMNOCK-DOON	33	23.7	17.5	3.11	42
50	WIGTOWN	33	38.2	23.8	4.27	16
51	STEWARTRY	20	30.6	18.2	4.25	39
52	NITHSDALE	52	31.8	22.3	3.17	21
53	ANNANDALE	25	24.2	16.3	3.37	50
54	ORKNEY	11	20.7	10.8--	3.34	56
55	SHETLAND	17	27.2	22.6	5.62	20
56	WESTERN ISLES	25	26.5	15.4	3.48	52
	ALL SCOTLAND	4353	29.1	20.9	0.32	

SCOTTISH CITIES

Colon — Females

		CASES	C.R.	S.R.	S.E.	RK
1	CAITHNESS	33	38.9	23.8	4.44	9
2	SUTHERLAND	15	39.1	20.0	5.91	25
3	ROSS-CROMARTY	62	48.6	26.8+	3.98	1
4	SKYE-LOCHALSH	15	47.9	27.5	8.70	2
5	LOCHABER	25	41.1	26.4	5.58	4
6	INVERNESS	55	32.7	18.3	2.69	36
7	BADENOCH	10	33.7	16.0	5.66	49
8	NAIRN	11	38.0	25.9	8.52	6
9	MORAY	109	44.7	23.9	2.49	8
10	BANFF-BUCHAN	95	40.3	20.8	2.33	19
11	GORDON	71	43.1	20.4	2.81	23
12	ABERDEEN	351	52.2	24.7++	1.48	7
13	KINCARDINE	61	53.8	26.3	3.76	5
14	ANGUS	154	55.0	23.8+	2.13	10
15	DUNDEE	254	41.9	20.7	1.46	20
16	PERTH-KINROSS	175	47.6	22.0	1.87	13
17	KIRKCALDY	174	37.7	20.6	1.67	22
18	N-E FIFE	91	44.3	17.9	2.22	37
19	DUNFERMLINE	105	28.1	15.6⁻	1.64	53
20	WEST LOTHIAN	105	27.2	18.9	1.91	30
21	EDINBURGH	640	43.3	18.9	0.85	28
22	MIDLOTHIAN	86	25.6	18.9	2.41	29
23	EAST LOTHIAN	89	40.9	18.9	2.07	27
24	TWEEDDALE	18	35.9	11.0--	3.20	55
25	ETTRICK	48	47.0	20.7	3.46	21
26	ROXBURGH	63	58.7	21.4	3.07	17
27	BERWICKSHIRE	21	38.5	18.8	4.03	46
28	CLACKMANNAN	53	36.2	21.6	3.12	16
29	STIRLING	83	34.2	19.9	2.33	26
30	FALKIRK	145	33.3	18.6	1.62	33
31	ARGYLL-BUTE	70	35.4	15.9	2.16	50
32	DUMBARTON	91	36.6	21.9	2.46	14
33	GLASGOW	1080	40.9	18.5	0.62	35
34	CLYDEBANK	59	34.8	21.1	2.86	18
35	BEARSDEN	38	31.7	16.9	2.80	42
36	STRATHKELVIN	61	24.0	18.8	2.22	43
37	CUMBERNAULD	30	16.8	17.6	3.28	39
38	MONKLANDS	96	26.5	18.7	2.00	32
39	MOTHERWELL	131	27.3	15.8⁻	1.44	52
40	HAMILTON	86	25.9	16.3	1.84	48
41	EAST KILBRIDE	62	24.6	20.2	2.82	24
42	EASTWOOD	60	37.4	16.3	2.29	47
43	CLYDESDALE	53	31.0	14.8⁻	2.16	54
44	RENFREW	196	30.0	17.6	1.32	38
45	INVERCLYDE	118	36.5	18.6	1.82	34
46	CUNNINGHAME	131	31.6	17.0	1.60	41
47	KILMARNOCK	82	32.3	17.5	2.05	40
48	KYLE-CARRICK	174	49.0	21.6	1.79	15
49	CUMNOCK-DOON	40	26.0	18.8	3.13	31
50	WIGTOWN	33	35.9	18.8	3.14	44
51	STEWARTRY	24	34.9	18.6	3.75	45
52	NITHSDALE	83	54.0	27.1+	3.08	3
53	ANNANDALE	48	44.6	22.5	3.43	12
54	ORKNEY	9	16.5	5.3--	1.86	56
55	SHETLAND	31	50.8	23.5	4.83	11
56	WESTERN ISLES	37	41.0	15.8	3.00	51
	ALL SCOTLAND	6138	38.1	19.3	0.27	

Legend:
- 25.0
- 22.5
- 20.0
- 17.5
- 15.0
- 12.5
- 10.0

Bar chart: SCOTTISH CITIES — A, D, E, G

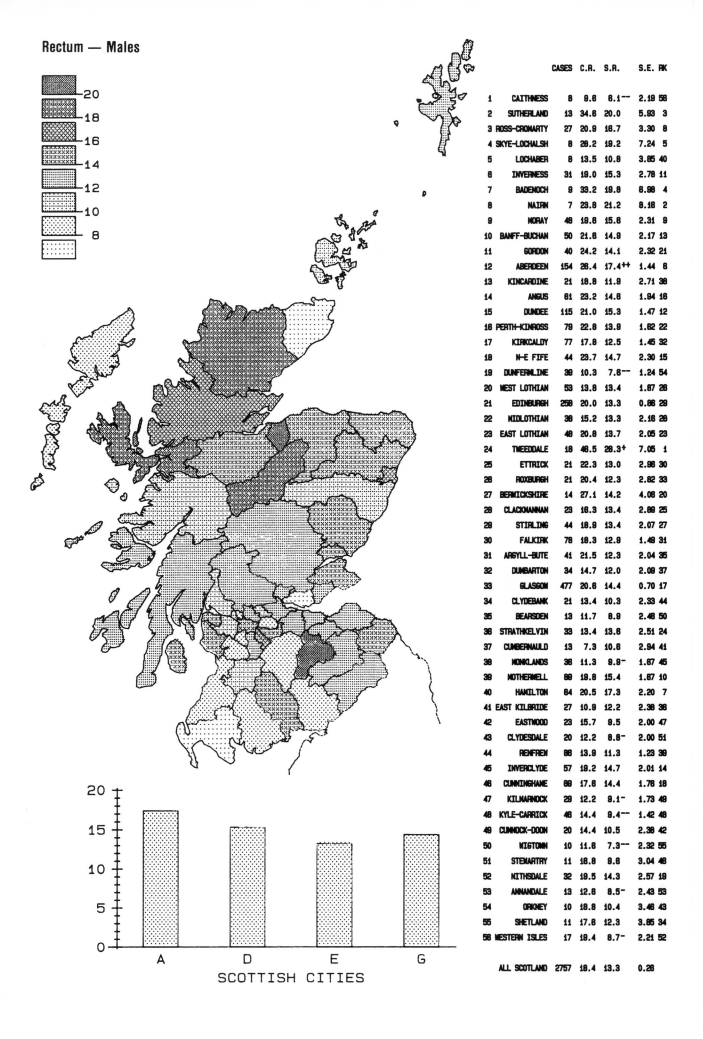

Rectum — Males

		CASES	C.R.	S.R.	S.E.	RK
1	CAITHNESS	8	9.6	6.1--	2.19	56
2	SUTHERLAND	13	34.6	20.0	5.93	3
3	ROSS-CROMARTY	27	20.9	16.7	3.30	8
4	SKYE-LOCHALSH	8	28.2	19.2	7.24	5
5	LOCHABER	8	13.5	10.6	3.85	40
6	INVERNESS	31	19.0	15.3	2.78	11
7	BADENOCH	9	33.2	19.8	6.98	4
8	NAIRN	7	23.8	21.2	8.16	2
9	MORAY	48	19.8	15.6	2.31	9
10	BANFF-BUCHAN	50	21.6	14.9	2.17	13
11	GORDON	40	24.2	14.1	2.32	21
12	ABERDEEN	154	26.4	17.4++	1.44	6
13	KINCARDINE	21	18.8	11.9	2.71	38
14	ANGUS	61	23.2	14.6	1.94	16
15	DUNDEE	115	21.0	15.3	1.47	12
16	PERTH-KINROSS	79	22.8	13.9	1.62	22
17	KIRKCALDY	77	17.8	12.5	1.45	32
18	N-E FIFE	44	23.7	14.7	2.30	15
19	DUNFERMLINE	38	10.3	7.6--	1.24	54
20	WEST LOTHIAN	53	13.8	13.4	1.87	26
21	EDINBURGH	256	20.0	13.3	0.66	29
22	MIDLOTHIAN	36	15.2	13.3	2.16	28
23	EAST LOTHIAN	46	20.8	13.7	2.05	23
24	TWEEDDALE	18	46.5	26.3+	7.05	1
25	ETTRICK	21	22.3	13.0	2.98	30
26	ROXBURGH	21	20.4	12.3	2.82	33
27	BERWICKSHIRE	14	27.1	14.2	4.06	20
28	CLACKMANNAN	23	16.3	13.4	2.89	25
29	STIRLING	44	18.9	13.4	2.07	27
30	FALKIRK	78	18.3	12.9	1.49	31
31	ARGYLL-BUTE	41	21.5	12.3	2.04	35
32	DUMBARTON	34	14.7	12.0	2.09	37
33	GLASGOW	477	20.6	14.4	0.70	17
34	CLYDEBANK	21	13.4	10.3	2.33	44
35	BEARSDEN	13	11.7	8.9	2.46	50
36	STRATHKELVIN	33	13.4	13.6	2.51	24
37	CUMBERNAULD	13	7.3	10.6	2.84	41
38	MONKLANDS	36	11.3	9.9-	1.67	45
39	MOTHERWELL	89	19.6	15.4	1.87	10
40	HAMILTON	64	20.5	17.3	2.20	7
41	EAST KILBRIDE	27	10.9	12.2	2.38	36
42	EASTWOOD	23	15.7	9.5	2.00	47
43	CLYDESDALE	20	12.2	8.8-	2.00	51
44	RENFREW	86	13.9	11.3	1.23	39
45	INVERCLYDE	57	19.2	14.7	2.01	14
46	CUNNINGHAME	89	17.6	14.4	1.76	18
47	KILMARNOCK	29	12.2	9.1-	1.73	49
48	KYLE-CARRICK	46	14.4	9.4--	1.42	48
49	CUMNOCK-DOON	20	14.4	10.5	2.38	42
50	WIGTOWN	10	11.6	7.3--	2.32	55
51	STEWARTRY	11	16.8	9.6	3.04	46
52	NITHSDALE	32	19.5	14.3	2.57	19
53	ANNANDALE	13	12.6	8.5-	2.43	53
54	ORKNEY	10	16.8	10.4	3.46	43
55	SHETLAND	11	17.6	12.3	3.85	34
56	WESTERN ISLES	17	19.4	8.7-	2.21	52
	ALL SCOTLAND	2757	18.4	13.3	0.26	

SCOTTISH CITIES

Rectum — Females

		CASES	C.R.	S.R.	S.E.	RK
1	CAITHNESS	17	20.0	11.0	2.81	9
2	SUTHERLAND	4	10.4	4.3	2.37	52
3	ROSS-CROMARTY	23	18.0	11.5	2.53	7
4	SKYE-LOCHALSH	10	31.9	11.0	3.85	8
5	LOCHABER	7	11.5	7.8	3.18	33
6	INVERNESS	25	14.8	8.5	1.86	22
7	BADENOCH	11	37.1	22.1	7.23	1
8	NAIRN	8	19.6	11.6	4.94	5
9	MORAY	51	20.9	12.1+	1.83	4
10	BANFF-BUCHAN	38	16.1	8.9	1.57	20
11	GORDON	30	18.2	9.1	1.83	19
12	ABERDEEN	140	20.8	10.6++	0.99	10
13	KINCARDINE	27	23.8	13.3	2.78	3
14	ANGUS	51	18.2	8.5	1.36	23
15	DUNDEE	99	16.2	7.3	0.84	36
16	PERTH-KINROSS	81	16.6	6.9	0.99	42
17	KIRKCALDY	65	14.1	7.9	1.04	31
18	N-E FIFE	49	23.8	11.5	1.85	6
19	DUNFERMLINE	50	13.4	7.5	1.11	35
20	WEST LOTHIAN	46	12.4	9.7	1.45	14
21	EDINBURGH	279	16.9	8.4	0.56	24
22	MIDLOTHIAN	31	12.0	8.3	1.52	26
23	EAST LOTHIAN	47	19.4	9.8	1.53	13
24	TWEEDDALE	5	11.2	3.3--	1.61	55
25	ETTRICK	19	16.6	8.0	2.06	29
26	ROXBURGH	17	15.3	5.9	1.60	47
27	BERWICKSHIRE	4	7.3	2.0--	1.04	56
28	CLACKMANNAN	25	17.1	10.4	2.19	11
29	STIRLING	38	15.7	8.5	1.65	17
30	FALKIRK	68	15.6	8.7	1.12	21
31	ARGYLL-BUTE	38	19.2	7.8	1.50	32
32	DUMBARTON	38	14.5	8.4	1.47	25
33	GLASGOW	415	15.7	7.6	0.41	34
34	CLYDEBANK	29	17.1	10.1	1.91	12
35	BEARSDEN	26	23.3	14.0+	2.90	2
36	STRATHKELVIN	35	13.7	9.4	1.66	16
37	CUMBERNAULD	18	10.1	8.6	2.32	18
38	MONKLANDS	29	8.3	6.0	1.18	46
39	MOTHERWELL	54	11.3	7.1	1.01	41
40	HAMILTON	35	10.6	7.2	1.27	39
41	EAST KILBRIDE	18	7.1	6.4	1.54	44
42	EASTWOOD	21	13.1	6.1	1.45	45
43	CLYDESDALE	13	7.6	4.3--	1.27	53
44	RENFREW	76	11.9	7.2	0.88	39
45	INVERCLYDE	43	13.3	7.2	1.15	38
46	CUNNINGHAME	51	12.3	6.6	1.03	43
47	KILMARNOCK	39	15.4	8.4	1.43	26
48	KYLE-CARRICK	53	14.9	7.3	1.08	37
49	CUMNOCK-DOON	19	13.3	8.4	2.01	27
50	WIGTOWN	10	10.9	5.3	1.87	49
51	STEWARTRY	9	13.1	5.2	1.90	51
52	NITHSDALE	26	16.3	8.0	1.64	30
53	ANNANDALE	13	12.1	5.3	1.59	50
54	ORKNEY	4	7.3	3.6-	1.86	54
55	SHETLAND	7	11.5	5.8	2.36	48
56	WESTERN ISLES	16	19.8	9.7	2.69	15
	ALL SCOTLAND	2464	15.4	8.1	0.18	

SCOTTISH CITIES

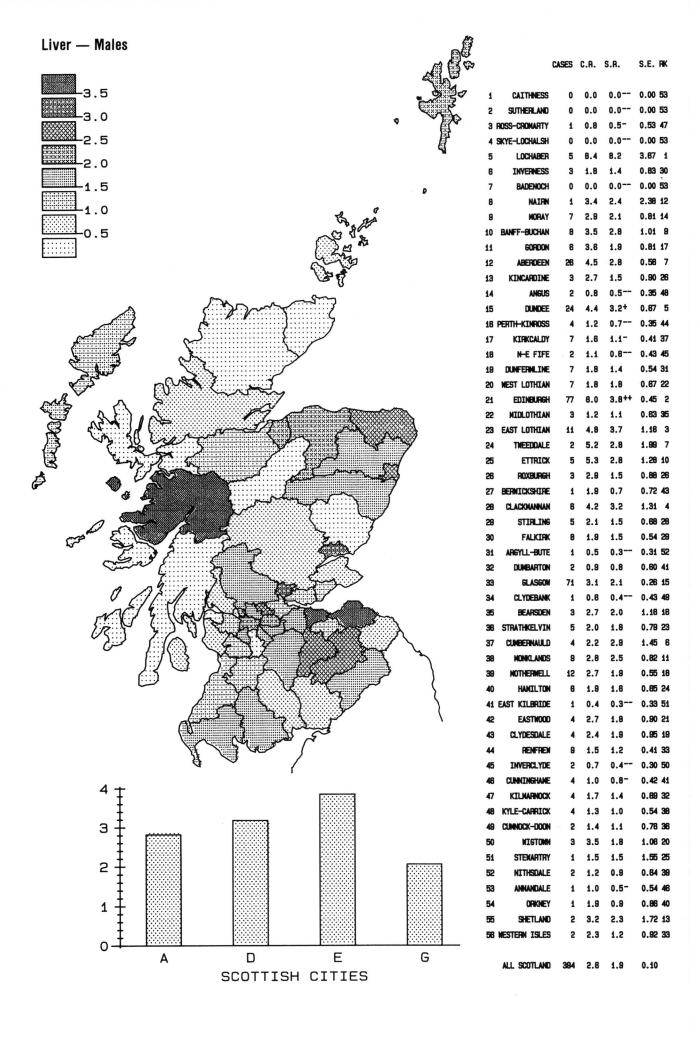

Liver — Males

		CASES	C.R.	S.R.	S.E.	RK
1	CAITHNESS	0	0.0	0.0--	0.00	53
2	SUTHERLAND	0	0.0	0.0--	0.00	53
3	ROSS-CROMARTY	1	0.8	0.5-	0.53	47
4	SKYE-LOCHALSH	0	0.0	0.0--	0.00	53
5	LOCHABER	5	8.4	8.2	3.67	1
6	INVERNESS	3	1.8	1.4	0.83	30
7	BADENOCH	0	0.0	0.0--	0.00	53
8	NAIRN	1	3.4	2.4	2.38	12
9	MORAY	7	2.9	2.1	0.81	14
10	BANFF-BUCHAN	8	3.5	2.8	1.01	9
11	GORDON	6	3.6	1.9	0.81	17
12	ABERDEEN	26	4.5	2.8	0.56	7
13	KINCARDINE	3	2.7	1.5	0.90	26
14	ANGUS	2	0.8	0.5--	0.35	48
15	DUNDEE	24	4.4	3.2+	0.87	5
16	PERTH-KINROSS	4	1.2	0.7-	0.35	44
17	KIRKCALDY	7	1.6	1.1-	0.41	37
18	N-E FIFE	2	1.1	0.8--	0.43	45
19	DUNFERMLINE	7	1.8	1.4	0.54	31
20	WEST LOTHIAN	7	1.8	1.8	0.67	22
21	EDINBURGH	77	6.0	3.8++	0.45	2
22	MIDLOTHIAN	3	1.2	1.1	0.63	35
23	EAST LOTHIAN	11	4.8	3.7	1.16	3
24	TWEEDDALE	2	5.2	2.8	1.99	7
25	ETTRICK	5	5.3	2.8	1.28	10
26	ROXBURGH	3	2.9	1.5	0.88	26
27	BERWICKSHIRE	1	1.9	0.7	0.72	43
28	CLACKMANNAN	8	4.2	3.2	1.31	4
29	STIRLING	5	2.1	1.5	0.68	28
30	FALKIRK	8	1.9	1.5	0.54	29
31	ARGYLL-BUTE	1	0.5	0.3--	0.31	52
32	DUMBARTON	2	0.9	0.8	0.60	41
33	GLASGOW	71	3.1	2.1	0.26	15
34	CLYDEBANK	1	0.8	0.4--	0.43	49
35	BEARSDEN	3	2.7	2.0	1.16	16
36	STRATHKELVIN	5	2.0	1.8	0.78	23
37	CUMBERNAULD	4	2.2	2.9	1.45	6
38	MONKLANDS	9	2.8	2.5	0.82	11
39	MOTHERWELL	12	2.7	1.9	0.55	18
40	HAMILTON	6	1.9	1.6	0.65	24
41	EAST KILBRIDE	1	0.4	0.3--	0.33	51
42	EASTWOOD	4	2.7	1.8	0.90	21
43	CLYDESDALE	4	2.4	1.8	0.95	19
44	RENFREW	9	1.5	1.2	0.41	33
45	INVERCLYDE	2	0.7	0.4--	0.30	50
46	CUNNINGHAME	4	1.0	0.8-	0.42	41
47	KILMARNOCK	4	1.7	1.4	0.69	32
48	KYLE-CARRICK	4	1.3	1.0	0.54	38
49	CUMNOCK-DOON	2	1.4	1.1	0.76	36
50	WIGTOWN	3	3.5	1.8	1.06	20
51	STEWARTRY	1	1.5	1.5	1.55	25
52	NITHSDALE	2	1.2	0.9	0.64	39
53	ANNANDALE	1	1.0	0.5-	0.54	46
54	ORKNEY	1	1.9	0.9	0.86	40
55	SHETLAND	2	3.2	2.3	1.72	13
56	WESTERN ISLES	2	2.3	1.2	0.92	33
	ALL SCOTLAND	384	2.6	1.9	0.10	

SCOTTISH CITIES

Liver — Females

		CASES	C.R.	S.R.	S.E.	RK
1	CAITHNESS	0	0.0	0.0--	0.00	44
2	SUTHERLAND	2	5.2	3.0	2.18	1
3	ROSS-CROMARTY	0	0.0	0.0--	0.00	44
4	SKYE-LOCHALSH	1	3.2	1.0	0.98	19
5	LOCHABER	0	0.0	0.0--	0.00	44
6	INVERNESS	0	0.0	0.0--	0.00	44
7	BADENOCH	2	6.7	2.3	1.69	2
8	NAIRN	0	0.0	0.0--	0.00	44
9	MORAY	3	1.2	1.6	0.84	6
10	BANFF-BUCHAN	4	1.7	0.9	0.46	24
11	GORDON	0	0.0	0.0--	0.00	44
12	ABERDEEN	12	1.6	1.0	0.33	16
13	KINCARDINE	1	0.9	0.4	0.37	38
14	ANGUS	4	1.4	1.0	0.53	21
15	DUNDEE	9	1.5	0.8	0.27	28
16	PERTH-KINROSS	4	1.1	0.8	0.43	25
17	KIRKCALDY	7	1.5	1.2	0.50	9
18	N-E FIFE	3	1.5	0.8	0.46	27
19	DUNFERMLINE	2	0.5	0.3	0.24	40
20	WEST LOTHIAN	10	2.6	2.1	0.70	3
21	EDINBURGH	33	2.2	1.0	0.23	17
22	MIDLOTHIAN	3	1.2	0.8	0.34	34
23	EAST LOTHIAN	5	2.1	1.1	0.52	14
24	TWEEDDALE	0	0.0	0.0--	0.00	44
25	ETTRICK	0	0.0	0.0--	0.00	44
26	ROXBURGH	2	1.6	1.1	0.89	13
27	BERWICKSHIRE	0	0.0	0.0--	0.00	44
28	CLACKMANNAN	2	1.4	1.2	0.90	11
29	STIRLING	2	0.8	0.4	0.33	37
30	FALKIRK	4	0.9	0.7	0.35	31
31	ARGYLL-BUTE	4	2.0	1.4	0.69	8
32	DUMBARTON	4	1.6	1.0	0.53	15
33	GLASGOW	42	1.6	0.7	0.12	30
34	CLYDEBANK	1	0.8	0.4	0.35	39
35	BEARSDEN	3	2.5	1.7	1.06	5
36	STRATHKELVIN	1	0.4	0.3	0.29	41
37	CUMBERNAULD	0	0.0	0.0--	0.00	44
38	MONKLANDS	4	1.2	1.0	0.50	20
39	MOTHERWELL	8	1.7	0.8	0.30	26
40	HAMILTON	1	0.3	0.2--	0.20	43
41	EAST KILBRIDE	5	2.0	1.6	0.81	4
42	EASTWOOD	3	1.9	0.9	0.56	22
43	CLYDESDALE	3	1.6	1.0	0.61	17
44	RENFREW	7	1.1	0.8	0.24	33
45	INVERCLYDE	1	0.3	0.6	0.62	32
46	CUNNINGHAME	4	1.0	0.5	0.27	36
47	KILMARNOCK	6	2.4	1.5	0.67	7
48	KYLE-CARRICK	9	2.5	1.2	0.42	12
49	CUMNOCK-DOON	2	1.4	0.7	0.52	29
50	WIGTOWN	0	0.0	0.0--	0.00	44
51	STEWARTRY	0	0.0	0.0--	0.00	44
52	NITHSDALE	2	1.2	0.5	0.36	35
53	ANNANDALE	0	0.0	0.0--	0.00	44
54	ORKNEY	1	1.8	1.2	1.22	9
55	SHETLAND	2	3.3	0.9	0.64	23
56	WESTERN ISLES	1	1.1	0.2--	0.22	42
	ALL SCOTLAND	229	1.4	0.8	0.06	

SCOTTISH CITIES

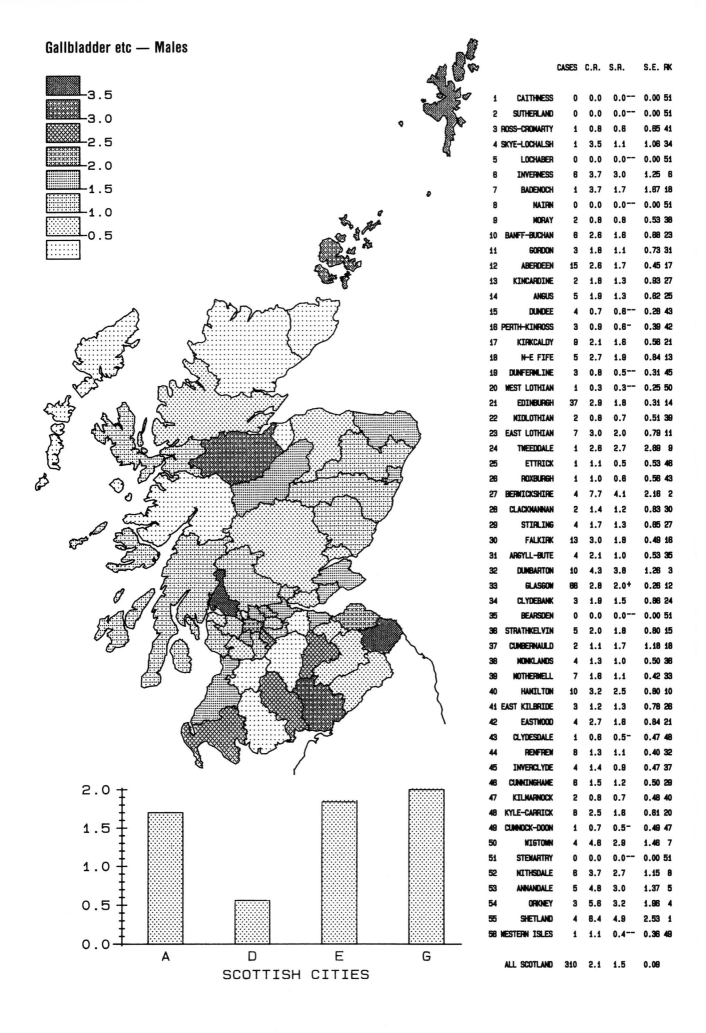

Gallbladder etc — Males

		CASES	C.R.	S.R.	S.E.	RK
1	CAITHNESS	0	0.0	0.0--	0.00	51
2	SUTHERLAND	0	0.0	0.0--	0.00	51
3	ROSS-CROMARTY	1	0.8	0.6	0.65	41
4	SKYE-LOCHALSH	1	3.5	1.1	1.06	34
5	LOCHABER	0	0.0	0.0--	0.00	51
6	INVERNESS	6	3.7	3.0	1.25	6
7	BADENOCH	1	3.7	1.7	1.67	19
8	NAIRN	0	0.0	0.0--	0.00	51
9	MORAY	2	0.8	0.8	0.53	38
10	BANFF-BUCHAN	6	2.6	1.6	0.68	23
11	GORDON	3	1.6	1.1	0.73	31
12	ABERDEEN	15	2.6	1.7	0.45	17
13	KINCARDINE	2	1.6	1.3	0.93	27
14	ANGUS	5	1.9	1.3	0.62	25
15	DUNDEE	4	0.7	0.6--	0.28	43
16	PERTH-KINROSS	3	0.9	0.6-	0.39	42
17	KIRKCALDY	9	2.1	1.6	0.56	21
18	N-E FIFE	5	2.7	1.9	0.84	13
19	DUNFERMLINE	3	0.8	0.5--	0.31	45
20	WEST LOTHIAN	1	0.3	0.3--	0.25	50
21	EDINBURGH	37	2.9	1.8	0.31	14
22	MIDLOTHIAN	2	0.8	0.7	0.51	39
23	EAST LOTHIAN	7	3.0	2.0	0.79	11
24	TWEEDDALE	1	2.8	2.7	2.69	9
25	ETTRICK	1	1.1	0.5	0.53	46
26	ROXBURGH	1	1.0	0.6	0.56	43
27	BERWICKSHIRE	4	7.7	4.1	2.16	2
28	CLACKMANNAN	2	1.4	1.2	0.83	30
29	STIRLING	4	1.7	1.3	0.65	27
30	FALKIRK	13	3.0	1.8	0.49	16
31	ARGYLL-BUTE	4	2.1	1.0	0.53	35
32	DUMBARTON	10	4.3	3.6	1.26	3
33	GLASGOW	66	2.8	2.0+	0.26	12
34	CLYDEBANK	3	1.9	1.5	0.86	24
35	BEARSDEN	0	0.0	0.0--	0.00	51
36	STRATHKELVIN	5	2.0	1.6	0.80	15
37	CUMBERNAULD	2	1.1	1.7	1.18	18
38	MONKLANDS	4	1.3	1.0	0.50	36
39	MOTHERWELL	7	1.6	1.1	0.42	33
40	HAMILTON	10	3.2	2.5	0.80	10
41	EAST KILBRIDE	3	1.2	1.3	0.78	26
42	EASTWOOD	4	2.7	1.6	0.84	21
43	CLYDESDALE	1	0.6	0.5-	0.47	48
44	RENFREW	8	1.3	1.1	0.40	32
45	INVERCLYDE	4	1.4	0.9	0.47	37
46	CUNNINGHAME	6	1.5	1.2	0.50	29
47	KILMARNOCK	2	0.8	0.7	0.48	40
48	KYLE-CARRICK	8	2.5	1.6	0.61	20
49	CUMNOCK-DOON	1	0.7	0.5-	0.49	47
50	WIGTOWN	4	4.8	2.9	1.46	7
51	STEWARTRY	0	0.0	0.0--	0.00	51
52	NITHSDALE	6	3.7	2.7	1.15	8
53	ANNANDALE	5	4.8	3.0	1.37	5
54	ORKNEY	3	5.6	3.2	1.96	4
55	SHETLAND	4	8.4	4.9	2.53	1
56	WESTERN ISLES	1	1.1	0.4--	0.36	49
	ALL SCOTLAND	310	2.1	1.5	0.09	

SCOTTISH CITIES

Gallbladder etc — Females

		CASES	C.R.	S.R.	S.E.	RK
1	CAITHNESS	0	0.0	0.0--	0.00	56
2	SUTHERLAND	2	5.2	3.0	2.16	8
3	ROSS-CROMARTY	1	0.8	0.7	0.71	49
4	SKYE-LOCHALSH	2	6.4	3.2	2.62	5
5	LOCHABER	1	1.6	0.6	0.60	51
6	INVERNESS	1	0.6	0.3--	0.28	55
7	BADENOCH	1	3.4	0.9	0.90	44
8	NAIRN	2	6.5	2.6	1.81	11
9	MORAY	8	2.5	1.6	0.68	31
10	BANFF-BUCHAN	14	5.9	3.2	0.84	6
11	GORDON	7	4.2	1.6	0.71	29
12	ABERDEEN	30	4.5	2.0	0.39	18
13	KINCARDINE	5	4.4	1.1	0.51	42
14	ANGUS	14	5.0	2.2	0.68	15
15	DUNDEE	21	3.5	1.6	0.38	30
16	PERTH-KINROSS	8	2.2	1.1	0.47	40
17	KIRKCALDY	21	4.6	2.3	0.54	14
18	N-E FIFE	7	3.4	0.9-	0.38	46
19	DUNFERMLINE	14	3.7	2.5	0.68	12
20	WEST LOTHIAN	10	2.6	1.9	0.62	21
21	EDINBURGH	51	3.5	1.4	0.23	33
22	MIDLOTHIAN	10	3.9	2.4	0.78	13
23	EAST LOTHIAN	8	3.3	1.4	0.54	32
24	TWEEDDALE	1	2.2	0.3--	0.34	54
25	ETTRICK	2	2.0	1.3	0.92	38
26	ROXBURGH	6	5.4	1.4	0.59	34
27	BERWICKSHIRE	1	1.6	0.4--	0.40	53
28	CLACKMANNAN	4	2.7	1.7	0.88	27
29	STIRLING	8	2.5	1.3	0.56	36
30	FALKIRK	15	3.4	1.9	0.52	20
31	ARGYLL-BUTE	12	6.1	2.6	0.98	10
32	DUMBARTON	7	2.6	1.3	0.49	37
33	GLASGOW	100	3.8	1.7	0.18	27
34	CLYDEBANK	5	2.9	1.7	0.78	25
35	BEARSDEN	4	3.3	1.2	0.62	39
36	STRATHKELVIN	16	6.3	3.5+	0.90	4
37	CUMBERNAULD	2	1.1	1.4	0.99	35
38	MONKLANDS	10	3.0	1.7	0.56	23
39	MOTHERWELL	15	3.1	1.7	0.45	25
40	HAMILTON	18	5.4	3.6+	0.87	3
41	EAST KILBRIDE	7	2.8	2.2	0.84	17
42	EASTWOOD	8	3.7	2.0	0.83	19
43	CLYDESDALE	12	7.0	4.3+	1.30	1
44	RENFREW	21	3.2	1.8	0.42	22
45	INVERCLYDE	9	2.8	1.7	0.62	23
46	CUNNINGHAME	7	1.7	0.8--	0.25	50
47	KILMARNOCK	5	2.0	0.9	0.43	43
48	KYLE-CARRICK	5	1.4	0.9-	0.41	46
49	CUMNOCK-DOON	2	1.4	0.8	0.55	48
50	WIGTOWN	5	5.4	2.9	1.48	9
51	STEWARTRY	3	4.4	1.1	0.68	40
52	NITHSDALE	8	4.8	3.1	1.20	7
53	ANNANDALE	9	8.4	4.0	1.50	2
54	ORKNEY	2	3.7	2.2	1.56	16
55	SHETLAND	2	3.3	0.9	0.64	44
56	WESTERN ISLES	2	2.2	0.5--	0.39	52
	ALL SCOTLAND	585	3.5	1.7	0.08	

SCOTTISH CITIES

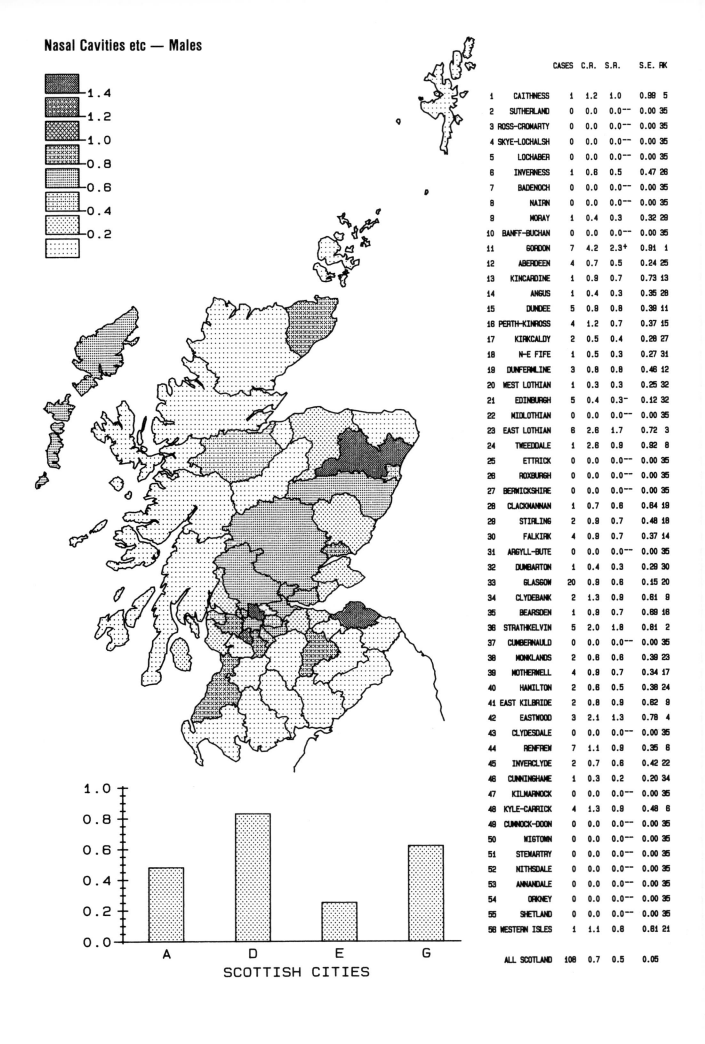

Nasal Cavities etc — Males

	CASES	C.R.	S.R.	S.E.	RK
1 CAITHNESS	1	1.2	1.0	0.99	5
2 SUTHERLAND	0	0.0	0.0--	0.00	35
3 ROSS-CROMARTY	0	0.0	0.0--	0.00	35
4 SKYE-LOCHALSH	0	0.0	0.0--	0.00	35
5 LOCHABER	0	0.0	0.0--	0.00	35
6 INVERNESS	1	0.6	0.5	0.47	26
7 BADENOCH	0	0.0	0.0--	0.00	35
8 NAIRN	0	0.0	0.0--	0.00	35
9 MORAY	1	0.4	0.3	0.32	29
10 BANFF-BUCHAN	0	0.0	0.0--	0.00	35
11 GORDON	7	4.2	2.3+	0.91	1
12 ABERDEEN	4	0.7	0.5	0.24	25
13 KINCARDINE	1	0.9	0.7	0.73	13
14 ANGUS	1	0.4	0.3	0.35	28
15 DUNDEE	5	0.9	0.8	0.39	11
16 PERTH-KINROSS	4	1.2	0.7	0.37	15
17 KIRKCALDY	2	0.5	0.4	0.28	27
18 N-E FIFE	1	0.5	0.3	0.27	31
19 DUNFERMLINE	3	0.8	0.8	0.46	12
20 WEST LOTHIAN	1	0.3	0.3	0.25	32
21 EDINBURGH	5	0.4	0.3-	0.12	32
22 MIDLOTHIAN	0	0.0	0.0--	0.00	35
23 EAST LOTHIAN	6	2.6	1.7	0.72	3
24 TWEEDDALE	1	2.6	0.9	0.92	8
25 ETTRICK	0	0.0	0.0--	0.00	35
26 ROXBURGH	0	0.0	0.0--	0.00	35
27 BERWICKSHIRE	0	0.0	0.0--	0.00	35
28 CLACKMANNAN	1	0.7	0.6	0.64	19
29 STIRLING	2	0.9	0.7	0.48	18
30 FALKIRK	4	0.9	0.7	0.37	14
31 ARGYLL-BUTE	0	0.0	0.0--	0.00	35
32 DUMBARTON	1	0.4	0.3	0.29	30
33 GLASGOW	20	0.9	0.6	0.15	20
34 CLYDEBANK	2	1.3	0.9	0.61	9
35 BEARSDEN	1	0.9	0.7	0.69	16
36 STRATHKELVIN	5	2.0	1.6	0.61	2
37 CUMBERNAULD	0	0.0	0.0--	0.00	35
38 MONKLANDS	2	0.6	0.6	0.39	23
39 MOTHERWELL	4	0.9	0.7	0.34	17
40 HAMILTON	2	0.6	0.5	0.38	24
41 EAST KILBRIDE	2	0.8	0.9	0.62	9
42 EASTWOOD	3	2.1	1.3	0.78	4
43 CLYDESDALE	0	0.0	0.0--	0.00	35
44 RENFREW	7	1.1	0.9	0.35	6
45 INVERCLYDE	2	0.7	0.6	0.42	22
46 CUNNINGHAME	1	0.3	0.2	0.20	34
47 KILMARNOCK	0	0.0	0.0--	0.00	35
48 KYLE-CARRICK	4	1.3	0.9	0.48	6
49 CUMNOCK-DOON	0	0.0	0.0--	0.00	35
50 WIGTOWN	0	0.0	0.0--	0.00	35
51 STEWARTRY	0	0.0	0.0--	0.00	35
52 NITHSDALE	0	0.0	0.0--	0.00	35
53 ANNANDALE	0	0.0	0.0--	0.00	35
54 ORKNEY	0	0.0	0.0--	0.00	35
55 SHETLAND	0	0.0	0.0--	0.00	35
56 WESTERN ISLES	1	1.1	0.8	0.61	21
ALL SCOTLAND	108	0.7	0.5	0.05	

SCOTTISH CITIES

Nasal Cavities etc — Females

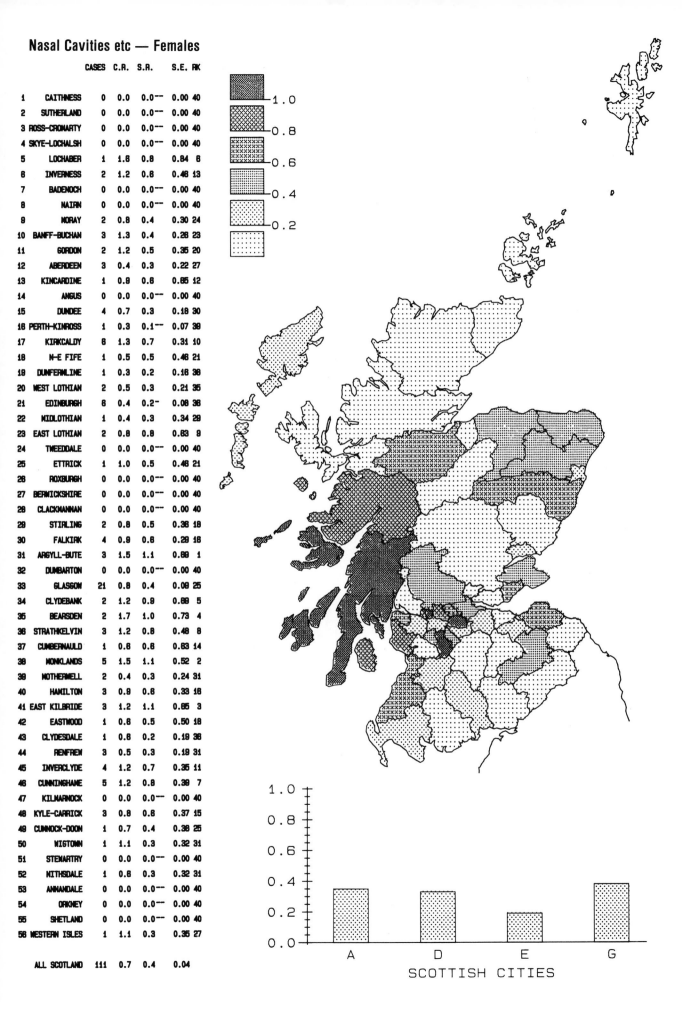

		CASES	C.R.	S.R.	S.E.	RK
1	CAITHNESS	0	0.0	0.0 --	0.00	40
2	SUTHERLAND	0	0.0	0.0 --	0.00	40
3	ROSS-CROMARTY	0	0.0	0.0 --	0.00	40
4	SKYE-LOCHALSH	0	0.0	0.0 --	0.00	40
5	LOCHABER	1	1.6	0.8	0.84	6
6	INVERNESS	2	1.2	0.6	0.46	13
7	BADENOCH	0	0.0	0.0 --	0.00	40
8	NAIRN	0	0.0	0.0 --	0.00	40
9	MORAY	2	0.8	0.4	0.30	24
10	BANFF-BUCHAN	3	1.3	0.4	0.26	23
11	GORDON	2	1.2	0.5	0.35	20
12	ABERDEEN	3	0.4	0.3	0.22	27
13	KINCARDINE	1	0.9	0.6	0.65	12
14	ANGUS	0	0.0	0.0 --	0.00	40
15	DUNDEE	4	0.7	0.3	0.18	30
16	PERTH-KINROSS	1	0.3	0.1 --	0.07	39
17	KIRKCALDY	6	1.3	0.7	0.31	10
18	N-E FIFE	1	0.5	0.5	0.46	21
19	DUNFERMLINE	1	0.3	0.2	0.16	38
20	WEST LOTHIAN	2	0.5	0.3	0.21	35
21	EDINBURGH	6	0.4	0.2 -	0.08	36
22	MIDLOTHIAN	1	0.4	0.3	0.34	29
23	EAST LOTHIAN	2	0.8	0.8	0.63	8
24	TWEEDDALE	0	0.0	0.0 --	0.00	40
25	ETTRICK	1	1.0	0.5	0.46	21
26	ROXBURGH	0	0.0	0.0 --	0.00	40
27	BERWICKSHIRE	0	0.0	0.0 --	0.00	40
28	CLACKMANNAN	0	0.0	0.0 --	0.00	40
29	STIRLING	2	0.8	0.5	0.36	18
30	FALKIRK	4	0.9	0.6	0.29	16
31	ARGYLL-BUTE	3	1.5	1.1	0.69	1
32	DUMBARTON	0	0.0	0.0 --	0.00	40
33	GLASGOW	21	0.8	0.4	0.09	25
34	CLYDEBANK	2	1.2	0.9	0.69	5
35	BEARSDEN	2	1.7	1.0	0.73	4
36	STRATHKELVIN	3	1.2	0.8	0.46	8
37	CUMBERNAULD	1	0.8	0.6	0.63	14
38	MONKLANDS	5	1.5	1.1	0.52	2
39	MOTHERWELL	2	0.4	0.3	0.24	31
40	HAMILTON	3	0.9	0.6	0.33	16
41	EAST KILBRIDE	3	1.2	1.1	0.65	3
42	EASTWOOD	1	0.8	0.5	0.50	18
43	CLYDESDALE	1	0.6	0.2	0.19	36
44	RENFREW	3	0.5	0.3	0.19	31
45	INVERCLYDE	4	1.2	0.7	0.35	11
46	CUNNINGHAME	5	1.2	0.8	0.39	7
47	KILMARNOCK	0	0.0	0.0 --	0.00	40
48	KYLE-CARRICK	3	0.8	0.6	0.37	15
49	CUMNOCK-DOON	1	0.7	0.4	0.38	25
50	WIGTOWN	1	1.1	0.3	0.32	31
51	STEWARTRY	0	0.0	0.0 --	0.00	40
52	NITHSDALE	1	0.6	0.3	0.32	31
53	ANNANDALE	0	0.0	0.0 --	0.00	40
54	ORKNEY	0	0.0	0.0 --	0.00	40
55	SHETLAND	0	0.0	0.0 --	0.00	40
56	WESTERN ISLES	1	1.1	0.3	0.35	27
	ALL SCOTLAND	111	0.7	0.4	0.04	

SCOTTISH CITIES

Bone — Males

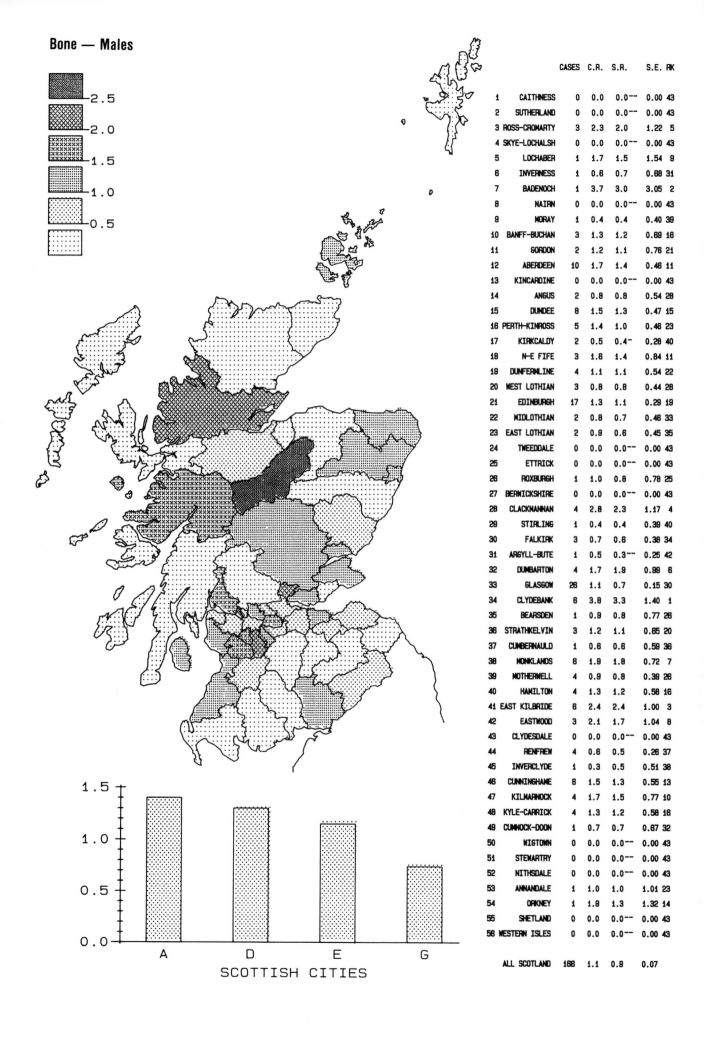

		CASES	C.R.	S.R.	S.E.	RK
1	CAITHNESS	0	0.0	0.0--	0.00	43
2	SUTHERLAND	0	0.0	0.0--	0.00	43
3	ROSS-CROMARTY	3	2.3	2.0	1.22	5
4	SKYE-LOCHALSH	0	0.0	0.0--	0.00	43
5	LOCHABER	1	1.7	1.5	1.54	9
6	INVERNESS	1	0.6	0.7	0.88	31
7	BADENOCH	1	3.7	3.0	3.05	2
8	NAIRN	0	0.0	0.0--	0.00	43
9	MORAY	1	0.4	0.4	0.40	39
10	BANFF-BUCHAN	3	1.3	1.2	0.69	16
11	GORDON	2	1.2	1.1	0.76	21
12	ABERDEEN	10	1.7	1.4	0.46	11
13	KINCARDINE	0	0.0	0.0--	0.00	43
14	ANGUS	2	0.8	0.8	0.54	28
15	DUNDEE	8	1.5	1.3	0.47	15
16	PERTH-KINROSS	5	1.4	1.0	0.46	23
17	KIRKCALDY	2	0.5	0.4-	0.28	40
18	N-E FIFE	3	1.6	1.4	0.84	11
19	DUNFERMLINE	4	1.1	1.1	0.54	22
20	WEST LOTHIAN	3	0.8	0.8	0.44	28
21	EDINBURGH	17	1.3	1.1	0.29	19
22	MIDLOTHIAN	2	0.8	0.7	0.46	33
23	EAST LOTHIAN	2	0.9	0.6	0.45	35
24	TWEEDDALE	0	0.0	0.0--	0.00	43
25	ETTRICK	0	0.0	0.0--	0.00	43
26	ROXBURGH	1	1.0	0.8	0.76	25
27	BERWICKSHIRE	0	0.0	0.0--	0.00	43
28	CLACKMANNAN	4	2.8	2.3	1.17	4
29	STIRLING	1	0.4	0.4	0.39	40
30	FALKIRK	3	0.7	0.6	0.38	34
31	ARGYLL-BUTE	1	0.5	0.3-	0.25	42
32	DUMBARTON	4	1.7	1.9	0.99	6
33	GLASGOW	26	1.1	0.7	0.15	30
34	CLYDEBANK	6	3.8	3.3	1.40	1
35	BEARSDEN	1	0.9	0.8	0.77	26
36	STRATHKELVIN	3	1.2	1.1	0.65	20
37	CUMBERNAULD	1	0.6	0.6	0.59	36
38	MONKLANDS	6	1.9	1.8	0.72	7
39	MOTHERWELL	4	0.9	0.8	0.39	26
40	HAMILTON	4	1.3	1.2	0.58	16
41	EAST KILBRIDE	6	2.4	2.4	1.00	3
42	EASTWOOD	3	2.1	1.7	1.04	8
43	CLYDESDALE	0	0.0	0.0--	0.00	43
44	RENFREW	4	0.6	0.5	0.26	37
45	INVERCLYDE	1	0.3	0.5	0.51	38
46	CUNNINGHAME	6	1.5	1.3	0.55	13
47	KILMARNOCK	4	1.7	1.5	0.77	10
48	KYLE-CARRICK	4	1.3	1.2	0.58	16
49	CUMNOCK-DOON	1	0.7	0.7	0.67	32
50	WIGTOWN	0	0.0	0.0--	0.00	43
51	STEWARTRY	0	0.0	0.0--	0.00	43
52	NITHSDALE	0	0.0	0.0--	0.00	43
53	ANNANDALE	1	1.0	1.0	1.01	23
54	ORKNEY	1	1.8	1.3	1.32	14
55	SHETLAND	0	0.0	0.0--	0.00	43
56	WESTERN ISLES	0	0.0	0.0--	0.00	43
	ALL SCOTLAND	168	1.1	0.9	0.07	

SCOTTISH CITIES

Bone — Females

		CASES	C.R.	S.R.	S.E.	RK
1	CAITHNESS	1	1.2	0.6	0.58	33
2	SUTHERLAND	0	0.0	0.0--	0.00	43
3	ROSS-CROMARTY	1	0.8	0.7	0.71	27
4	SKYE-LOCHALSH	0	0.0	0.0--	0.00	43
5	LOCHABER	0	0.0	0.0--	0.00	43
6	INVERNESS	2	1.2	0.9	0.64	18
7	BADENOCH	0	0.0	0.0--	0.00	43
8	NAIRN	0	0.0	0.0--	0.00	43
9	MORAY	2	0.8	0.9	0.63	18
10	BANFF-BUCHAN	2	0.8	1.0	0.71	16
11	GORDON	1	0.6	0.2-	0.18	42
12	ABERDEEN	9	1.3	1.2	0.44	9
13	KINCARDINE	1	0.9	1.0	1.02	15
14	ANGUS	3	1.1	0.8	0.52	22
15	DUNDEE	9	1.5	1.2	0.44	10
16	PERTH-KINROSS	3	0.8	0.7	0.44	30
17	KIRKCALDY	8	1.3	1.5	0.65	6
18	N-E FIFE	5	2.4	1.8	0.83	2
19	DUNFERMLINE	2	0.5	0.3	0.26	37
20	WEST LOTHIAN	7	1.8	1.5	0.60	5
21	EDINBURGH	14	0.9	0.8	0.24	22
22	MIDLOTHIAN	3	1.2	1.1	0.65	11
23	EAST LOTHIAN	2	0.8	0.8	0.55	25
24	TWEEDDALE	0	0.0	0.0--	0.00	43
25	ETTRICK	3	2.8	1.4	1.02	8
26	ROXBURGH	0	0.0	0.0--	0.00	43
27	BERWICKSHIRE	0	0.0	0.0--	0.00	43
28	CLACKMANNAN	1	0.7	0.8	0.83	21
29	STIRLING	2	0.8	0.8	0.60	20
30	FALKIRK	4	0.9	0.8	0.41	26
31	ARGYLL-BUTE	2	1.0	0.8	0.71	22
32	DUMBARTON	0	0.0	0.0--	0.00	43
33	GLASGOW	26	1.0	0.8	0.14	32
34	CLYDEBANK	1	0.8	0.2-	0.21	41
35	BEARSDEN	1	0.8	0.9	0.91	17
36	STRATHKELVIN	3	1.2	1.1	0.66	14
37	CUMBERNAULD	2	1.1	1.1	0.76	11
38	MONKLANDS	1	0.3	0.3-	0.25	40
39	MOTHERWELL	2	0.4	0.3-	0.23	38
40	HAMILTON	4	1.2	1.1	0.56	13
41	EAST KILBRIDE	4	1.8	1.5	0.73	7
42	EASTWOOD	1	0.8	0.4	0.41	36
43	CLYDESDALE	1	0.8	0.8	0.85	31
44	RENFREW	7	1.1	0.7	0.28	29
45	INVERCLYDE	3	0.9	0.7	0.42	28
46	CUNNINGHAME	3	0.7	0.4	0.26	35
47	KILMARNOCK	7	2.8	1.8	0.67	3
48	KYLE-CARRICK	3	0.8	0.3--	0.18	39
49	CUMNOCK-DOON	0	0.0	0.0--	0.00	43
50	WIGTOWN	1	1.1	0.4	0.45	34
51	STEWARTRY	0	0.0	0.0--	0.00	43
52	NITHSDALE	4	2.3	1.9	1.06	1
53	ANNANDALE	3	2.8	1.6	1.10	4
54	ORKNEY	0	0.0	0.0--	0.00	43
55	SHETLAND	0	0.0	0.0--	0.00	43
56	WESTERN ISLES	0	0.0	0.0--	0.00	43
	ALL SCOTLAND	182	1.0	0.8	0.07	

SCOTTISH CITIES

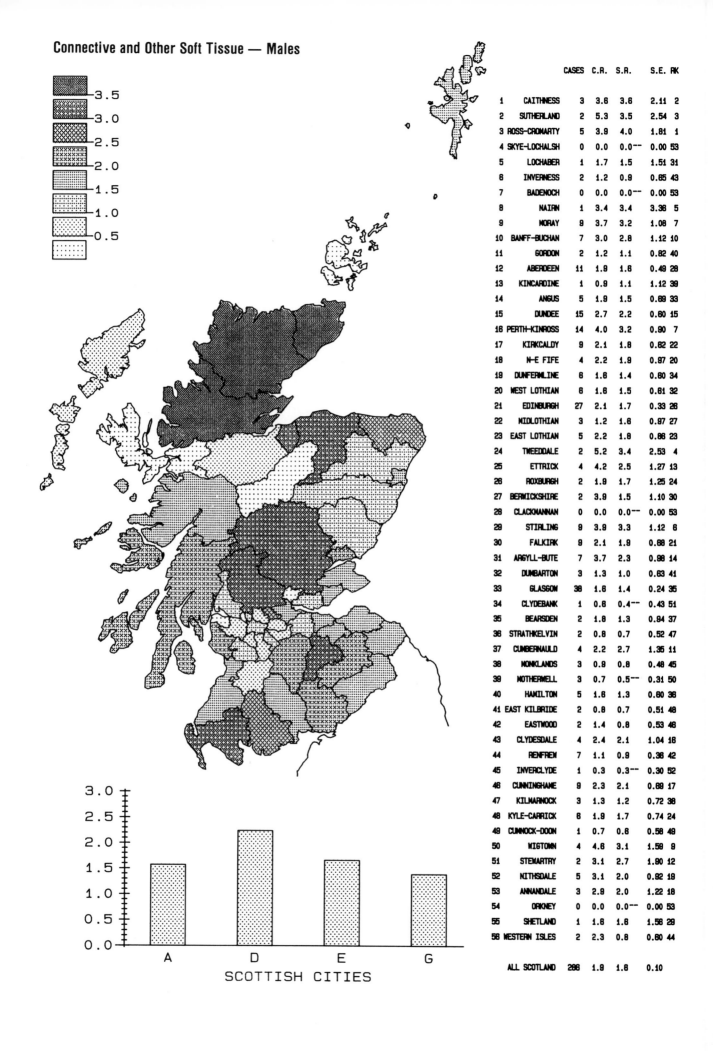

Connective and Other Soft Tissue — Males

		CASES	C.R.	S.R.	S.E.	RK
1	CAITHNESS	3	3.6	3.6	2.11	2
2	SUTHERLAND	2	5.3	3.5	2.54	3
3	ROSS-CROMARTY	5	3.9	4.0	1.81	1
4	SKYE-LOCHALSH	0	0.0	0.0 --	0.00	53
5	LOCHABER	1	1.7	1.5	1.51	31
6	INVERNESS	2	1.2	0.9	0.65	43
7	BADENOCH	0	0.0	0.0 --	0.00	53
8	NAIRN	1	3.4	3.4	3.38	5
9	MORAY	9	3.7	3.2	1.08	7
10	BANFF-BUCHAN	7	3.0	2.8	1.12	10
11	GORDON	2	1.2	1.1	0.82	40
12	ABERDEEN	11	1.9	1.6	0.49	26
13	KINCARDINE	1	0.9	1.1	1.12	39
14	ANGUS	5	1.9	1.5	0.69	33
15	DUNDEE	15	2.7	2.2	0.60	15
16	PERTH-KINROSS	14	4.0	3.2	0.90	7
17	KIRKCALDY	9	2.1	1.8	0.62	22
18	N-E FIFE	4	2.2	1.9	0.97	20
19	DUNFERMLINE	6	1.6	1.4	0.60	34
20	WEST LOTHIAN	8	1.6	1.5	0.61	32
21	EDINBURGH	27	2.1	1.7	0.33	26
22	MIDLOTHIAN	3	1.2	1.6	0.97	27
23	EAST LOTHIAN	5	2.2	1.8	0.86	23
24	TWEEDDALE	2	5.2	3.4	2.53	4
25	ETTRICK	4	4.2	2.5	1.27	13
26	ROXBURGH	2	1.9	1.7	1.25	24
27	BERWICKSHIRE	2	3.9	1.5	1.10	30
28	CLACKMANNAN	0	0.0	0.0 --	0.00	53
29	STIRLING	9	3.9	3.3	1.12	6
30	FALKIRK	9	2.1	1.6	0.66	21
31	ARGYLL-BUTE	7	3.7	2.3	0.98	14
32	DUMBARTON	3	1.3	1.0	0.63	41
33	GLASGOW	36	1.6	1.4	0.24	35
34	CLYDEBANK	1	0.6	0.4 --	0.43	51
35	BEARSDEN	2	1.6	1.3	0.94	37
36	STRATHKELVIN	2	0.8	0.7	0.52	47
37	CUMBERNAULD	4	2.2	2.7	1.35	11
38	MONKLANDS	3	0.9	0.8	0.46	45
39	MOTHERWELL	3	0.7	0.5 --	0.31	50
40	HAMILTON	5	1.6	1.3	0.60	38
41	EAST KILBRIDE	2	0.8	0.7	0.51	48
42	EASTWOOD	2	1.4	0.8	0.53	46
43	CLYDESDALE	4	2.4	2.1	1.04	16
44	RENFREW	7	1.1	0.9	0.36	42
45	INVERCLYDE	1	0.3	0.3 --	0.30	52
46	CUNNINGHAME	9	2.3	2.1	0.69	17
47	KILMARNOCK	3	1.3	1.2	0.72	36
48	KYLE-CARRICK	8	1.9	1.7	0.74	24
49	CUMNOCK-DOON	1	0.7	0.6	0.58	49
50	WIGTOWN	4	4.6	3.1	1.59	9
51	STEWARTRY	2	3.1	2.7	1.90	12
52	NITHSDALE	5	3.1	2.0	0.92	19
53	ANNANDALE	3	2.9	2.0	1.22	18
54	ORKNEY	0	0.0	0.0 --	0.00	53
55	SHETLAND	1	1.6	1.6	1.56	29
56	WESTERN ISLES	2	2.3	0.8	0.60	44
	ALL SCOTLAND	266	1.9	1.6	0.10	

SCOTTISH CITIES

Connective and Other Soft Tissue — Females

		CASES	C.R.	S.R.	S.E.	RK
1	CAITHNESS	3	3.5	2.2	1.37	11
2	SUTHERLAND	1	2.6	1.2	1.22	29
3	ROSS-CROMARTY	8	6.3	4.2	1.62	2
4	SKYE-LOCHALSH	1	3.2	3.6	3.58	3
5	LOCHABER	2	3.3	2.1	1.49	13
6	INVERNESS	3	1.0	0.7	0.39	42
7	BADENOCH	1	3.4	2.4	2.44	8
8	NAIRN	1	3.3	1.4	1.42	25
9	MORAY	4	1.6	1.0	0.56	35
10	BANFF-BUCHAN	1	0.4	0.2--	0.19	51
11	GORDON	6	3.6	3.0	1.36	4
12	ABERDEEN	24	3.6	2.6+	0.60	5
13	KINCARDINE	3	2.6	1.7	1.05	18
14	ANGUS	5	1.8	1.4	0.69	27
15	DUNDEE	12	2.0	1.5	0.46	23
16	PERTH-KINROSS	8	2.2	1.4	0.56	27
17	KIRKCALDY	14	3.0	1.9	0.56	15
18	N-E FIFE	7	3.4	2.2	0.90	10
19	DUNFERMLINE	7	1.9	1.5	0.59	22
20	WEST LOTHIAN	8	2.1	1.9	0.68	16
21	EDINBURGH	43	2.9	2.1+	0.37	14
22	MIDLOTHIAN	5	1.9	1.5	0.68	24
23	EAST LOTHIAN	3	1.2	0.8	0.51	40
24	TWEEDDALE	1	2.2	0.7	0.71	41
25	ETTRICK	3	2.9	2.1	1.30	12
26	ROXBURGH	3	2.7	1.4	1.02	26
27	BERWICKSHIRE	2	3.7	2.5	2.15	7
28	CLACKMANNAN	2	1.4	1.6	1.35	17
29	STIRLING	1	0.4	0.1--	0.14	52
30	FALKIRK	3	0.7	0.4--	0.27	49
31	ARGYLL-BUTE	5	2.5	1.5	0.77	21
32	DUMBARTON	3	1.2	0.8	0.46	38
33	GLASGOW	42	1.6	1.2	0.20	30
34	CLYDEBANK	1	0.8	0.7	0.66	43
35	BEARSDEN	2	1.7	1.5	1.14	20
36	STRATHKELVIN	1	0.4	0.4	0.45	46
37	CUMBERNAULD	2	1.1	1.0	0.74	33
38	MONKLANDS	1	0.3	0.1--	0.13	53
39	MOTHERWELL	4	0.8	0.6-	0.33	46
40	HAMILTON	2	0.6	0.6	0.46	44
41	EAST KILBRIDE	2	0.8	0.6	0.41	47
42	EASTWOOD	2	1.2	0.8	0.47	45
43	CLYDESDALE	0	0.0	0.0--	0.00	55
44	RENFREW	2	0.3	0.1--	0.08	54
45	INVERCLYDE	4	1.2	1.0	0.51	36
46	CUNNINGHAME	6	1.4	1.0	0.46	31
47	KILMARNOCK	3	1.2	0.8	0.52	38
48	KYLE-CARRICK	5	1.4	1.0	0.49	31
49	CUMNOCK-DOON	3	2.1	0.9	0.55	37
50	WIGTOWN	2	2.2	1.0	0.78	33
51	STEWARTRY	0	0.0	0.0--	0.00	55
52	NITHSDALE	4	2.3	1.7	0.88	18
53	ANNANDALE	2	1.8	2.7	2.13	6
54	ORKNEY	1	1.6	0.4-	0.39	50
55	SHETLAND	4	6.5	4.8	2.60	1
56	WESTERN ISLES	3	3.3	2.4	1.58	9
	ALL SCOTLAND	281	1.8	1.3	0.08	

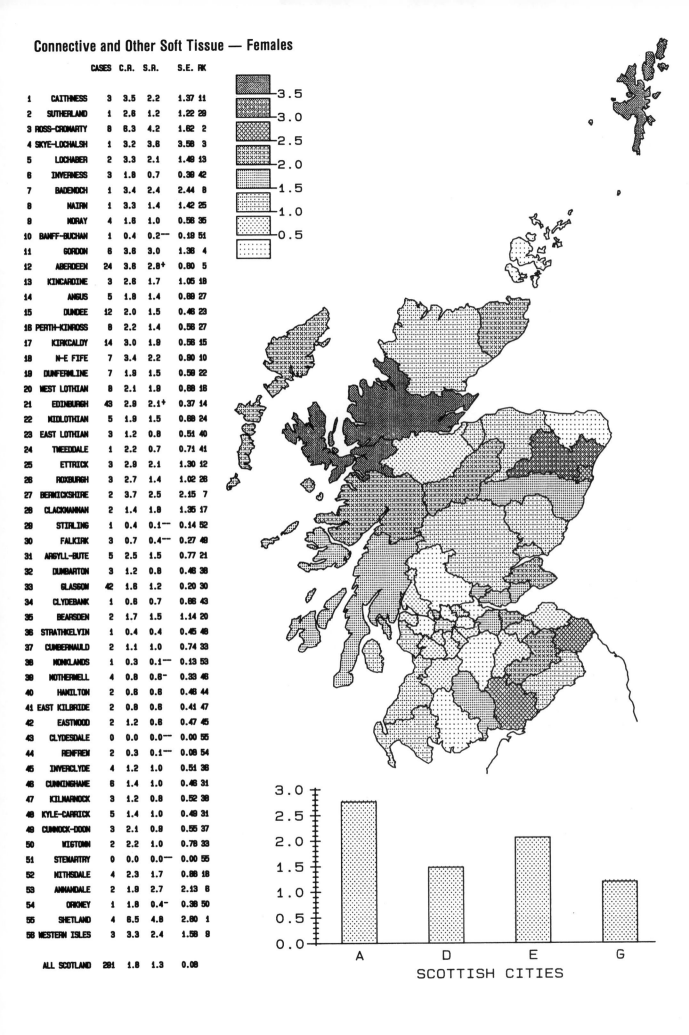

SCOTTISH CITIES

Other and Unspecified Female Genital Organs

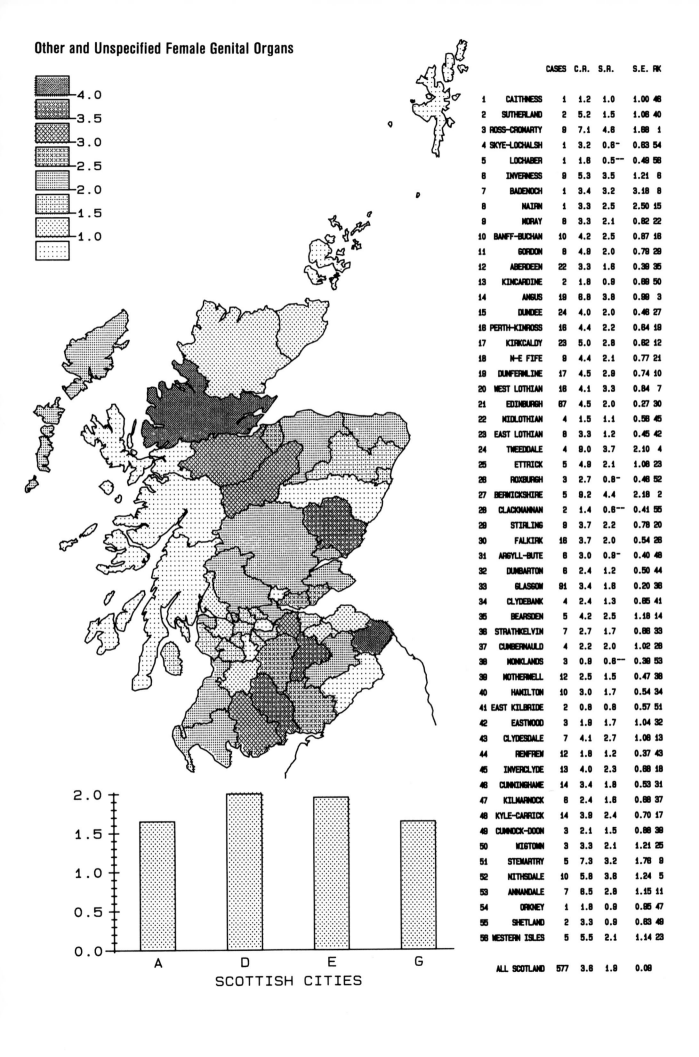

		CASES	C.R.	S.R.	S.E.	RK
1	CAITHNESS	1	1.2	1.0	1.00	46
2	SUTHERLAND	2	5.2	1.5	1.06	40
3	ROSS-CROMARTY	9	7.1	4.6	1.88	1
4	SKYE-LOCHALSH	1	3.2	0.6⁻	0.63	54
5	LOCHABER	1	1.6	0.5⁻⁻	0.49	56
6	INVERNESS	9	5.3	3.5	1.21	6
7	BADENOCH	1	3.4	3.2	3.18	8
8	NAIRN	1	3.3	2.5	2.50	15
9	MORAY	8	3.3	2.1	0.82	22
10	BANFF-BUCHAN	10	4.2	2.5	0.87	16
11	GORDON	8	4.9	2.0	0.79	29
12	ABERDEEN	22	3.3	1.6	0.39	35
13	KINCARDINE	2	1.6	0.9	0.69	50
14	ANGUS	19	6.8	3.6	0.99	3
15	DUNDEE	24	4.0	2.0	0.46	27
16	PERTH-KINROSS	16	4.4	2.2	0.64	19
17	KIRKCALDY	23	5.0	2.6	0.62	12
18	N-E FIFE	9	4.4	2.1	0.77	21
19	DUNFERMLINE	17	4.5	2.9	0.74	10
20	WEST LOTHIAN	16	4.1	3.3	0.84	7
21	EDINBURGH	67	4.5	2.0	0.27	30
22	MIDLOTHIAN	4	1.5	1.1	0.56	45
23	EAST LOTHIAN	8	3.3	1.2	0.45	42
24	TWEEDDALE	4	8.0	3.7	2.10	4
25	ETTRICK	5	4.9	2.1	1.06	23
26	ROXBURGH	3	2.7	0.8⁻	0.46	52
27	BERWICKSHIRE	5	8.2	4.4	2.18	2
28	CLACKMANNAN	2	1.4	0.6⁻⁻	0.41	55
29	STIRLING	8	3.7	2.2	0.78	20
30	FALKIRK	16	3.7	2.0	0.54	26
31	ARGYLL-BUTE	6	3.0	0.8⁻	0.40	48
32	DUMBARTON	8	2.4	1.2	0.50	44
33	GLASGOW	91	3.4	1.6	0.20	36
34	CLYDEBANK	4	2.4	1.3	0.65	41
35	BEARSDEN	5	4.2	2.5	1.18	14
36	STRATHKELVIN	7	2.7	1.7	0.66	33
37	CUMBERNAULD	4	2.2	2.0	1.02	26
38	MONKLANDS	3	0.9	0.6⁻⁻	0.39	53
39	MOTHERWELL	12	2.5	1.5	0.47	38
40	HAMILTON	10	3.0	1.7	0.54	34
41	EAST KILBRIDE	2	0.8	0.8	0.57	51
42	EASTWOOD	3	1.8	1.7	1.04	32
43	CLYDESDALE	7	4.1	2.7	1.06	13
44	RENFREW	12	1.8	1.2	0.37	43
45	INVERCLYDE	13	4.0	2.3	0.66	16
46	CUNNINGHAME	14	3.4	1.8	0.53	31
47	KILMARNOCK	8	2.4	1.6	0.66	37
48	KYLE-CARRICK	14	3.9	2.4	0.70	17
49	CUMNOCK-DOON	3	2.1	1.5	0.66	39
50	WIGTOWN	3	3.3	2.1	1.21	25
51	STEWARTRY	5	7.3	3.2	1.76	9
52	NITHSDALE	10	5.6	3.6	1.24	5
53	ANNANDALE	7	6.5	2.8	1.15	11
54	ORKNEY	1	1.6	0.9	0.85	47
55	SHETLAND	2	3.3	0.9	0.63	49
56	WESTERN ISLES	5	5.5	2.1	1.14	23
	ALL SCOTLAND	577	3.6	1.9	0.09	

Kidney — Males

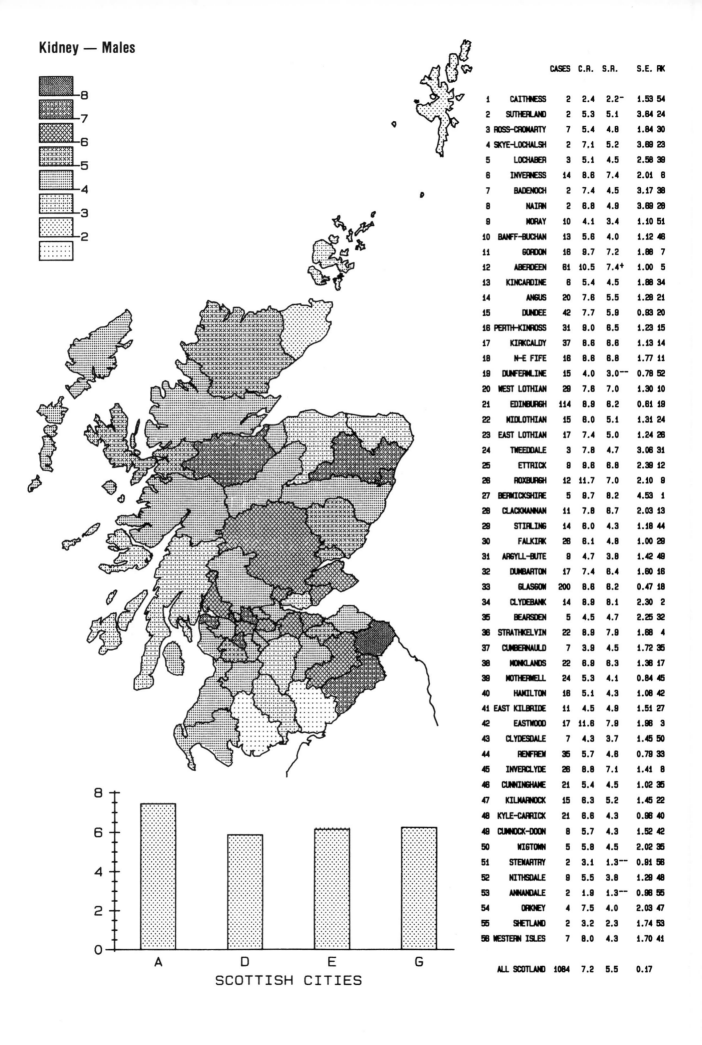

		CASES	C.R.	S.R.	S.E.	RK
1	CAITHNESS	2	2.4	2.2⁻	1.53	54
2	SUTHERLAND	2	5.3	5.1	3.64	24
3	ROSS-CROMARTY	7	5.4	4.8	1.84	30
4	SKYE-LOCHALSH	2	7.1	5.2	3.89	23
5	LOCHABER	3	5.1	4.5	2.58	39
6	INVERNESS	14	8.6	7.4	2.01	6
7	BADENOCH	2	7.4	4.5	3.17	38
8	NAIRN	2	6.8	4.9	3.69	26
9	MORAY	10	4.1	3.4	1.10	51
10	BANFF-BUCHAN	13	5.6	4.0	1.12	46
11	GORDON	16	9.7	7.2	1.88	7
12	ABERDEEN	61	10.5	7.4+	1.00	5
13	KINCARDINE	6	5.4	4.5	1.88	34
14	ANGUS	20	7.6	5.5	1.28	21
15	DUNDEE	42	7.7	5.9	0.93	20
16	PERTH-KINROSS	31	9.0	6.5	1.23	15
17	KIRKCALDY	37	8.6	6.6	1.13	14
18	N-E FIFE	16	8.6	6.6	1.77	11
19	DUNFERMLINE	15	4.0	3.0⁻⁻	0.78	52
20	WEST LOTHIAN	29	7.6	7.0	1.30	10
21	EDINBURGH	114	8.9	6.2	0.61	19
22	MIDLOTHIAN	15	8.0	5.1	1.31	24
23	EAST LOTHIAN	17	7.4	5.0	1.24	26
24	TWEEDDALE	3	7.8	4.7	3.06	31
25	ETTRICK	9	9.6	6.6	2.39	12
26	ROXBURGH	12	11.7	7.0	2.10	9
27	BERWICKSHIRE	5	9.7	8.2	4.53	1
28	CLACKMANNAN	11	7.6	6.7	2.03	13
29	STIRLING	14	6.0	4.3	1.18	44
30	FALKIRK	26	6.1	4.8	1.00	29
31	ARGYLL-BUTE	9	4.7	3.8	1.42	49
32	DUMBARTON	17	7.4	6.4	1.60	16
33	GLASGOW	200	6.6	6.2	0.47	18
34	CLYDEBANK	14	8.9	8.1	2.30	2
35	BEARSDEN	5	4.5	4.7	2.25	32
36	STRATHKELVIN	22	8.9	7.9	1.68	4
37	CUMBERNAULD	7	3.9	4.5	1.72	35
38	MONKLANDS	22	8.9	8.3	1.36	17
39	MOTHERWELL	24	5.3	4.1	0.84	45
40	HAMILTON	18	5.1	4.3	1.06	42
41	EAST KILBRIDE	11	4.5	4.9	1.51	27
42	EASTWOOD	17	11.8	7.9	1.96	3
43	CLYDESDALE	7	4.3	3.7	1.45	50
44	RENFREW	35	5.7	4.8	0.79	33
45	INVERCLYDE	26	6.6	7.1	1.41	8
46	CUNNINGHAME	21	5.4	4.5	1.02	36
47	KILMARNOCK	15	6.3	5.2	1.45	22
48	KYLE-CARRICK	21	6.6	4.3	0.96	40
49	CUMNOCK-DOON	8	5.7	4.3	1.52	42
50	WIGTOWN	5	5.6	4.5	2.02	36
51	STEWARTRY	2	3.1	1.3⁻⁻	0.91	56
52	NITHSDALE	9	5.5	3.8	1.29	48
53	ANNANDALE	2	1.9	1.3⁻⁻	0.96	55
54	ORKNEY	4	7.5	4.0	2.03	47
55	SHETLAND	2	3.2	2.3	1.74	53
56	WESTERN ISLES	7	8.0	4.3	1.70	41
	ALL SCOTLAND	1084	7.2	5.5	0.17	

SCOTTISH CITIES

Kidney - Females

		CASES	C.R.	S.R.		S.E.	RK
1	CAITHNESS	3	3.5	2.2		1.30	39
2	SUTHERLAND	0	0.0	0.0	--	0.00	53
3	ROSS-CROMARTY	3	2.4	1.1	--	0.68	50
4	SKYE-LOCHALSH	0	0.0	0.0	--	0.00	53
5	LOCHABER	4	6.6	5.5		3.28	4
6	INVERNESS	13	7.7	5.1		1.51	5
7	BADENOCH	0	0.0	0.0	--	0.00	53
8	NAIRN	2	6.5	5.0		3.53	6
9	MORAY	6	2.5	1.0	--	0.42	51
10	BANFF-BUCHAN	8	3.4	1.8		0.67	46
11	GORDON	8	4.9	3.3		1.22	22
12	ABERDEEN	55	6.2	4.4	+	0.70	7
13	KINCARDINE	5	4.4	1.8		0.88	45
14	ANGUS	9	3.2	1.2	--	0.42	48
15	DUNDEE	31	5.1	3.0		0.64	26
16	PERTH-KINROSS	22	6.0	3.8		0.84	16
17	KIRKCALDY	23	5.0	3.2		0.70	23
18	N-E FIFE	8	3.9	2.0		0.75	42
19	DUNFERMLINE	13	3.5	2.6		0.79	33
20	WEST LOTHIAN	20	5.2	4.2		0.95	9
21	EDINBURGH	107	7.2	3.5		0.37	19
22	MIDLOTHIAN	18	6.2	4.2		1.07	9
23	EAST LOTHIAN	14	5.8	3.0		0.84	26
24	TWEEDDALE	2	4.5	3.1		2.29	25
25	ETTRICK	5	4.9	4.0		1.91	12
26	ROXBURGH	8	5.4	3.6		1.57	16
27	BERWICKSHIRE	0	0.0	0.0	--	0.00	53
28	CLACKMANNAN	5	3.4	2.5		1.11	36
29	STIRLING	8	3.7	2.3		0.84	37
30	FALKIRK	25	5.7	3.2		0.67	24
31	ARGYLL-BUTE	8	4.0	1.4	--	0.54	47
32	DUMBARTON	14	5.6	3.6		1.04	16
33	GLASGOW	137	5.2	2.7		0.27	30
34	CLYDEBANK	8	3.5	2.0		0.82	44
35	BEARSDEN	4	3.3	3.4		1.88	21
36	STRATHKELVIN	11	4.3	4.0		1.32	13
37	CUMBERNAULD	5	2.8	2.5		1.13	34
38	MONKLANDS	10	3.0	2.0		0.68	41
39	MOTHERWELL	18	3.8	2.8		0.74	29
40	HAMILTON	14	4.2	3.8		1.11	14
41	EAST KILBRIDE	8	3.2	2.7		0.98	30
42	EASTWOOD	5	3.1	1.1	--	0.51	49
43	CLYDESDALE	11	6.4	4.2		1.38	8
44	RENFREW	31	4.7	3.0		0.56	27
45	INVERCLYDE	15	4.6	2.5		0.68	34
46	CUNNINGHAME	20	4.6	3.4		0.85	20
47	KILMARNOCK	8	2.4	2.0		0.83	43
48	KYLE-CARRICK	10	2.8	2.3		0.84	38
49	CUMNOCK-DOON	4	2.8	2.1		1.05	40
50	WIGTOWN	8	9.6	7.1		2.53	1
51	STEWARTRY	5	7.3	3.8		1.82	15
52	NITHSDALE	12	7.0	4.1		1.23	11
53	ANNANDALE	2	1.9	0.9	--	0.63	52
54	ORKNEY	7	12.6	6.8		2.76	2
55	SHETLAND	2	3.3	2.7		1.88	32
56	WESTERN ISLES	7	7.8	6.1		2.47	3
	ALL SCOTLAND	803	5.0	3.0		0.12	

SCOTTISH CITIES

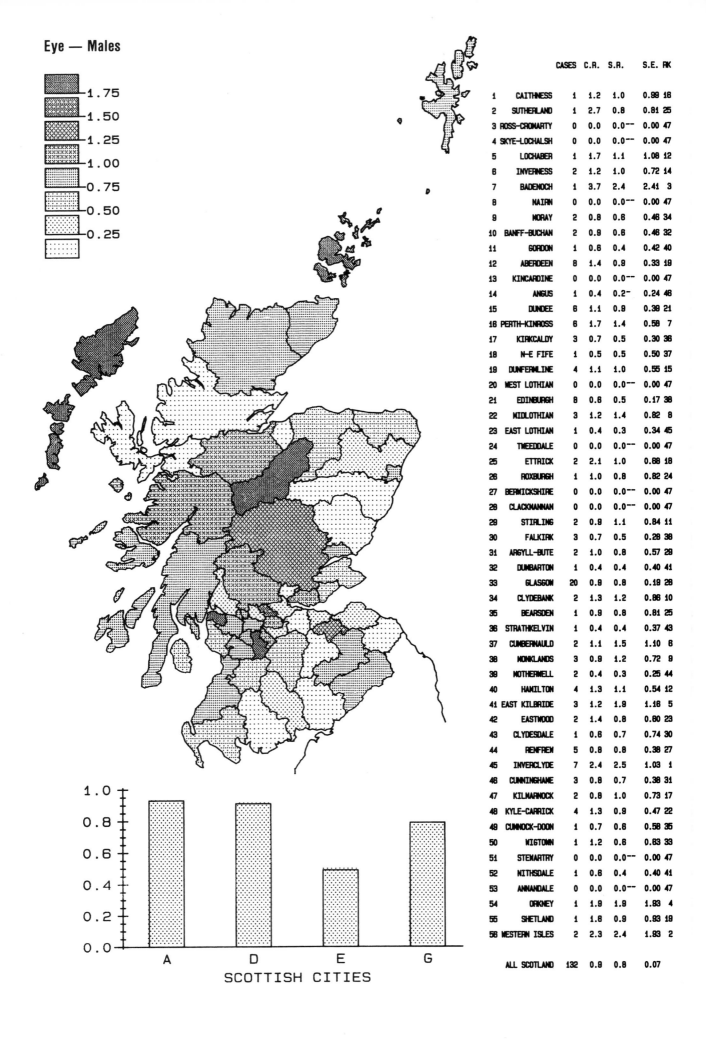

Eye — Males

		CASES	C.R.	S.R.	S.E.	RK
1	CAITHNESS	1	1.2	1.0	0.99	16
2	SUTHERLAND	1	2.7	0.8	0.81	25
3	ROSS-CROMARTY	0	0.0	0.0 --	0.00	47
4	SKYE-LOCHALSH	0	0.0	0.0 --	0.00	47
5	LOCHABER	1	1.7	1.1	1.08	12
6	INVERNESS	2	1.2	1.0	0.72	14
7	BADENOCH	1	3.7	2.4	2.41	3
8	NAIRN	0	0.0	0.0 --	0.00	47
9	MORAY	2	0.8	0.6	0.46	34
10	BANFF-BUCHAN	2	0.9	0.6	0.46	32
11	GORDON	1	0.6	0.4	0.42	40
12	ABERDEEN	8	1.4	0.9	0.33	19
13	KINCARDINE	0	0.0	0.0 --	0.00	47
14	ANGUS	1	0.4	0.2 -	0.24	46
15	DUNDEE	6	1.1	0.9	0.39	21
16	PERTH-KINROSS	6	1.7	1.4	0.58	7
17	KIRKCALDY	3	0.7	0.5	0.30	36
18	N-E FIFE	1	0.5	0.5	0.50	37
19	DUNFERMLINE	4	1.1	1.0	0.55	15
20	WEST LOTHIAN	0	0.0	0.0 --	0.00	47
21	EDINBURGH	8	0.6	0.5	0.17	38
22	MIDLOTHIAN	3	1.2	1.4	0.82	8
23	EAST LOTHIAN	1	0.4	0.3	0.34	45
24	TWEEDDALE	0	0.0	0.0 --	0.00	47
25	ETTRICK	2	2.1	1.0	0.68	18
26	ROXBURGH	1	1.0	0.8	0.82	24
27	BERWICKSHIRE	0	0.0	0.0 --	0.00	47
28	CLACKMANNAN	0	0.0	0.0 --	0.00	47
29	STIRLING	2	0.9	1.1	0.84	11
30	FALKIRK	3	0.7	0.5	0.28	38
31	ARGYLL-BUTE	2	1.0	0.8	0.57	29
32	DUMBARTON	1	0.4	0.4	0.40	41
33	GLASGOW	20	0.9	0.8	0.19	26
34	CLYDEBANK	2	1.3	1.2	0.86	10
35	BEARSDEN	1	0.9	0.8	0.81	25
36	STRATHKELVIN	1	0.4	0.4	0.37	43
37	CUMBERNAULD	2	1.1	1.5	1.10	6
38	MONKLANDS	3	0.9	1.2	0.72	9
39	MOTHERWELL	2	0.4	0.3	0.25	44
40	HAMILTON	4	1.3	1.1	0.54	12
41	EAST KILBRIDE	3	1.2	1.9	1.16	5
42	EASTWOOD	2	1.4	0.8	0.60	23
43	CLYDESDALE	1	0.6	0.7	0.74	30
44	RENFREW	5	0.8	0.8	0.38	27
45	INVERCLYDE	7	2.4	2.5	1.03	1
46	CUNNINGHAME	3	0.8	0.7	0.38	31
47	KILMARNOCK	2	0.8	1.0	0.73	17
48	KYLE-CARRICK	4	1.3	0.9	0.47	22
49	CUMNOCK-DOON	1	0.7	0.6	0.58	35
50	WIGTOWN	1	1.2	0.6	0.63	33
51	STEWARTRY	0	0.0	0.0 --	0.00	47
52	NITHSDALE	1	0.6	0.4	0.40	41
53	ANNANDALE	0	0.0	0.0 --	0.00	47
54	ORKNEY	1	1.9	1.9	1.93	4
55	SHETLAND	1	1.6	0.9	0.93	19
56	WESTERN ISLES	2	2.3	2.4	1.83	2
	ALL SCOTLAND	132	0.9	0.8	0.07	

SCOTTISH CITIES

Eye — Females

		CASES	C.R.	S.R.	S.E.	RK
1	CAITHNESS	2	2.4	2.2	1.55	1
2	SUTHERLAND	1	2.6	0.6	0.58	28
3	ROSS-CROMARTY	1	0.8	0.4	0.39	39
4	SKYE-LOCHALSH	0	0.0	0.0--	0.00	46
5	LOCHABER	2	3.3	1.9	1.45	4
6	INVERNESS	0	0.0	0.0--	0.00	46
7	BADENOCH	0	0.0	0.0--	0.00	46
8	NAIRN	1	3.3	1.7	1.68	7
9	MORAY	2	0.8	0.4	0.32	35
10	BANFF-BUCHAN	6	2.5	1.9	0.92	6
11	GORDON	2	1.2	0.8	0.57	22
12	ABERDEEN	4	0.6	0.6	0.38	27
13	KINCARDINE	3	2.6	1.4	0.85	8
14	ANGUS	5	1.8	1.1	0.56	14
15	DUNDEE	4	0.7	0.3-	0.19	41
16	PERTH-KINROSS	3	0.8	0.6	0.38	31
17	KIRKCALDY	8	2.0	1.9	0.72	4
18	N-E FIFE	1	0.5	0.2-	0.18	45
19	DUNFERMLINE	1	0.3	0.2-	0.20	43
20	WEST LOTHIAN	3	0.8	0.7	0.39	25
21	EDINBURGH	16	1.1	0.8	0.23	23
22	MIDLOTHIAN	0	0.0	0.0--	0.00	46
23	EAST LOTHIAN	4	1.7	0.8	0.51	20
24	TWEEDDALE	2	4.5	1.3	1.06	10
25	ETTRICK	1	1.0	0.6	0.62	28
26	ROXBURGH	1	0.9	1.0	1.02	16
27	BERWICKSHIRE	1	1.8	0.9	0.85	18
28	CLACKMANNAN	1	0.7	0.6	0.58	28
29	STIRLING	2	0.8	0.7	0.47	25
30	FALKIRK	4	0.9	1.3	0.69	11
31	ARGYLL-BUTE	2	1.0	0.6	0.41	31
32	DUMBARTON	1	0.4	0.4	0.42	38
33	GLASGOW	32	1.2	0.9	0.21	17
34	CLYDEBANK	3	1.6	1.2	0.76	12
35	BEARSDEN	0	0.0	0.0--	0.00	46
36	STRATHKELVIN	1	0.4	0.3	0.35	40
37	CUMBERNAULD	3	1.7	2.1	1.24	2
38	MONKLANDS	2	0.6	0.5	0.32	34
39	MOTHERWELL	3	0.6	0.5	0.30	33
40	HAMILTON	1	0.3	0.2-	0.20	43
41	EAST KILBRIDE	3	1.2	1.1	0.66	15
42	EASTWOOD	0	0.0	0.0--	0.00	46
43	CLYDESDALE	1	0.6	0.2-	0.24	42
44	RENFREW	5	0.8	0.8	0.39	19
45	INVERCLYDE	6	1.9	1.2	0.54	12
46	CUNNINGHAME	3	0.7	0.7	0.41	24
47	KILMARNOCK	0	0.0	0.0--	0.00	46
48	KYLE-CARRICK	4	1.1	1.4	0.87	9
49	CUMNOCK-DOON	1	0.7	0.4	0.44	37
50	WIGTOWN	1	1.1	0.8	0.79	20
51	STEWARTRY	0	0.0	0.0--	0.00	46
52	NITHSDALE	0	0.0	0.0--	0.00	46
53	ANNANDALE	0	0.0	0.0--	0.00	46
54	ORKNEY	1	1.6	2.0	2.04	3
55	SHETLAND	1	1.6	0.4	0.45	35
56	WESTERN ISLES	0	0.0	0.0--	0.00	46
	ALL SCOTLAND	156	1.0	0.8	0.07	

SCOTTISH CITIES

Brain and Nervous System — Males

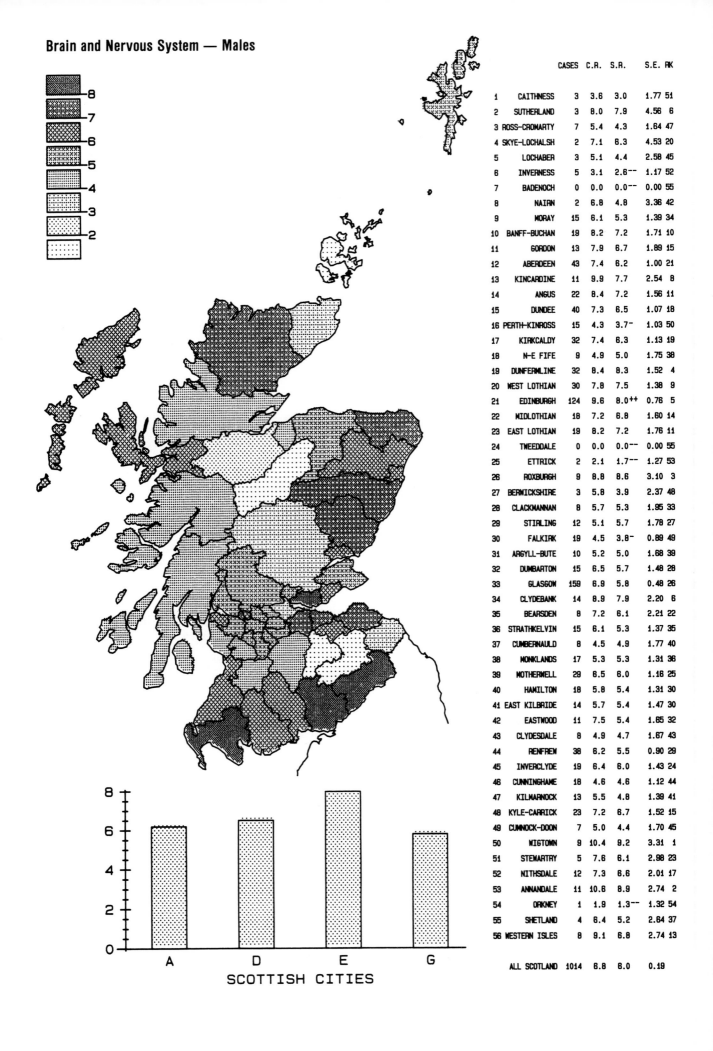

		CASES	C.R.	S.R.	S.E.	RK
1	CAITHNESS	3	3.6	3.0	1.77	51
2	SUTHERLAND	3	8.0	7.9	4.56	6
3	ROSS-CROMARTY	7	5.4	4.3	1.64	47
4	SKYE-LOCHALSH	2	7.1	6.3	4.53	20
5	LOCHABER	3	5.1	4.4	2.58	45
6	INVERNESS	5	3.1	2.6--	1.17	52
7	BADENOCH	0	0.0	0.0--	0.00	55
8	NAIRN	2	6.8	4.8	3.38	42
9	MORAY	15	6.1	5.3	1.39	34
10	BANFF-BUCHAN	19	8.2	7.2	1.71	10
11	GORDON	13	7.9	6.7	1.89	15
12	ABERDEEN	43	7.4	6.2	1.00	21
13	KINCARDINE	11	9.9	7.7	2.54	8
14	ANGUS	22	8.4	7.2	1.56	11
15	DUNDEE	40	7.3	6.5	1.07	18
16	PERTH-KINROSS	15	4.3	3.7-	1.03	50
17	KIRKCALDY	32	7.4	6.3	1.13	19
18	N-E FIFE	9	4.9	5.0	1.75	38
19	DUNFERMLINE	32	8.4	8.3	1.52	4
20	WEST LOTHIAN	30	7.8	7.5	1.38	9
21	EDINBURGH	124	9.6	8.0++	0.78	5
22	MIDLOTHIAN	18	7.2	6.8	1.60	14
23	EAST LOTHIAN	19	8.2	7.2	1.76	11
24	TWEEDDALE	0	0.0	0.0--	0.00	55
25	ETTRICK	2	2.1	1.7--	1.27	53
26	ROXBURGH	9	8.8	8.6	3.10	3
27	BERWICKSHIRE	3	5.8	3.9	2.37	48
28	CLACKMANNAN	8	5.7	5.3	1.95	33
29	STIRLING	12	5.1	5.7	1.78	27
30	FALKIRK	19	4.5	3.8-	0.89	49
31	ARGYLL-BUTE	10	5.2	5.0	1.68	39
32	DUMBARTON	15	6.5	5.7	1.48	28
33	GLASGOW	159	6.9	5.8	0.48	26
34	CLYDEBANK	14	8.9	7.9	2.20	6
35	BEARSDEN	8	7.2	6.1	2.21	22
36	STRATHKELVIN	15	6.1	5.3	1.37	35
37	CUMBERNAULD	8	4.5	4.9	1.77	40
38	MONKLANDS	17	5.3	5.3	1.31	36
39	MOTHERWELL	29	6.5	6.0	1.16	25
40	HAMILTON	18	5.8	5.4	1.31	30
41	EAST KILBRIDE	14	5.7	5.4	1.47	30
42	EASTWOOD	11	7.5	5.4	1.65	32
43	CLYDESDALE	8	4.9	4.7	1.67	43
44	RENFREW	38	6.2	5.5	0.90	29
45	INVERCLYDE	19	6.4	6.0	1.43	24
46	CUNNINGHAME	18	4.6	4.6	1.12	44
47	KILMARNOCK	13	5.5	4.8	1.39	41
48	KYLE-CARRICK	23	7.2	6.7	1.52	15
49	CUMNOCK-DOON	7	5.0	4.4	1.70	45
50	WIGTOWN	9	10.4	9.2	3.31	1
51	STEWARTRY	5	7.6	6.1	2.98	23
52	NITHSDALE	12	7.3	6.6	2.01	17
53	ANNANDALE	11	10.6	8.9	2.74	2
54	ORKNEY	1	1.9	1.3--	1.32	54
55	SHETLAND	4	6.4	5.2	2.64	37
56	WESTERN ISLES	8	9.1	6.8	2.74	13
	ALL SCOTLAND	1014	6.8	6.0	0.19	

SCOTTISH CITIES

Brain and Nervous System — Females

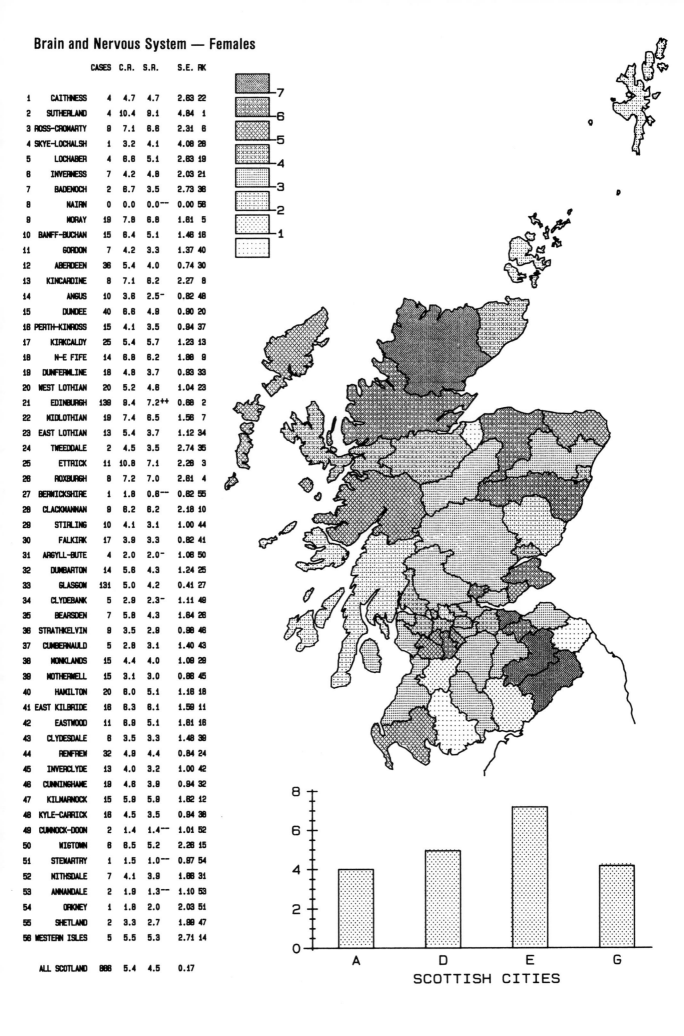

		CASES	C.R.	S.R.	S.E.	RK
1	CAITHNESS	4	4.7	4.7	2.63	22
2	SUTHERLAND	4	10.4	8.1	4.64	1
3	ROSS-CROMARTY	9	7.1	6.6	2.31	6
4	SKYE-LOCHALSH	1	3.2	4.1	4.06	28
5	LOCHABER	4	6.6	5.1	2.63	19
6	INVERNESS	7	4.2	4.6	2.03	21
7	BADENOCH	2	8.7	3.5	2.73	36
8	NAIRN	0	0.0	0.0--	0.00	56
9	MORAY	19	7.8	6.6	1.61	5
10	BANFF-BUCHAN	15	6.4	5.1	1.46	16
11	GORDON	7	4.2	3.3	1.37	40
12	ABERDEEN	36	5.4	4.0	0.74	30
13	KINCARDINE	8	7.1	6.2	2.27	8
14	ANGUS	10	3.6	2.5-	0.82	48
15	DUNDEE	40	6.6	4.9	0.90	20
16	PERTH-KINROSS	15	4.1	3.5	0.84	37
17	KIRKCALDY	25	5.4	5.7	1.23	13
18	N-E FIFE	14	6.8	6.2	1.88	9
19	DUNFERMLINE	18	4.6	3.7	0.93	33
20	WEST LOTHIAN	20	5.2	4.6	1.04	23
21	EDINBURGH	138	8.4	7.2++	0.66	2
22	MIDLOTHIAN	19	7.4	6.5	1.56	7
23	EAST LOTHIAN	13	5.4	3.7	1.12	34
24	TWEEDDALE	2	4.5	3.5	2.74	35
25	ETTRICK	11	10.8	7.1	2.28	3
26	ROXBURGH	8	7.2	7.0	2.61	4
27	BERWICKSHIRE	1	1.6	0.6--	0.62	55
28	CLACKMANNAN	9	6.2	6.2	2.16	10
29	STIRLING	10	4.1	3.1	1.00	44
30	FALKIRK	17	3.9	3.3	0.82	41
31	ARGYLL-BUTE	4	2.0	2.0-	1.06	50
32	DUMBARTON	14	5.6	4.3	1.24	25
33	GLASGOW	131	5.0	4.2	0.41	27
34	CLYDEBANK	5	2.9	2.3-	1.11	49
35	BEARSDEN	7	5.8	4.3	1.64	26
36	STRATHKELVIN	9	3.5	2.8	0.98	46
37	CUMBERNAULD	5	2.8	3.1	1.40	43
38	MONKLANDS	15	4.4	4.0	1.09	29
39	MOTHERWELL	15	3.1	3.0	0.86	45
40	HAMILTON	20	6.0	5.1	1.16	16
41	EAST KILBRIDE	16	6.3	6.1	1.59	11
42	EASTWOOD	11	6.9	5.1	1.61	16
43	CLYDESDALE	8	3.5	3.3	1.46	39
44	RENFREW	32	4.9	4.4	0.84	24
45	INVERCLYDE	13	4.0	3.2	1.00	42
46	CUNNINGHAME	19	4.6	3.9	0.94	32
47	KILMARNOCK	15	5.9	5.9	1.62	12
48	KYLE-CARRICK	16	4.5	3.5	0.94	38
49	CUMNOCK-DOON	2	1.4	1.4--	1.01	52
50	WIGTOWN	8	6.5	5.2	2.26	15
51	STEWARTRY	1	1.5	1.0--	0.97	54
52	NITHSDALE	7	4.1	3.9	1.86	31
53	ANNANDALE	2	1.9	1.3--	1.10	53
54	ORKNEY	1	1.8	2.0	2.03	51
55	SHETLAND	2	3.3	2.7	1.99	47
56	WESTERN ISLES	5	5.5	5.3	2.71	14
	ALL SCOTLAND	866	5.4	4.5	0.17	

Lymphatic Leukaemia — Males

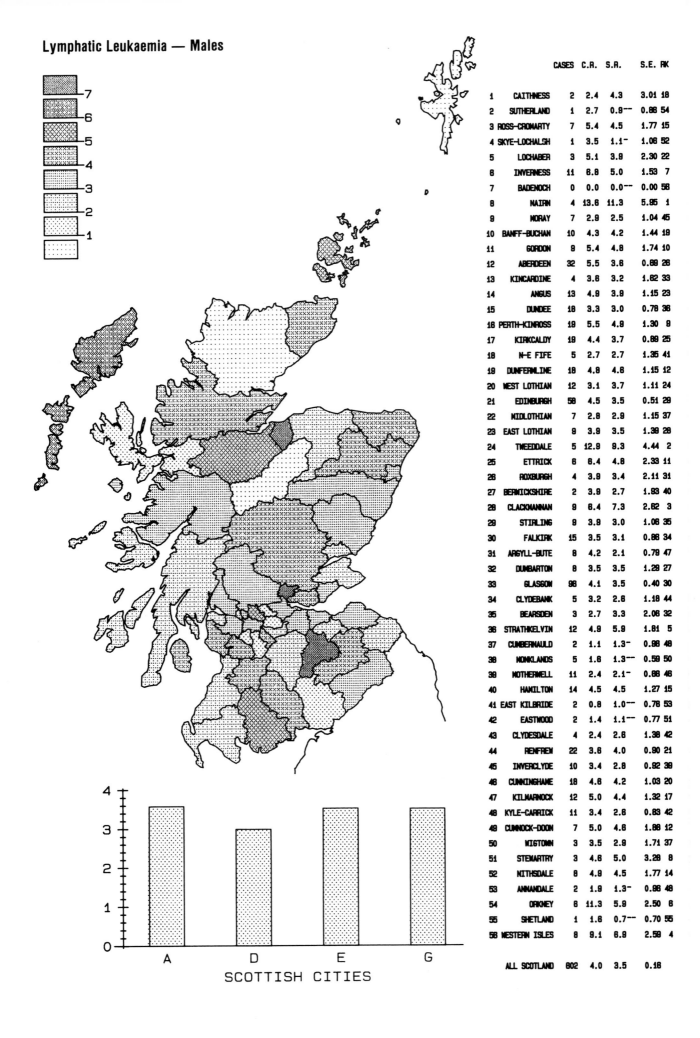

		CASES	C.R.	S.R.	S.E.	RK
1	CAITHNESS	2	2.4	4.3	3.01	18
2	SUTHERLAND	1	2.7	0.9--	0.88	54
3	ROSS-CROMARTY	7	5.4	4.5	1.77	15
4	SKYE-LOCHALSH	1	3.5	1.1-	1.06	52
5	LOCHABER	3	5.1	3.9	2.30	22
8	INVERNESS	11	6.8	5.0	1.53	7
7	BADENOCH	0	0.0	0.0--	0.00	56
8	NAIRN	4	13.6	11.3	5.95	1
9	MORAY	7	2.9	2.5	1.04	45
10	BANFF-BUCHAN	10	4.3	4.2	1.44	19
11	GORDON	9	5.4	4.8	1.74	10
12	ABERDEEN	32	5.5	3.8	0.69	26
13	KINCARDINE	4	3.8	3.2	1.62	33
14	ANGUS	13	4.9	3.9	1.15	23
15	DUNDEE	18	3.3	3.0	0.78	36
16	PERTH-KINROSS	19	5.5	4.8	1.30	8
17	KIRKCALDY	19	4.4	3.7	0.89	25
18	N-E FIFE	5	2.7	2.7	1.35	41
19	DUNFERMLINE	18	4.8	4.8	1.15	12
20	WEST LOTHIAN	12	3.1	3.7	1.11	24
21	EDINBURGH	58	4.5	3.5	0.51	29
22	MIDLOTHIAN	7	2.8	2.9	1.15	37
23	EAST LOTHIAN	9	3.9	3.5	1.39	28
24	TWEEDDALE	5	12.9	8.3	4.44	2
25	ETTRICK	6	6.4	4.8	2.33	11
26	ROXBURGH	4	3.9	3.4	2.11	31
27	BERWICKSHIRE	2	3.9	2.7	1.93	40
28	CLACKMANNAN	9	8.4	7.3	2.62	3
29	STIRLING	9	3.9	3.0	1.06	35
30	FALKIRK	15	3.5	3.1	0.86	34
31	ARGYLL-BUTE	8	4.2	2.1	0.79	47
32	DUMBARTON	8	3.5	3.5	1.29	27
33	GLASGOW	98	4.1	3.5	0.40	30
34	CLYDEBANK	5	3.2	2.8	1.18	44
35	BEARSDEN	3	2.7	3.3	2.06	32
36	STRATHKELVIN	12	4.9	5.9	1.81	5
37	CUMBERNAULD	2	1.1	1.3-	0.96	48
38	MONKLANDS	5	1.6	1.3--	0.59	50
39	MOTHERWELL	11	2.4	2.1-	0.66	46
40	HAMILTON	14	4.5	4.5	1.27	15
41	EAST KILBRIDE	2	0.8	1.0--	0.76	53
42	EASTWOOD	2	1.4	1.1--	0.77	51
43	CLYDESDALE	4	2.4	2.6	1.36	42
44	RENFREW	22	3.6	4.0	0.90	21
45	INVERCLYDE	10	3.4	2.8	0.92	39
46	CUNNINGHAME	18	4.6	4.2	1.03	20
47	KILMARNOCK	12	5.0	4.4	1.32	17
48	KYLE-CARRICK	11	3.4	2.6	0.83	42
49	CUMNOCK-DOON	7	5.0	4.8	1.86	12
50	WIGTOWN	3	3.5	2.9	1.71	37
51	STEWARTRY	3	4.8	5.0	3.28	8
52	NITHSDALE	8	4.9	4.5	1.77	14
53	ANNANDALE	2	1.9	1.3-	0.98	48
54	ORKNEY	6	11.3	5.9	2.50	6
55	SHETLAND	1	1.6	0.7--	0.70	55
56	WESTERN ISLES	8	9.1	6.9	2.59	4
	ALL SCOTLAND	602	4.0	3.5	0.16	

Lymphatic Leukaemia — Females

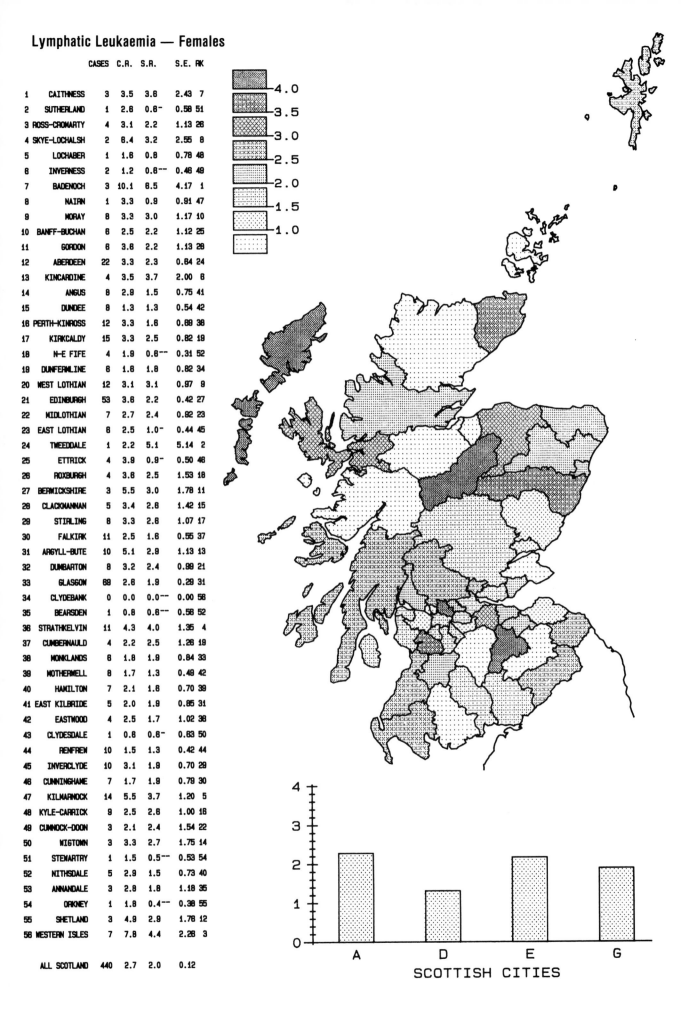

		CASES	C.R.	S.R.	S.E.	RK
1	CAITHNESS	3	3.5	3.6	2.43	7
2	SUTHERLAND	1	2.6	0.6⁻	0.58	51
3	ROSS-CROMARTY	4	3.1	2.2	1.13	26
4	SKYE-LOCHALSH	2	6.4	3.2	2.55	8
5	LOCHABER	1	1.6	0.8	0.78	48
6	INVERNESS	2	1.2	0.6⁻⁻	0.46	49
7	BADENOCH	3	10.1	8.5	4.17	1
8	NAIRN	1	3.3	0.9	0.91	47
9	MORAY	8	3.3	3.0	1.17	10
10	BANFF-BUCHAN	6	2.5	2.2	1.12	25
11	GORDON	6	3.6	2.2	1.13	26
12	ABERDEEN	22	3.3	2.3	0.64	24
13	KINCARDINE	4	3.5	3.7	2.00	6
14	ANGUS	8	2.9	1.5	0.75	41
15	DUNDEE	8	1.3	1.3	0.54	42
16	PERTH-KINROSS	12	3.3	1.6	0.69	38
17	KIRKCALDY	15	3.3	2.5	0.82	19
18	N-E FIFE	4	1.9	0.8⁻⁻	0.31	52
19	DUNFERMLINE	8	1.8	1.8	0.82	34
20	WEST LOTHIAN	12	3.1	3.1	0.97	9
21	EDINBURGH	53	3.6	2.2	0.42	27
22	MIDLOTHIAN	7	2.7	2.4	0.82	23
23	EAST LOTHIAN	6	2.5	1.0⁻	0.44	45
24	TWEEDDALE	1	2.2	5.1	5.14	2
25	ETTRICK	4	3.9	0.9⁻	0.50	46
26	ROXBURGH	4	3.6	2.5	1.53	18
27	BERWICKSHIRE	3	5.5	3.0	1.78	11
28	CLACKMANNAN	5	3.4	2.6	1.42	15
29	STIRLING	8	3.3	2.6	1.07	17
30	FALKIRK	11	2.5	1.6	0.55	37
31	ARGYLL-BUTE	10	5.1	2.9	1.13	13
32	DUMBARTON	8	3.2	2.4	0.99	21
33	GLASGOW	69	2.6	1.9	0.29	31
34	CLYDEBANK	0	0.0	0.0⁻⁻	0.00	56
35	BEARSDEN	1	0.8	0.6⁻	0.58	52
36	STRATHKELVIN	11	4.3	4.0	1.35	4
37	CUMBERNAULD	4	2.2	2.5	1.26	19
38	MONKLANDS	6	1.8	1.9	0.84	33
39	MOTHERWELL	8	1.7	1.3	0.48	42
40	HAMILTON	7	2.1	1.8	0.70	39
41	EAST KILBRIDE	5	2.0	1.9	0.85	31
42	EASTWOOD	4	2.5	1.7	1.02	36
43	CLYDESDALE	1	0.8	0.6⁻	0.63	50
44	RENFREW	10	1.5	1.3	0.42	44
45	INVERCLYDE	10	3.1	1.9	0.70	29
46	CUNNINGHAME	7	1.7	1.9	0.79	30
47	KILMARNOCK	14	5.5	3.7	1.20	5
48	KYLE-CARRICK	9	2.5	2.6	1.00	16
49	CUMNOCK-DOON	3	2.1	2.4	1.54	22
50	WIGTOWN	3	3.3	2.7	1.75	14
51	STEWARTRY	1	1.5	0.5⁻⁻	0.53	54
52	NITHSDALE	5	2.9	1.5	0.73	40
53	ANNANDALE	3	2.8	1.8	1.18	35
54	ORKNEY	1	1.8	0.4⁻⁻	0.38	55
55	SHETLAND	3	4.9	2.9	1.78	12
56	WESTERN ISLES	7	7.8	4.4	2.26	3
	ALL SCOTLAND	440	2.7	2.0	0.12	

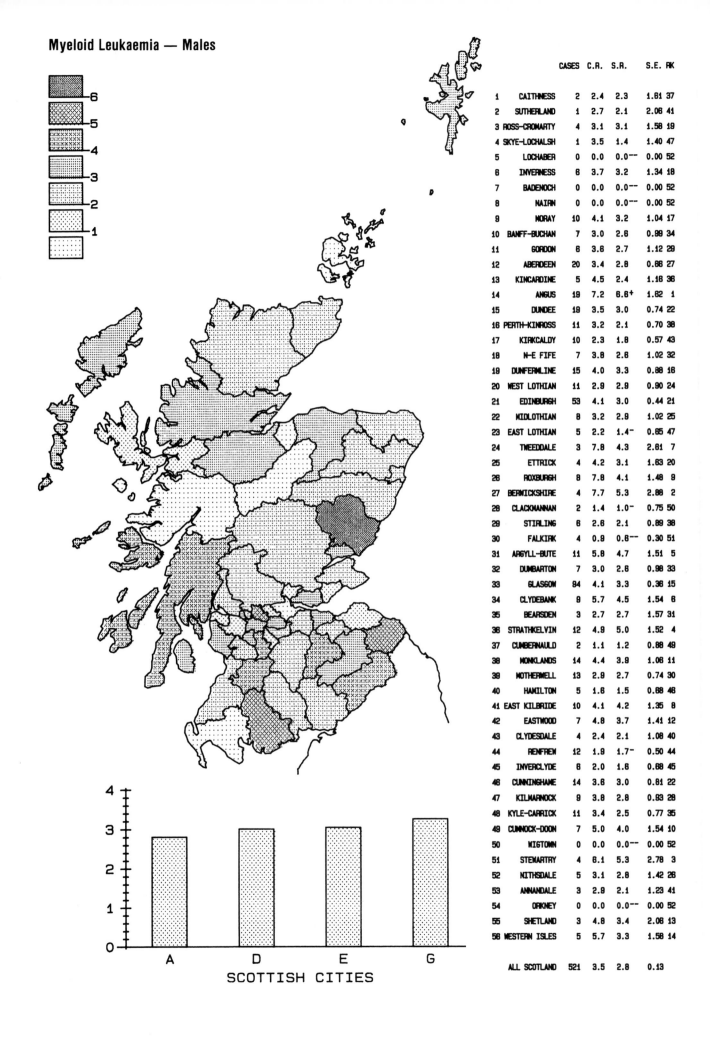

Myeloid Leukaemia — Males

				CASES	C.R.	S.R.	S.E.	RK
1	CAITHNESS			2	2.4	2.3	1.61	37
2	SUTHERLAND			1	2.7	2.1	2.06	41
3	ROSS-CROMARTY			4	3.1	3.1	1.58	19
4	SKYE-LOCHALSH			1	3.5	1.4	1.40	47
5	LOCHABER			0	0.0	0.0 --	0.00	52
6	INVERNESS			8	3.7	3.2	1.34	16
7	BADENOCH			0	0.0	0.0 --	0.00	52
8	NAIRN			0	0.0	0.0 --	0.00	52
9	MORAY			10	4.1	3.2	1.04	17
10	BANFF-BUCHAN			7	3.0	2.6	0.99	34
11	GORDON			6	3.6	2.7	1.12	29
12	ABERDEEN			20	3.4	2.6	0.66	27
13	KINCARDINE			5	4.5	2.4	1.16	36
14	ANGUS			19	7.2	6.6 +	1.62	1
15	DUNDEE			19	3.5	3.0	0.74	22
16	PERTH-KINROSS			11	3.2	2.1	0.70	38
17	KIRKCALDY			10	2.3	1.8	0.57	43
18	N-E FIFE			7	3.8	2.6	1.02	32
19	DUNFERMLINE			15	4.0	3.3	0.88	16
20	WEST LOTHIAN			11	2.9	2.9	0.90	24
21	EDINBURGH			53	4.1	3.0	0.44	21
22	MIDLOTHIAN			8	3.2	2.9	1.02	25
23	EAST LOTHIAN			5	2.2	1.4 -	0.65	47
24	TWEEDDALE			3	7.8	4.3	2.81	7
25	ETTRICK			4	4.2	3.1	1.63	20
26	ROXBURGH			8	7.8	4.1	1.48	9
27	BERWICKSHIRE			4	7.7	5.3	2.86	2
28	CLACKMANNAN			2	1.4	1.0 -	0.75	50
29	STIRLING			8	2.6	2.1	0.89	38
30	FALKIRK			4	0.9	0.8 --	0.30	51
31	ARGYLL-BUTE			11	5.8	4.7	1.51	5
32	DUMBARTON			7	3.0	2.6	0.98	33
33	GLASGOW			94	4.1	3.3	0.36	15
34	CLYDEBANK			9	5.7	4.5	1.54	6
35	BEARSDEN			3	2.7	2.7	1.57	31
36	STRATHKELVIN			12	4.9	5.0	1.52	4
37	CUMBERNAULD			2	1.1	1.2	0.88	49
38	MONKLANDS			14	4.4	3.9	1.06	11
39	MOTHERWELL			13	2.9	2.7	0.74	30
40	HAMILTON			5	1.6	1.5	0.68	46
41	EAST KILBRIDE			10	4.1	4.2	1.35	8
42	EASTWOOD			7	4.8	3.7	1.41	12
43	CLYDESDALE			4	2.4	2.1	1.08	40
44	RENFREW			12	1.9	1.7 -	0.50	44
45	INVERCLYDE			8	2.0	1.6	0.68	45
46	CUNNINGHAME			14	3.6	3.0	0.81	22
47	KILMARNOCK			9	3.8	2.6	0.93	28
48	KYLE-CARRICK			11	3.4	2.5	0.77	35
49	CUMNOCK-DOON			7	5.0	4.0	1.54	10
50	WIGTOWN			0	0.0	0.0 --	0.00	52
51	STEWARTRY			4	6.1	5.3	2.78	3
52	NITHSDALE			5	3.1	2.6	1.42	26
53	ANNANDALE			3	2.9	2.1	1.23	41
54	ORKNEY			0	0.0	0.0 --	0.00	52
55	SHETLAND			3	4.6	3.4	2.06	13
56	WESTERN ISLES			5	5.7	3.3	1.58	14
	ALL SCOTLAND			521	3.5	2.8	0.13	

SCOTTISH CITIES

A D E G

Myeloid Leukaemia — Females

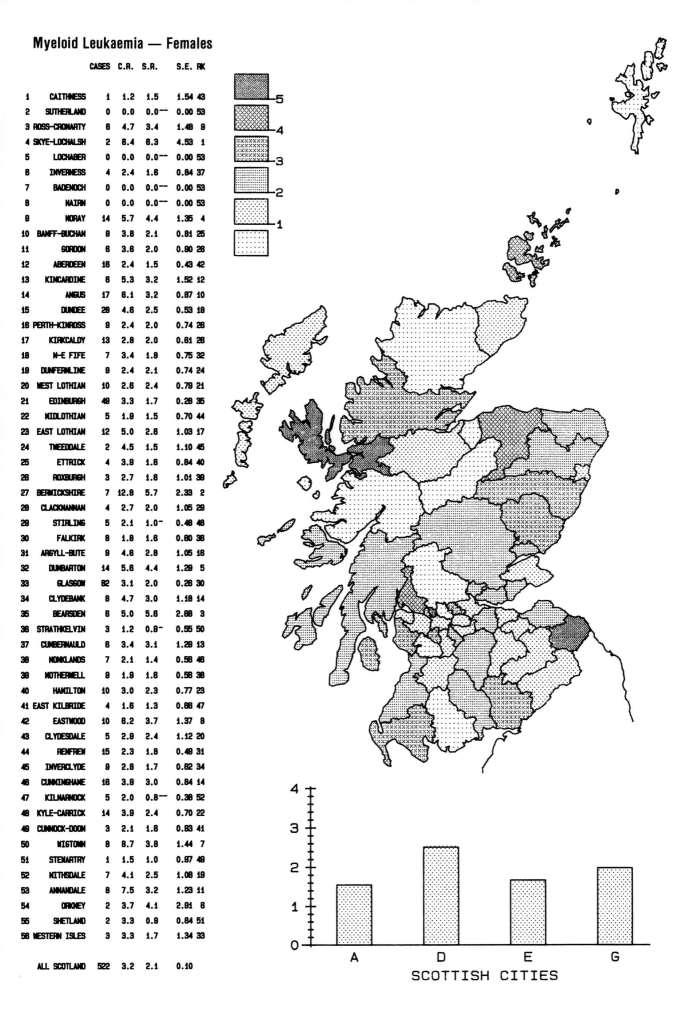

		CASES	C.R.	S.R.	S.E.	RK
1	CAITHNESS	1	1.2	1.5	1.54	43
2	SUTHERLAND	0	0.0	0.0--	0.00	53
3	ROSS-CROMARTY	8	4.7	3.4	1.48	9
4	SKYE-LOCHALSH	2	8.4	6.3	4.53	1
5	LOCHABER	0	0.0	0.0--	0.00	53
6	INVERNESS	4	2.4	1.6	0.84	37
7	BADENOCH	0	0.0	0.0--	0.00	53
8	NAIRN	0	0.0	0.0--	0.00	53
9	MORAY	14	5.7	4.4	1.35	4
10	BANFF-BUCHAN	9	3.8	2.1	0.81	25
11	GORDON	8	3.8	2.0	0.90	26
12	ABERDEEN	18	2.4	1.5	0.43	42
13	KINCARDINE	8	5.3	3.2	1.52	12
14	ANGUS	17	6.1	3.2	0.87	10
15	DUNDEE	28	4.6	2.5	0.53	18
16	PERTH-KINROSS	9	2.4	2.0	0.74	26
17	KIRKCALDY	13	2.8	2.0	0.61	28
18	N-E FIFE	7	3.4	1.8	0.75	32
19	DUNFERMLINE	9	2.4	2.1	0.74	24
20	WEST LOTHIAN	10	2.8	2.4	0.79	21
21	EDINBURGH	49	3.3	1.7	0.28	35
22	MIDLOTHIAN	5	1.9	1.5	0.70	44
23	EAST LOTHIAN	12	5.0	2.6	1.03	17
24	TWEEDDALE	2	4.5	1.5	1.10	45
25	ETTRICK	4	3.9	1.6	0.84	40
26	ROXBURGH	3	2.7	1.6	1.01	39
27	BERWICKSHIRE	7	12.8	5.7	2.33	2
28	CLACKMANNAN	4	2.7	2.0	1.05	29
29	STIRLING	5	2.1	1.0-	0.48	48
30	FALKIRK	8	1.9	1.6	0.60	38
31	ARGYLL-BUTE	9	4.8	2.8	1.05	16
32	DUMBARTON	14	5.6	4.4	1.29	5
33	GLASGOW	82	3.1	2.0	0.26	30
34	CLYDEBANK	8	4.7	3.0	1.18	14
35	BEARSDEN	6	5.0	5.6	2.66	3
36	STRATHKELVIN	3	1.2	0.9-	0.55	50
37	CUMBERNAULD	6	3.4	3.1	1.29	13
38	MONKLANDS	7	2.1	1.4	0.56	46
39	MOTHERWELL	9	1.9	1.6	0.56	38
40	HAMILTON	10	3.0	2.3	0.77	23
41	EAST KILBRIDE	4	1.6	1.3	0.68	47
42	EASTWOOD	10	6.2	3.7	1.37	8
43	CLYDESDALE	5	2.9	2.4	1.12	20
44	RENFREW	15	2.3	1.8	0.49	31
45	INVERCLYDE	9	2.8	1.7	0.62	34
46	CUNNINGHAME	18	3.9	3.0	0.84	14
47	KILMARNOCK	5	2.0	0.8--	0.36	52
48	KYLE-CARRICK	14	3.9	2.4	0.70	22
49	CUMNOCK-DOON	3	2.1	1.6	0.93	41
50	WIGTOWN	8	8.7	3.8	1.44	7
51	STEWARTRY	1	1.5	1.0	0.97	49
52	NITHSDALE	7	4.1	2.5	1.08	19
53	ANNANDALE	8	7.5	3.2	1.23	11
54	ORKNEY	2	3.7	4.1	2.91	6
55	SHETLAND	2	3.3	0.9	0.84	51
56	WESTERN ISLES	3	3.3	1.7	1.34	33
	ALL SCOTLAND	522	3.2	2.1	0.10	

SCOTTISH CITIES

Chapter 8

OVERVIEW: THE WAY AHEAD

This chapter summarizes the major geographical features of the distribution of cancer in Scotland. To set the findings in perspective, this synopsis is followed by a consideration of the major categories of cancer cause likely to be operating in Scotland. Priorities for future research are put forward, and the techniques used by epidemiologists to investigate risk factors are discussed.

SUMMARY OF FINDINGS

A summary in any form inevitably leads to loss of information, and the reader is urged to examine the more detailed text in Chapter 7.

Earlier chapters have illustrated the variation in the characterization of the Scottish districts. It is dangerous to assume that differences in the rates of cancer reflect these characteristics directly. We repeat that a significantly high level of a given cancer in a district is not necessarily due to a universally unavoidable exposure to a particular risk, but may result from the personal habits, diet or workplace exposures of a segment of the population. A district with higher levels of one cancer may have lower levels of another. In the present state of our knowledge, some differences will remain unexplained. Further, the suggestions made in this volume about possible explanations for observed differences are suggestions only, not proof. Hence, areas for future research in Scotland into the causes of several sites of cancers are proposed below, using the observed geographical differences.

The numbers in round brackets after the name of a site of cancer refer to the 8th revision of the International Classification of Diseases (ICD) (WHO, 1967), and the percentages in square brackets give the proportion of all cancer in Scottish males (M) and females (F) represented by the site in question (Fig. 8.1).

Lip (ICD8 140) [M 0.8%; F 0.1%]

The falling north-south gradient of incidence rates of lip cancer among males is one of the most noticeable features of all the maps produced. The marked aggregation of high rates in the north-west and north-east of Scotland accords well with the distribution of persons employed in farming, fishing and forestry. Persons in these occupations, which involve mainly outdoor work, have a greater exposure to sunlight, which is thought to be the major risk factor for lip cancer. It would be worthwhile to study the etiology of lip cancer in Scotland, although the small numbers of cases scattered over remote areas may present logistic difficulties — perhaps suitable for postal questionnaire. Information on occupation and smoking habits would be essential. Equally rewarding may be the investment in 'health awareness' schemes in areas of high incidence, to encourage people to come for diagnosis and treatment of early lesions. The feasibility, and acceptability, of introducing barrier cream protection for the lips of outdoor workers may also prove an interesting area of research.

Oral cavity and pharynx (ICD8 141, 143-145, 146, 148) [M 1.3%; F 0.8%]

Because cancer occurs relatively infrequently on the tongue, gums, the roof and floor of the mouth, the inner aspect of the cheeks, the oropharynx and hypopharynx, all these sites were combined as 'oral cavity and pharynx'. No striking pattern is evident in the geographical distribution of these cancers. In males, both the highest and the lowest rates are found in rural areas in the north and south of the country. For females, there appears to be a small pocket of districts with high incidence rates south-east and south-west of Glasgow.

Oesophagus (ICD8 150) M 2.5%; F 2.2%]

There are three foci of higher than average rates of cancer of the oesophagus in males: the Grampian region, the Western Isles and the Firth of Clyde.

For females, high rates are seen in the southern part of Scotland, notably in the eastern Borders, south Lothian, south Ayrshire and Clydesdale. The all-Scottish rate for females (4.2) is the highest currently recorded among any European female population with published cancer incidence data. Not only are the rates high, but substantial increases have taken place over time, particularly among younger age groups. These increases are also seen in males.

Several factors may serve to increase the risk of oesophageal cancer — cigarette smoking, consumption of spirits and a diet low in the protective factors believed to occur in fresh fruit and vegetables. Studies to show whether these factors are, indeed, responsible for oesophageal cancer in Scotland should be carried out with some priority, in view of the rapidly

Figure 8.1. The common cancers in Scottish males and females, 1975-1980, percentage of all forms of cancer

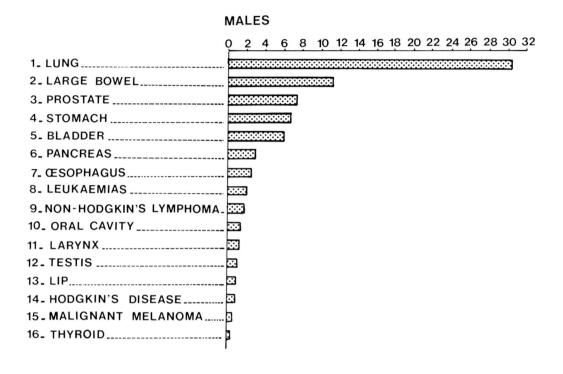

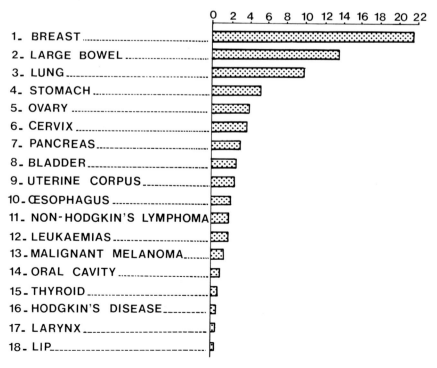

increasing incidence in young people, the unusual level of this cancer in females and the extremely high rate of mortality from such tumours.

Stomach (ICD8 151) [M 6.8%; F 5.4%]

The main features of the geographical distribution of stomach cancer in people of each sex are the consistent finding of low rates in the rural north and south of Scotland, and the concentrations of high rates in and around the major industrial centres of Dundee and Glasgow. Although stomach cancer incidence is decreasing in Scotland, it remains the third commonest cancer in males, accounting for 6.8% of all reported malignancies. In females, this cancer is the fourth commonest, comprising 5.4% of all cancers.

A series of studies of stomach cancer in Scotland appears justified, in order to determine the reasons underlying these distinctive patterns. The studies should investigate occupational factors, such as metalworking, as well as dietary consumption of items believed to be protective, such as fresh fruit and vegetables.

Large bowel (ICD8 153,154) [M 11.2%; F 13.7%]

The outstanding feature of the distribution of large-bowel cancer is the aggregation of landward areas with similar incidence. High levels are found in people of each sex in northern Scotland, and low levels in the south; rates in the central belt are intermediate. Further, the three-fold range of variation between highest and lowest incidence rates within Scotland is the same as that found over the continent of Europe.

Biochemical studies of diet, including estimations of fat and fibre, measurements of faecal bile-acid concentration and determinations of serum cholesterol and vitamin levels, should be conducted in areas of Scotland of contrasting risk of large-bowel cancer. These studies should have priority in view of the high and increasing incidence in Scotland and the large within-country variation in incidence rates.

Pancreas (ICD8 157) [M 2.9%; F 3.0%]

Pancreatic cancer is an increasingly common form of malignancy in Scotland, with very poor survival rates. Several striking features of the geographical pattern found in Scotland are: consistently low rates in people of each sex in the western Highland districts; high rates in people of each sex in and around the four Scottish cities; high rates in males in both urban and rural districts of south-east Scotland; and a belt of high incidence rates in females in southern-central Scotland.

Little is known about the risk factors for most cases of pancreatic cancer, and no explanation of the Scottish pattern can yet be proposed. The variations in incidence within Scotland should be used to improve our understanding of the causes of this form of cancer.

Larynx (ICD8 161) [M 1.3%; F 0.4%]

The one important feature of laryngeal cancer in Scotland is the similarity of the rates found in the Scottish cities. In each city, the rates were higher than the all-Scottish rate for people of each sex. This may be due to an association with cigarette smoking, the single most important known risk factor.

Trachea, bronchus and lung (ICD8 162) [M 30.4%; F 9.8%]

Cancer of the trachea, bronchus and lung (lung cancer) is tragically common in Scotland, and the incidence continues to increase. Rates are high in people of each sex in the districts of central Scotland and in the cities and their surrounding districts.

The problem presented by lung cancer in Scotland is clearly illustrated by the rates found in Glasgow males, among whom the age-standardized incidence (130.7) is the highest recorded anywhere in the world. In terms of crude rates, the numbers are even more startling — nearly one new case of lung cancer for every 500 males each year. If the level in Glasgow were reduced to that of the next highest rate (that of New Orleans blacks), the toll among Glasgow males would be reduced by over 100 cases each year.

The high rates of lung cancer found in densely populated urban areas should not be interpreted as implying an association between urban living *per se* and an increased risk. Most likely, they reflect the pattern of tobacco usage in large industrial conurbations; as much as 85% of lung cancer is caused by smoking cigarettes. Unless women take action to reduce smoking, the regrettable truth is that the incidence rates in females will continue their dramatic rise over the coming decades. The recently launched campaign to make Glasgow a non-smoking city by the year 2000 is to be thoroughly applauded; if successful, some 400 Glasgow men would be spared from lung cancer each year.

Malignant melanoma of the skin (ICD8 172) [M 0.7%; F 1.3%]

The incidence of malignant melanoma is increasing in people of each sex in Scotland. No striking pattern is evident for the rates among the

Scottish districts, other than a tendency that more districts with high rates are found in the north of the country. Rates in Glasgow City appear to be low in people of each sex.

Breast (ICD8 174) [F 21.6%]

Breast cancer is the commonest form of cancer in women, and the incidence is increasing in Scottish females in all but the youngest (20-34) age groups. Higher incidence rates occur in less densely populated areas of the country, while lower rates occur generally in areas of essentially urban and industrial character. The regional variation observed may be due to regional dietary practices, although little, if any, information is available about dietary patterns within Scotland. Other worthwhile investigations include an assessment of the age at first full-time pregnancy by district and the degree of protection conferred by such pregnancies throughout the country.

Uterine cervix (ICD8 180) [F 3.8%], uterine corpus (ICD8 182.0) [F 2.6%], ovary (ICD8 183) [F 4.1%]

The most important feature of the geographical distribution of these three forms of cancer in Scotland is the lower-than-average rates in the districts surrounding Glasgow in western central Scotland. This finding is particularly intriguing, as risk factors for cervical and corpus cancer act in opposite directions. For example, childbearing increases the risk for cervical cancer but decreases the risk for corpus cancer. A series of case-control studies (see below) is needed, to follow confirmation of the unexpectedly low rates and isolated high rates in some parts of the country.

Prostate (ICD8 185) [M 7.6%]

The incidence of cancer of the prostate gland has increased in Scotland since 1960. The rate is higher in north-east Scotland, and there is an aggregation of districts with lower rates in and around the City of Glasgow. The incidence in Glasgow (17.8) is considerably lower than that in Aberdeen (26.5), Edinburgh (25.5) and Dundee (23.0). While little is known about the risk factors for prostatic cancer, study of the differences in incidence between Glasgow and other Scottish cities could produce some interesting clues to the etiology of this disease.

Testis (ICD8 186) [M 0.9%]

Testicular cancer is an increasingly common form of malignant disease, especially among young men in

Scotland. There is no obvious geographical pattern, although statistically there is weak evidence that the pattern is non-random.

Bladder (ICD8 188) [M 6.0%; F 2.7%]

Bladder cancer is unusually common in Scotland, and the incidence is rising in people of each sex. High rates are seen in the Stirling and Kilmarnock districts, and higher-than-average rates occur in each of the cities in people of both sexes. Case-control studies could be rewarding, perhaps, to uncover risk factors other than smoking. While textile industry workers have been shown historically to have an increased risk of bladder cancer, no excess risk is seen in those Border and Highland districts where weaving, spinning and textile manufacture are concentrated.

Thyroid (ICD8 193) [M 0.2%; F 0.7%]

The incidence of cancer of the thyroid gland is three times higher in females than in males and has remained essentially unchanged for 20 years. Two patterns are apparent among females: first, districts with high rates are found on either side of the Firth of Forth, extending northwards through Fife and Perth; and, second, a belt of lower-than-average rates extends across the south of Scotland from east to west. It is difficult to explain this pattern on the basis of current knowledge about cause.

Hodgkin's disease (ICD8 201) [M 0.8%; F 0.6%], non-Hodgkin's lymphoma (ICD8 200, 202) [M 1.9%; F 1.9%]

Both of these forms of cancer are occurring with increasing incidence in Scotland. The outstanding feature of both is the virtual absence of cases among residents of north-west Scotland and the western Highlands.

The incidence of Hodgkin's disease is elevated in the adjacent districts of Clydesdale and Tweeddale in people of each sex; since a similar finding is apparent for non-Hodgkin's lymphoma, these districts are appropriate for research. A similar situation is found in north-eastern Scotland. The incidence of Hodgkin's disease is also high in males in East Lothian and Berwickshire, but female rates are low in these districts, indicating that there may be some local environmental factor to which males but not females are exposed. While Hodgkin's disease is equally common in the four cities in people of each sex, the incidence of non-Hodgkin's lymphoma is substantially higher in Aberdeen and Edinburgh than in Dundee and

Glasgow, again in people of each sex. Although little is known that could explain these variations, analysis of the differences could prove useful in increasing our understanding.

The leukaemias (ICD8 204-207) [M 2.0%; F 1.8%]

The incidence of leukaemia is increasing in people of each sex, but no geographical pattern is apparent. A notable feature is the seven-fold range of variation seen in males and females within Scotland, which is much larger than the 1.5- to two-fold variation reported internationally.

MAJOR CAUSES OF CANCER

Following the descriptions of the distribution of the commoner cancers in Scotland, suggestions were made in Chapter 7 (and to a lesser extent above) about possible causes, on the basis of work performed, for the most part, in other countries. Given the shortage of resources, both in trained manpower and in money, the choice of priorities for investigation in Scotland, as elsewhere, is critical.

Over the past five years, a degree of agreement has arisen concerning the relative importance of the various components of the environment to the causation of cancer (Doll, 1977; Wynder & Gori, 1977; Higginson & Muir, 1979; Doll & Peto, 1981). In this context, the word 'environment' means the total environment, not just man-made chemicals; it includes not only air, food and water and exposures in the workplace, but also personal habits, such as use of tobacco and alcohol, type and amount of food eaten, and culturally influenced habits, such as age at first coitus and at first full-term pregnancy.

While the proportions that can be ascribed to any given group of risk factors will vary from country to country, the likely percentages for different environmental factors in Scotland are listed in Table 8.1.

This table is simplistic in that most cancers are likely to have more than one contributory determinant. Nonetheless, in broad terms, the figures

Table 8.1. Proportion of cancers in Scotland, suspected to be due to environmental factors, by sex

Factor	Males (%)	Females (%)
Tobacco	30	7
Tobacco/alcohol	5	3
Sunlight	10	10
Occupation	5	2
Radiation	1	1
Drugs	1	1
Lifestyle	30	63
Congenital	2	2
Unknown	16	11

indicate the relative importance of the risk factors. By far the largest categories in males are tobacco and lifestyle; in females, lifestyle predominates. Lifestyle consists of two major elements - diet and cultural habits, which influence the risk of cancers of the digestive tract (stomach, large bowel) and reproductive system (breast, uterus, ovary, prostate). The cultural habits in question are mainly those linked to reproduction and include age at first full-term pregnancy and at first sexual intercourse (Chapter 7). However, while there is good evidence that diet and cultural habits do influence risk — in the form of geographical differences and changes in risk on migration from one country to another and over time — it is not known for the most part how they exert their effect. Hence, it is difficult to propose rational preventive measures. As far as diet is concerned, there seems to be a consensus today that reduction of the amount of fat and an increase in the amount of fresh fruit, vegetables and fibre in the diet can only be beneficial. The third of cancers in Scottish men and the tenth in women that are caused by tobacco and excess alcohol can be prevented on the basis of current knowledge.

EPIDEMIOLOGICAL STUDIES

As noted above, the varying distributions of cancers within Scotland have suggested possible causes and areas for further epidemiological research.

Epidemiologists use four major methods to identify risk factors:

(1) Case characteristics

With this method, a series of patients with the cancer under study are asked about various aspects of their life. The questions could cover place of birth, residence, occupation, marital status, obstetric history, personal habits and so on, in an attempt to uncover a common thread. In the Scottish context, cancer of the lip is an obvious choice for this approach. If the investigators suspect the existence of common exposures and risk factors, the next step is to mount a case-control study (see below).

The case-characteristic type of investigation is most effective when the cancer is uncommon and the cause unusual. Study of a rare cancer means that a system must be established for interviewing these persons. In Scotland, it would take several years to interview, for example, 50 persons with cancer of the eye. If clinical records of good quality are available, however, these studies can sometimes be carried out by review of the case-sheets of patients diagnosed in the past.

(2) Correlation study

Statistical comparison of the rate of a cancer with the distribution of a series of exposures is called a correlation study.

As noted above, there is some danger in assuming that differences in the character of Scottish districts are related directly to the variations in the rates of cancer between different districts, as depicted on the maps. For example, there is less rainfall in the north-east than in the south-west. The maps show that the incidence of cancer of the large bowel is higher in the north-east than in the south-west. This does not necessarily mean that there is any direct or, indeed, indirect relationship between the incidence of colorectal cancer and rainfall, although there might be some subtle secondary factor to which both phenomena are related. Many other apparent 'correlations' of this kind could be demonstrated, most of which would prove, on further study, to be spurious. Although it may be worthwhile to relate cancer distribution to diet, dietary information frequently does not exist at the level of the local authority district.

For statistical reasons, positive correlations may emerge by chance alone, while those that are real may fail to be detected, again by chance.

Case-characteristic and correlation studies are 'weak' in that conclusions are at best tenuous, surrounded by a large measure of uncertainty, and can permit no conclusion regarding the causal nature of any observed association or correlation. The studies may, however, help in deciding whether it would be worth proceeding with the 'stronger' forms of epidemiological enquiry outlined below.— case-control and cohort studies.

(3) Case-control study

With this method, patients with the cancer under study are asked a series of questions about exposures that are believed to be linked to the cancer; and the same questions are put to controls, i.e., persons of the same age and sex without the cancer in question. The answers are then compared. A case-control study would find that a greater proportion of lung cancer patients have smoked than controls. This type of study is very popular among epidemiologists in that it is relatively quick to undertake (two to three years) and fairly cheap (usually costing £50 000 to £ 100 000). For some kinds of exposure, however, there are problems in measurement, notably for diet. The presence of cancer may have altered the diet, and it is very difficult to remember what was eaten 10 or more years ago. Nonetheless, in Scotland, case-control studies of breast and large-bowel cancer would seem very worthwhile. Given the very high rates of lung cancer in and around Glasgow, it would be of greatest interest to determine, for a given number of cigarettes smoked, the degree to which risk is influenced by diet, a question readily examined by a case-control study. More complex case-control studies may need laboratory support, e.g., for measurements of blood and urine.

The mounting of a case-control study and interviewing adequate numbers of people in a reasonably short time imply a major collaborative effort and willingness on the part of those who diagnose and treat cancer patients to allow them to be interviewed. As Scotland is a small country, the opportunities for undertaking case-control studies should be good.

(4) Cohort study

In this type of study, a group of people with a common exposure is identified and followed over time to see whether they develop more cancer than the population in general. A typical example is a record of names of persons employed in a particular industry or factory. This type of study can be mounted very cheaply by cancer registries, but the results are frequently difficult to interpret because information is also needed about the personal habits of cohort members, such as smoking and alcohol consumption. Here, trade unions can be of great help by providing membership lists [as done so successfully for the Danish breweries (Jensen, 1980]: employers' lists of those working in specific areas of a particular enterprise may help, also. Scotland again, is very favourably placed for this type of study.

The scope of cohort studies can be greatly extended for the study of digestive tract cancer if samples of blood, urine, etc, can be taken and stored until a cohort member develops cancer. The levels of various body chemicals in cancer patients and in fellow cohort members who do not develop cancer could then be compared.

In adition, there is clearly much more to be done with the existing data. To cite but one example, to investigate whether certain sites of cancer have the same pattern of distribution, e.g., large bowel and breast, as this could suggest common determinants.

CONCLUSION

Time is the epidemiologist's greatest enemy. Normally, it takes 20-30 years' exposure to a risk factor before cancer appears. During this time, the future cancer patient is exposed to many other factors and it can be very difficult to determine which of these are really significant. Fortunately, not everybody who is exposed develops cancer, but this, too, makes the epidemiologist's task more difficult.

Although treatment for some cancers is improving (e.g., enormous advances have been made in the treatment of Hodgkin's disease, cancer of the testis and acute lymphatic leukaemia in children), for the common cancers, such as lung, large bowel, stomach and breast, progress over the past 30 years has been disappointingly slow.

Each country devotes large sums of money to detection, diagnosis and treatment of cancer. Study of the causes, with a view to preventing cancer, has received little attention, and it is here that a major investment is needed. The establishment by the Cancer Research Campaign of a Cancer Epidemiology Unit at the University of Edinburgh is a small but welcome step in this direction.

ACKNOWLEDGEMENTS

An atlas of this nature depends on contributions from many sources. Among the many who have helped, the Editors would like to mention the following:

— In Scotland:

The Registrar General for Scotland

The Staff of the Scottish Office Computer, Saughton

The Graphics Group of the Scottish Development Department, directed by Mr C. Downie, who provided a great deal of useful information

Miss J. Forbes, Senior Lecturer, Town and Country Planning, University of Glasgow, who supplied helpful descriptions of the topography and demography in the west of Scotland and of the decline of industry in Scotland, as well as other information

The Librarian of the Information Services Division, Mrs M. Wellman, and the staffs of the Library of the Scottish Office and Central Public Library Edinburgh, who helped with much of the descriptive material

Dr D. Bruce, Industry Department for Scotland, who provided occupational statistics for Scottish cities

Mr A. Redfern, Assistant Public Relations Officer, City of Glasgow District Council, who provided information on Glasgow

Mr P. Trotter, of the Information Services Division, who was a most valuable source of information on Scottish cancer registries and other items

Miss A. Munro, Mr G. Foggo, Mr C. Jones and Mrs P. Robertson of that Division, who also provided much support

Mrs C. Mackie, Miss L. Porteous, Mrs L. Campbell of the same Division, who willingly typed many drafts

Mrs V. Carstairs, who supplied data for Figures 3.1 and 3.3 and Miss V. Howie, who provided data for Table 3.4

— In England:

The Office of Population Censuses and Surveys, which provided valuable information on population distribution (Dr M.A. Alderson)

— In Lyon:

Miss A.M. Corre, who achieved miracles on the word processor, ably supported by Mrs A. Romanoff and Miss O. Bouvy

Mrs M.-M. Courcier and Mrs J. Thévenoux who prepared the printed text and layout

The Editors are also indebted for many useful suggestions made by participants at a Working Party on Mapping of Cancer; Professor E. Schifflers of the Department of Mathematics, University of Namur, Belgium, was particularly fertile in ideas.

In compiling the text, the Editors used information from many sources, articles, reports and books. These are acknowledged in the list of references. However, some of the sources provided more extensive information than others. In this respect, special acknowledgement is made of information derived from Mr David Hamilton's book *The Healers, a History of Medicine in Scotland*, Canongate Publishing, Edinburgh, which helped considerably in the preparation of the description of early landmarks in medicine in Scotland in Chapter 5. For Chapter 2, The Scottish Habitat, much background information was gained from Mr G.H. Dury's book, *The British Isles*, fifth edition, published by Heinemann, London, 1973. Professor T.C. Smout's book, *A History of the Scottish People, 1560-1830*, published by Collins-Fontana, Glasgow, 1972, was valuable in preparing the section on the origins of the Scottish people in Chapter 3.

REFERENCES

Ackerman, L.V. & del Regato, J.A. (1962) *Cancer: Diagnosis, Treatment and Prognosis*, St Louis, MO, CV Mosby Co

Alderson, M. (1981) *International Mortality Staistics*, New York, Facts on File Inc.

Armstrong, B. & Doll, R. (1975) Environmental factors and cancer incidence and mortality in different countries with special references to dietary practices. *Int. J. Cancer*, **15**, 617-631

Austin, D.F. (1982) *Larynx*. In: Schottenfeld, D. & Fraumeni, J.F., eds, *Cancer Epidemiology and Prevention*, Philadelphia, PA, W.B. Saunders

Beatson, G.T. (1896) On the treatment of inoperable cases of carcinoma of the mamma, suggestions of a new method of treatment with illustrative cases. *Lancet*, **ii**, 104-107

Beral, V., Fraser, P. & Chilvers, C. (1978) Does pregnancy protect against ovarian cancer? *Lancet*, **i**, 1083-1087

Beral, V., Evans, S., Shaw, H. & Milton, G. (1982) Malignant melanoma and exposure to fluorescent lighting at work. *Lancet*, **ii**, 290-293

Berg, J.W. (1975) Can nutrition explain the pattern of international epidemiology of hormone-dependent cancers? *Cancer Res.*, **35**, 3345-3350

Berry, S. (1984) The greening of Glasgow. *The Scotsman*, Edinburgh

Bjelke, E. (1978) Dietary factors and the epidemiology of cancer of the stomach and large bowel. *Aktuel. Ernaehrungsmed. Klin. Prax.*, **2**, 10-17

Blaxter, K. (1977) *The Scottish diet*. In: Underwood R., ed., *The Future of Scotland*, London, Croom Helm, pp. 75-86

Blot, W.J., Fraumeni, J.F. & Stone, B.J. (1978) Geographic correlates of pancreas cancer in the United States. *Cancer*, **42**, 373-380

Boyle, P. (1985) Incidence versus mortality data: A time and a place for both. (submitted for publication)

Boyle, P. & Zaridze, D.G. (1984) Colorectal cancer as a disease of the environment. *Ecol. Dis.*, **2**, 241-248

Boyle, P., Scully, C., Smans, M. & Robertson, A.G. (1983a) Lip cancer mortality in Great Britain. *IRCS Med. Sci.*, **11**, 178-179

Boyle, P., Day, N.E. & Magnus, K. (1983b) Mathematical modelling of malignant melanoma time trends in Norway, 1953-1978. *Am. J. Epidemiol.*, **118**, 887-896

Boyle, P., Zaridze, D.G. & Smans, M. (1984) Descriptive epidemiology of colorectal cancer. *Int. J. Cancer* (in press)

Boyle, P., Smans, M. & Muir, C.S. (1985a) Recent advances in descriptive *epidemiology. The cancer atlas. (submitted for publication)*

Boyle, P., Kaye, S.B. & Gillis, C.R. (1985b) Improving prognosis of testicular cancer in Scotland. *IRCS Med. Sci.* (in press)

Brawn, P.N. (1983) The origin of germ cell tumours of the testis. *Cancer*, **51**, 1610-1614

Breslow, N.E. & Day, N.E. (1975) Indirect standardisation and multiplicative models for rates, with reference to the age-adjustment of cancer incidence and relative frequency data. *J. chron. Dis.*, **28**, 289-303

Broders, A.C. (1920) Squamous cell epithelioma of the lip. *J. Am. med. Assoc.*, **74**, 656-664

Brown, J. (1858) *Rab and His Friends, Horae Subsecivae*, London, Locke and Sydenham (Republished as *Rab and His Friends and Other Papers and Essays*, London, Dent, 1906)

Bull, A.W., Soullier, B.K., Wilson, P.S., Hayden, M.T. & Nigro, N.D. (1979) Promotion of azoxymethane-induced intestinal cancer by high-fat diet in rats. *Cancer Res.*, **39**, 4956-4959

Cairns, J. (1977) *Cancer, Science and Society*, San Francisco, CA, Freeman

Cameron, H.M. & McGoogan, E. (1981) A prospective study of 1152 hospital autopsies. I. Inaccuracies in death certification. *J. Pathol.*, **133**, 273-283

Cantor, K.P., Hoover, R., Mason, T.J. & McCabe, L.J. (1978) Associations of cancer mortality with halomethanes in drinking water. *J. natl Cancer Inst.*, **61**, 979-985

Casagrande, J.T., Louie, E.W., Pike, M.C., Roy, S., Ross, R.K. & Henderson, B.E. (1979) Incessant ovulation and ovarian cancer. *Lancet*, **ii**, 170-173

Case, R.A.M. & Hosker, M.E. (1954) Tumour of the urinary bladder as an occupational disease in the rubber industry in England and Wales. *Br. J. prev. soc. Med.*, **8**, 39-50

Case, R.A.M., Hosker, M.E., McDonald, D.B. & Pearson, J.T. (1954) Tumours of the urinary bladder in workmen engaged in the manufacture and use of certain dyestuff intermediates in the British chemical industry. I. The role of aniline, benzidine, *alpha*-naphthylamine and *beta*-naphthylamine. *Br. J. ind. Med.*, **11**, 75-104

Cederlof, R., Friberg, L., Hrubec Z. & Lorich, U. (1975) *The Relationship of Smoking and Some Social Covariables to Mortality and Cancer Morbidity, A 10 Year Follow-up in a Probability Sample of 55,000 Subjects Age 18-69. Parts 1 and 2*, Stockholm, Department of Environmental Hygiene, Karolinska Institute

Choi, N.W., Howe, G.R., Miller, A.B., Matthews, V., Morgan, R.W., Munan, L., Burch, J.D., Feather, J., Jain, M. & Kelly, A. (1978) An epidemiologic study of breast cancer. *Am. J. Epidemiol.*, **107,** 510-521

Clemmesen, J. (1965) Statistical studies in the aetiology of malignant neoplasms: I. Review and results. *Acta pathol. microbiol. scand.*, Suppl. 174

Clemmesen, J. (1969) Statistical studies in malignant neoplasms. III. Testis cancer. *Acta pathol. microbiol. scand.*, Suppl. 209

Cohart, E.M. (1954) Socioeconomic distribution of stomach cancer in New Haven. *Cancer*, **7**, 455-461

Coldman, A.J., Elwood, J.M. & Gallagher, R. (1982) Sports activities and risk of testicular cancer. *Br. J. Cancer*, **46**, 749-756

Cole, P., Monson, R.R., Haning, H. & Friedell, G.H. (1971) Smoking and cancer of the lower urinary tract. *New Engl. J. Med.*, **284**, 129-134

Colman, M., Kirsch, M. & Creditor, M. (1978) *Tumours associated with medical X-ray therapy exposure in childhood*. In: *Late Biological Effects of Ionizing Radiation*, Vienna, International Atomic Energy Agency, pp. 167-180

Cook-Mozaffari, P.J., Azordegan, F., Day, N.E., Ressicaud, A., Sabai, C. & Aramesh, B. (1979) Oesophageal cancer studies in the Caspian littoral of Iran: Results of a case-control study. *Br. J. Cancer*, **39**, 293-309

Correa, P., Cuello, C. & Eisenberg, M. (1969) *Epidemiology of different types of thyroid cancer*. In: Hedinger, C., ed., *Thyroid Cancer (UICC Monograph Series No. 12)*, New York, Springer Verlag, pp. 81-93

Correa, P., Sasano, N., Stemmermann, G.N. & Haenszel, W. (1973) Pathology of gastric carcinoma in Japanese populations: Comparisons between Miyagi Prefecture, Japan and Hawaii. *J. natl Cancer Inst.*, **51**, 1449-1459

Cosman, B., Heddle, S.B. & Crikelair, G.F. (1976) The increasing incidence of melanoma. *Plast. reconstr. Surg.*, **57**, 50-56

Court-Brown, W.M. & Doll, R. (1965) Mortality from cancer and other causes after radiotherapy for ankylosing spondylitis. *Br. med. J.*, **ii**, 1327-1332

Cramer, D.W. (1982) *Uterine cervix*. In: Schottenfeld, D. & Fraumeni, J.F., eds, *Cancer Epidemiology and Prevention*, Philadelphia, PA, W.B. Saunders, pp. 881-900

Craver, L.F. (1932) Clinical study of aetiology of gastric and esophageal cancer. *Am. J. Cancer*, **16**, 68-102

Crawford, A., Plant, M.A., Kreitman, N. & Latcham, R.W. (1984) Regional variations in British alcohol morbidity rates: A myth uncovered? II: Population surveys. *Br. med. J.*, **289**, 1343-1345

Cullen, W. (1777) *First Line in the Practice of Physic*, Vols I and II, Edinburgh

Cunningham, A.S. (1976) Lymphomas and animal-protein consumption. *Lancet*, **ii**, 1184-1186

Davies, J.M. (1981) Testicular cancer in England and Wales: Some epidemiological aspects. *Lancet*, **i**, 928-932

Day, N.E. & Munoz, N. (1982) *Esophagus*. In: Schottenfeld, D. & Fraumeni, J.F., eds, *Cancer Epidemiology and Prevention*, Philadelphia, PA, W.B. Saunders, pp. 596-623

Department of Agriculture and Fisheries for Scotland (1982) *Economic Report on Scottish Agriculture, 1981*, Edinburgh, HMSO

Department of Employment (1974) *Family Expenditure Survey 1973*, London, HMSO

Department of Employment (1981) *Family Expenditure Survey 1980*, London, HMSO

Doll, R. (1977) Strategy for detection of cancer hazards to man. *Nature*, **265**, 589-596

Doll, R. & Peto, R. (1976) Mortality in relation to smoking: 20 years observations on male British doctors. *Br. med. J.*, **ii**, 1525-1536

Doll, R. & Peto, R. (1981) The causes of cancer: quantitative estimates of avoidable risks of cancer in the United States today. *J. natl Cancer Inst.*, **66**, 1191-1308

Doll, R., Payne, P. & Waterhouse, J.A.H., eds (1966) *Cancer Incidence in Five Continents*, Vol. I, New York, Springer

Doll, R., Muir, C.S. & Waterhouse, J.A.H., eds (1970) *Cancer Incidence in Five Continents*, Vol. II (*A Technical Report, UICC*), Berlin, Springer

Dorn, H.F. (1959) *The mortality of smokers and non-smokers*. In: *Proceedings of Social Statistics Session of American Statistical Association, Chicago*, Washington DC, American Statistical Association, pp. 34-71

Drassar, B.S. & Irving, D. (1973) Environmental factors and cancer of the colon and breast. *Br. J. Cancer*, **27**, 167-172

Durst, M., Gissman, L., Ikenberg, H. & zur Hausen, H. (1983) A papillomavirus DNA from a cervical carcinoma and its prevalence in cancer biopsy samples from different geographic regions. *Proc. natl Acad. Sci. USA*, **80**, 3812-3815

Dury, G.H. (1973) *The British Isles*, 5th ed., London, Heinemann

Ebenius, B. (1943) Cancer of the lip. *Acta Radiol.*, Suppl. 48

Editorial Committee for the Atlas of Cancer Mortality in the People's Republic of China (1979) *Atlas of Cancer Mortality in the People's Republic of China*, Shanghai, China Map Press

Einhorn, L.H. (1981) Testicular cancer as a model for a curable neoplasm. *Cancer Res.*, **41**, 3275-3280

Eklund, G. & Malec, E. (1978) Sunlight and incidence of cutaneous malignant melanoma. Effect of latitude and domicile in Sweden. *Scand. J. plast. reconstr. Surg.*, **12**, 231-241

Elwood, J.M. & Lee, J.A.H. (1974) Trends in mortality from primary tumours of skin in Canada. *Can. med. Assoc. J.*, **110**, 913-915

Elwood, J.M., Cole, P., Rothman, K.J. & Kaplan, S.D. (1977) Epidemiology of endometrial cancer. *J. natl Cancer Inst.*, **59**, 1055-1060

Enstrom, J.E. (1980) *Health and dietary practices and cancer mortality among California Mormons.* In: Cairns, J., Lyons, J.L. & Skolnick, M., eds, *Cancer Incidence in Defined Populations (Banbury Report No. 4)*, Cold Spring Harbor, NY, Cold Spring Harbor Laboratory, pp. 163-186

Forestry Commission (1983) *62nd Annual Report 1981/82*, London, HMSO

Frentzel-Beyme, R., Leutner, R., Wagner, G. & Wiebelth, H. (1979) *Krebsatlas der Bundesrepublik Deutschland*, Berlin, Springer-Verlag

Gardner, M.J., Winter, P.D., Taylor, C.P. & Acheson, E.D. (1984) *Atlas of Cancer Mortality in England and Wales, 1968-1978*, Chichester, John Wiley and Sons

Gaskill, S.P., McGuire, W.L., Osborne, C.K. & Stern, M.P. (1979) Breast cancer mortality and diet in the United States. *Cancer Res.*, **39**, 3628-3637

Gatti, R.A. & Good, R.A. (1971) Occurrence of malignancy in immunodeficiency diseases. *Cancer*, **28**, 89-98

Gellin, G.A., Kopf, A.W. & Garfinkel, L. (1969) Malignant melanoma: A controlled study of possibly associated factors. *Arch. Dermatol.*, **9**, 43-48

General Register Office (1983a) *Census 1981. Scotland Summary* Vol. I., Edinburgh, HMSO

General Register Office (1983b) *Census 1981: Great Britain - Economic Activity, Part IV*, London, Office of Population Censuses and Surveys, HMSO

George, W.K., Miller, M. & George, W.D. (1965) Hodgkin's disease in room-mates. *J. Am. Coll. Health Assoc.*, **13**, 399-402

Goldstein, M.R. (1982) No association found between coffee and cancer of the pancreas. *New Engl. J. Med.*, **306**, 997

Graham, S., Schotz, W. & Martino, P. (1972) Alimentary factors in the epidemiology of gastric cancer. *Cancer*, **30**, 927-938

Graham, S., Dayal, H., Rohrer, T., Swanson, M., Sultz, H., Shedd, D. & Fischman, S. (1977) Dentition, diet, tobacco and alcohol in the epidemiology of oral cancer. *J. natl Cancer Inst.*, **59**, 1611-1618

Graham, S., Marshall, J., Mettlin, C., Rzepka, T., Nemoto, T. & Byers, T. (1982) Diet in the epidemiology of breast cancer. *Am. J. Epidemiol.*, **116**, 68-75

Greene, A., MacLennan, R. & Siskind, V. (1985) Common acquired naevi and the risk of malignant melanoma. *Int. J. Cancer*, **35**, 297-300

Greene, M.H. (1982) *Non-Hodgkin's lymphoma and mycosis fungoides.* In: Schottenfeld, D. & Fraumeni, J.F., eds, *Cancer Epidemiology and Prevention*, Philadelphia, PA, W.B. Saunders, pp. 754-778

Grufferman, S. (1977) Clustering and aggregation of exposures in Hodgkin's disease. *Cancer*, **39**, 1829-1833

Grufferman, S. (1982) *Hodgkin's disease.* In: Schottenfeld, D. & Fraumeni, J.F., eds, *Cancer Epidemiology and Prevention*, Philadelphia, PA, W.B. Saunders, pp. 739-753

Gutensohn, N. & Cole, P. (1980) *Epidemiology of Hodgkin's Disease. Seminars in Oncology*, New York, Grune and Stratton

Haddow, A.J. (1968) *Cancer in the Scottish Highlands.* In: *8th Annual Report of Regional Hospital Board*, Glasgow, pp. 31-44

Haenszel, W. (1982) *Migrant studies.* In: Schottenfeld, D. & Fraumeni, J.F., eds, *Cancer Epidemiology and Prevention*, Philadelphia, PA, W.B. Saunders, pp. 194-207

Haenszel, W., Kurihara, M., Locke, F.B., Shimuzu, K. & Segi, M. (1976) Stomach cancer in Japan. *J. natl Cancer Inst.*, **56**, 265-274

Hamalainen, M.J. (1955) Cancer of the lip. *Ann. Chir. Gynaecol. Fenn.*, **44** (Suppl. 6)

Hamilton, D.N.H. (1981) *The Healers, a History of Medicine in Scotland*, Edinburgh, Canongate Publishing

Hammond, E.C. & Horn, D. (1958) Smoking and death rates — Report on 44 months of follow-up of 187 783 men. II. Death rates by cause. *J. Am. med. Assoc.*, **166**, 1294-1308

Harsanyi, Z.P., Post, P.W., Brinkmann, J.P., Chedekel, M.R. & Deibel, R.M. (1980) Mutagenicity of melanin from human red hair. *Experientia*, **36**, 291-292

Haviland, A. (1855) *The Geographical Distribution of Diseases in Great Britain*, London, Smith Elder

Heasman, M.A. & Lipworth, L. (1968) *Accuracy of Certification of Cause of Death*, London, HMSO

Heasman, M.A., Kemp, I.W., McLaren, A.M., Trotter, P., Gillis, C.R. & Hole, D.J. (1984) Incidence of leukaemia in young persons in West of Scotland. *Lancet*, **i**, 1188-1189

Heath, C.W., Jr (1982) *The leukaemias.* In: Schottenfeld, D. & Fraumeni, J.F., eds, *Cancer Epidemiology and Prevention*, Philadelphia, W.B. Saunders, pp. 728-738

Heister, L. (1747) *Chirurgie*, Nurenberg, p. 325

Henderson, B.E., Benton, B., Jing, J., Yu, M.C. & Pike, M.C. (1979) Risk factors for cancer of the testis in young men. *Int. J. Cancer*, **23**, 598-602

Herity, B.A., O'Halloran, M.J., Bourke, G.J. & Wilson-Davis, K. (1975) A study of breast cancer in Irish women. *Br. J. prev. soc. Med.*, **29**, 178-181

Higginson, J. & Muir, C.S. (1979) Environmental carcinogenesis. Misconceptions and limitations to cancer control. *J. natl Cancer Inst.*, **63**, 1291-1298

Hirabayashi, R. & Lindsay, S. (1965) The relation of thyroid carcinoma and chronic thyroiditis. *Surg. Gynecol. Obstet.*, **121**, 243-252

Hirayama, T. (1971) Epidemiology of stomach cancer. *Gann*, **11**, 3-19

Hirayama, T. (1977) *Changing patterns of cancer in Japan with special reference to the decrease in stomach cancer mortality.* In: Hiatt, H.H., Watson, J.D. & Winsten, J.A., eds, *Origins of Human Cancer*, Book A, Cold Spring Harbor, NY, Cold Spring Harbor Laboratory, pp. 55-75

Holman, C.D. & Armstrong, B.K. (1984) Cutaneous malignant melanoma and indicators of total accumulated exposure to sun: An analysis separating histogenetic type. *J. natl Cancer Inst.*, **73**, 75-82

Holman, C.D., James, I.R., Gattey, P.H. & Armstrong, B.K. (1980) An analysis of trends of mortality from malignant melanoma of the skin in Australia. *Int. J. Cancer*, **26**, 703-709

Hoover, R. & Cole, P. (1973) Temporal aspects of occupational bladder carcinogenesis. *New Engl. J. Med.*, **288**, 1040-1043

Houghton, A., Flannery, J. & Viola, M.V. (1980) Malignant melanoma in Connecticut and Denmark. *Int. J. Cancer*, **25**, 95-104

Howe, G.M. (1970) *National Atlas of Disease. Mortality in the United Kingdom*, London, Nelson

Howe, G.R., Burch, J.D., Miller, A.B., Cook, G.M., Estève, J., Morrison, B., Gordon, P., Chambers, L.W., Fodor, G. & Winsor, G.M. (1980) Tobacco use, occupation, coffee, various nutrients and bladder cancer. *J. natl Cancer Inst.*, **64**, 701-713

IARC (1985) *IARC Monographs on the Evaluation of the Carcinogenic Risk of Chemicals to Humans*, Vol. 38, *Tobacco Smoking*, Lyon (in press)

Independent Advisory Group (1984) *Report: Investigation of Possible Increased Incidence of Cancer in West Cumbria*, London, HMSO

Information Services Division (1981) *Cancer Registration and Survival Statistics, Scotland, 1963-1977*, Edinburgh, Scottish Health Service

Information Services Division (1982) *Hospital Bed Resources, 1980*, Edinburgh, Scottish Health Service

James, W. & Stein, L. (1961) *Estimation with quadratic loss*. In: *Proceeding of the Fourth Berkeley Symposium*, Vol. 1, Berkeley, University of California, pp. 361-379

Jensen, O.M. (1980) *Cancer Morbidity and Causes of Death among Danish Brewery Workers*, Lyon, International Agency for Research on Cancer

Jensen, O.M. & Bolander, A.M. (1980) Trends in malignant melanoma of the skin. *World Health Stat. Q.*, **33**, 2-26

Jick, H., Watkins, R.N., Hunter, J.R., Dinan, B.J., Madsen, S., Rothman, K.J. & Walker, A.M. (1979) Replacement estrogens and endometrial cancer. *New Engl. J. Med.*, **300**, 218-222

Joint Iran-IARC Study Group (1977) Esophageal cancer studies in the Caspian littoral of Iran: Results of population studies — a prodrome. *J. natl Cancer Inst.*, **59**, 1127-1138

Kademian, M.T. & Caldwell, W.L. (1976) Testicular seminomas: A case report of two brothers with seminoma and a review of the literature of testicular malignancies occurring in closely related family members. *J. Urol.*, **116**, 380-381

Kaplan, S.D. & Acheson, R.M. (1966) A Single etiological hypothesis for breast cancer? *J. chron. Dis.*, **19**, 1221-1230

Kassan, S.S., Thomas, T.L., Moutsopoulos, H.M., Hoover, R., Kimberly, R.P., Budman, D.R., Costa, J., Decker, J.L. & Chused, T.M. (1978) Increased risk of lymphoma in Sicca syndrome. *Ann. intern. Med.*, **89**, 888-892

Keller, A.Z. (1970) Cellular types, survival, race, nativity, occupations, habits and associated diseases in the pathogenesis of lip cancers. *Am. J. Epidemiol.*, **91**, 486-499

Kelsey, J.L. (1979) A review of the epidemiology of human breast cancer. *Epidemiol. Rev.*, **1**, 74-109

Kemp, I.W. & Bledin, K.D. (1981) *Occupational Mortality, 1969-1973*. Edinburgh, Registrar General Scotland, HMSO

Kemp, I.W., Stein, G.J. & Heasman, M.A. (1980) Myeloid leukaemia in Scotland. *Lancet*, **ii**, 732-734

Kinlen, L.J., Sheil, A.G., Peto, J. & Doll, R. (1979) A collaborative study of cancer in patients who have received immunosuppressive therapy. *Br. med. J.*, **ii**, 1461-1466

Klepp, O. & Magnus, K. (1979) Some environmental and bodily characteristics of melanoma patients: A case-control study. *Int. J. Cancer*, **23**, 482-488

Kolonel, L.N., Hankin, J.H., Lee, J., Chu, S.Y., Nomura, A.M. & Hinds, M.W. (1981) Nutrient intakes in relation to cancer incidence in Hawaii. *Br. J. Cancer*, **44**, 332-339

Lancaster, H.O. (1956) Some geographical aspects of the mortality from melanoma in Europeans. *Med. J. Aust.*, **1**, 1082-1087

Last, J. (1983) *A Dictionary of Epidemiology*. Oxford, Oxford University Press

Latcham, R.W., Kreitman, N., Plant, M.A. & Crawford, A. (1984) Regional variations in British alcohol morbidity rates: A myth uncovered? I. Clinical surveys. *Br. med. J.*, **289**, 1341-1343

Leach, J.F., Beadle, P.C., Pingstone, A.R. & Aughton, P. (1979) Some effects of the variation of atmospheric particulates on disease in Great Britain. *Aviat. Space environ. Med.*, **50**, 72-79

Lee, J.A.H. & Carter, A.P. (1970) Secular trends in mortality from malignant melanoma. *J. natl Cancer Inst.*, **45**, 91-97

Levin, D.L., Devesa, S.S., Goodwin, J.D., II & Silverman, D.T. (1974) *Cancer Rates and Risks (DHEW Publication No. (NIH) 75-691)*, Washington DC, US Government Printing Office

Lilienfeld, A.M., Coombs, J., Bross, I.D. & Chamberlain, A. (1975) Marital and reproductive experience in a community-wide epidemiological study of breast cancer. *Johns Hopkins med. J.*, **136**, 157-162

Lindqvist, C. (1979) Risk factors in lip cancer. A questionnaire survey. *Am. J. Epidemiol.*, **109**, 521-530

Lockwood, K. (1961) On the etiology of bladder tumors in Kobenhavn-Frederiksberg: An inquiry of 369 patients and 369 controls. *Acta pathol. microbiol. scand.*, **51**, 1-166

Logan, W.P.D. (1982) *Cancer Mortality by Occupation and Social Class, 1851-1971 (IARC Scientific Publications No. 36)*, Lyon, International Agency for Research on Cancer

Lombard, H.L. & Potter, E.A. (1950) Epidemiologic aspects of cancer of the cervix. *Cancer*, **3**, 960-968

Lowe, C.R. & MacMahon, B. (1970) Breast cancer and reproductive history of women in South Wales. *Lancet*, **i**, 153-157

Lubin, J.H., Blot, W.J. & Burns, P.G. (1981) Breast cancer following high dietary fat and protein consumption. *Am. J. Epidemiol.*, *114*, 422

Lynch, G.A. (1967) Cancer of the lip. *Ulster med. J.*, *36*, 44-50

Mackay, G.A. (1983) *The Scottish fishing industry in 1982*. In: *Scottish Government Yearbook 1983*, Edinburgh, Research Centre for Social Sciences, University of Edinburgh

Macklin, M.T. (1959) Comparison of the number of breast cancer deaths observed in relatives of breast cancer patients, and the number expected on the basis of mortality rates. *J. natl Cancer Inst.*, **22**, 927-951

MacMahon, B. & Pugh, R.C.B. (1968) *Principles of Epidemiology*, Boston, MA, Little Brown

MacMahon, B., Cole, P. & Brown, J. (1973) Etiology of human breast cancer: A review. *J. natl Cancer Inst.*, **50**, 21-42

MacMahon, B., Yen, S., Trichopoulos, D., Warren, K. & Nardi, G. (1981) Coffee and cancer of the pancreas. *New Engl. J. Med.*, **304**, 630-633

Magnus, K. (1973) Incidence of malignant melanoma of the skin in Norway, 1955-1970. Variations in time and space and solar radiation. *Cancer*, **32**, 1275-1286

Magnus, K. (1977) Incidence of malignant melanoma of the skin in the five Nordic countries: Significance of solar radiation. *Int. J. Cancer*, **20**, 477-485

Magnus, K. (1981) Habits of sun exposure and risk of malignant melanoma: An analysis of incidence rates in Norway 1955-1977 by cohort, sex, age and primary tumor site. *Cancer*, **48**, 2329-2335

Mason, T.J., McKay, F.W., Hoover, R. & Fraumeni, J.F. (1975) *Atlas of Cancer Mortality for US Counties: 1950-1969 (DHEW Publication No. 75-780)*, Washington DC, US Government Printing Office

Miller, A.B., Kelly, A., Choi, N.W., Matthews, V., Morgan, R.W., Munan, L., Burch, J.D., Feather, J., Howe, G.R. & Jain, M. (1978) A study of diet and breast cancer. *Am. J. Epidemiol.*, **107**, 499-509

Mosbech, J. & Videback, A. (1955) On the aetiology of esophageal carcinoma. *J. natl Cancer Inst.*, **15**, 1665-1673

National Academy of Sciences (1982) *Diet, Nutrition and Cancer*, Washington DC

National Food Survey Committee (1976-1980) *Annual Reports: Household Food Consumption and Expenditure*, London, HMSO

Newhall, S.M., Nick, D. & Kudd, D.B. (1943) Final report of the OSA sub-committee on the spacing of the Munsell colors. *J. opt. Soc. Am.*, **33**, 385-396

Nickerson, D. (1940) History of the Munsell color system and its scientific application. *J. opt. Soc. Am.*, **30**, 575-586

Nigro, N.D., Singh, D.V., Campbell, R.L. & Pak, M.S. (1975) Effect of dietary beef fat on intestinal tumor formation by azoxymethane in rats. *J. natl Cancer Inst.*, **54**, 439-442

Nomura, A. (1982) *Stomach*. In: Schottenfeld, D. & Fraumeni, J.F., eds, *Cancer Epidemiology and Prevention*, Philadelphia, W.B. Saunders, pp. 624-637

Okada, S., Hamilton, H.B., Egami, N. *et al* (1975) A review of 30 years study of Hiroshima and Nagasaki atomic bomb survivors. *Jpn Radiat. Res.*, **16**, 1-164

Pack, G.T., Davis, J. & Oppenheim, A. (1963) The relation of race and complexion to the incidence of moles and melanomas. *Ann. N.Y. Acad. Sci.*, **100**, 719-742

Pederson, E., Hogetveit, A.C. & Anderson, A. (1973) Cancer of respiratory organs at a nickel refinery in Norway. *Int. J. Cancer*, **12**, 32-41

Percy, C., Stanek, E. & Gloeckler, L. (1981) Accuracy of cancer death certificates and its effect on cancer mortality statistics. *Am. J. publ. Health*, **71**, 242-250

Perry, R. (1844) *Facts and Observations on the Sanitary State of Glasgow During the Last Year; with Statistical Tables of the Late Epidemic Shewing the Connection Existing between Poverty, Disease and Crime*, Glasgow

Phillips, R.L. (1975) Role of life-style and dietary habits in risk of cancer among Seventh-Day Adventists. *Cancer Res.*, **35**, 3513-3522

Phillips, R.L., Kuzma, J.W. & Lotz, T.M. (1980) *Cancer mortality among comparable members versus non-members of the Seventh-Day Adventist Church*. In: Cairns, J., Lyon, J.L. & Skolnick, M., eds, *Cancer Incidence in Defined Populations (Banbury Report No. 4)*, Cold Spring Harbor, NY, Cold Spring Harbor Laboratory, pp. 3-30

Pugh, R.C.B. (1976) *Testicular tumours: introduction*. In: Pugh, R.C.B., ed., *Pathology of the Testis*, Oxford, Blackwell Scientific Publications

Raghavan, D., Jelihovsky, T. & Fox, R.M. (1980) Father-son testicular malignancy: Does genetic anticipation occur? *Cancer*, **45**, 1005-1009

Ramazzini, B. (1743) *De Morbis Artificum, Diatriba*, Venice, J. Corona

Ravnihar, B., MacMahon, B. & Lindtner, J. (1971) Epidemiologic features of breast cancer in Slovenia, 1965-1967. *Eur. J. Cancer*, **7**, 295-306

Registrar General for England and Wales (1978) *Occupational Mortality. Registrar General's Decennial Supplement for England and Wales, 1970-1972 (Series DS No 1)*, London, Office of Population Censuses and Surveys

Registrar General Scotland (1982) *Census 1981. Reports for Regions, Vol. I (Tables A and B)*, Edinburgh, HMSO

Rehn, L. (1895) Blasengeschwuelste bei Fuchsin-Arbeitern. *Arch. Klin. Chir.*, **50**, 588-600

Ron, E. & Modan, B. (1980) Benign and malignant thyroid neoplasms after childhood irradiation for tinea capitis. *J. natl Cancer Inst.*, **65**, 7-11

Rothman, K. & Keller, A. (1972) The effect of joint exposure to alcohol and tobacco on risk of cancer of the mouth and pharynx. *J. chron. Dis.*, **25**, 711-716

Rotkin, I.D. (1966) Further studies in cervical cancer inheritance. *Cancer*, **19**, 1251-1268

Rotkin, I.D. (1967) Epidemiology of cancer of the cervix. III. Sexual characteristics of a cervical cancer population. *Am. J. publ. Health*, **57**, 815-829

Sanghvi, L.D., Rao, K.C.M. & Khanolkar, V.R. (1955) Smoking and chewing of tobacco in relation to cancer of the upper alimentary tract. *Br. med. J.*, **i**, 1111-1114

Schottenfeld, D. & Warshauer, M.E. (1982) *Testis*. In: Schottenfeld, D. & Fraumeni, J.F., eds, *Cancer Epidemiology and Prevention*, Philadelphia, PA, W.B. Saunders, pp. 947-957

Schwartz, D., Lellouch, J., Flamant, R. & Denoix, P.F. (1962) Alcool et cancer. Resultats d'une enquête rétrospective. *Rev. fr. Etud. Clin. Biol.*, **7**, 590-604

Scottish Health Service Planning Council (1979) *Cancer Services in Scotland*, Edinburgh, Scottish Home and Health Department

Scottish Information Office (1974) *Scotland Today*, Edinburgh, Central Office of Information, HMSO

Scotto, J., Fraumeni, J.F. & Lee, J.A.H. (1976) Melanomas of the eye and other non-cutaneous sites: Epidemiologic aspects. *J. natl Cancer Inst.*, **56**, 489-491

Segi, M. (1977) *Atlas of Cancer Mortality in Japan by Cities and Counties, 1969-1971*, Tokyo, Daiwa Health Foundation

Shapiro, S. (1977) Evidence on screening for breast cancer from a randomised trial. *Cancer*, **39**, 2772-2782

Shedd, D.P., Von Essen, C.F., Connelly, R.R. & Eisenberg, H. (1970) Cancer of the lip in Connecticut, 1935-1959. *Cancer Bull.*, **22**, 116-120

Shillitoe, E.J. & Silverman, S. (1979) *Oral cancer and herpes simplex virus — A review*. In: Little, J.W., ed., *Oral Surgery*, St Louis, MO, C.V. Mosby Co.

Sigurjonsson, J. (1967) Occupational variations in mortality from gastric cancer in relation to dietary differences. *Br. J. Cancer*, **21**, 651-656

Silverberg, S.G., Mabowski, E.L. & Rocks, W.D. (1977) Endometrial cancer in women under 40 years of age. *Cancer*, **39**, 592-598

Smith, P.G. (1982) *Spatial and temporal clustering*. In: Schottenfeld, D. & Fraumeni, J.F., eds, *Cancer Epidemiology and Prevention*, Philadelphia, PA, W.B. Saunders, pp. 391-407

Smout, T.C. (1972) *A History of the Scottish People, 1560-1830*, Glasgow, Collins-Fontana

Snow, J. (1936) *On the mode of communication of cholera*. In: *Snow on Cholera*, New York, Commonwealth Fund, pp. 1-175

Society for Investigating the Nature and Cure of Cancer (1806) *Edinburgh Med. Surg. J.*, **2**, 382-389

Spitzer, W.O., Hill, G.B., Chambers, L.W., Helliwell, B.E. & Murphy, H.B. (1975) The occupation of fishing as a risk factor in cancer of the lip. *New Engl. J. Med.*, **293**, 419-424

Staszewski, J. (1971) Age at menarche and breast cancer. *J. natl Cancer Inst.*, **47**, 935-940

Steiner, P. (1956) Etiology and histogenesis of carcinoma of esophagus. *Cancer*, **9**, 436-452

Stell, P.M. & McGill, T. (1973) Asbestos and laryngeal cancer. *Lancet*, **ii**, 416-417

Stocks, P. (1928) *On the Evidence for a Regional Distribution of Cancer in England and Wales, Report of the International Conference on Cancer*, London, British Empire Cancer Campaign, pp. 508-519.

Stocks, P. (1936) *Distribution in England and Wales of Cancer of Various Organs, 13th Annual Report*, London, British Empire Cancer Campaign, pp. 239-280

Stocks, P. (1937) *Distribution in England and Wales of Cancer of Various Organs, 14th Annual Report*, London, British Empire Cancer Campaign, pp. 198-223

Stocks, P. (1939) *Distribution in England and Wales of Cancer in Various Organs, 16th Annual Report*, London, British Empire Cancer Campaign, pp. 308-343

Stukonis, M.K. (1982) *Cancer Incidence Cumulative Rates. International Comparison Based on Data from Cancer Incidence in Five Continents* (*IARC Technical Report 78/002*), Lyon, International Agency for Research on Cancer

Svejda, J. & Kosut, V. (1971) Epidemiology of malignant tumours with special regard to the orofacial region. *Neoplasma*, **18**, 193-196

Szpak, C.A., Stone, M.J. & Frenkel, E.P. (1977) Some observations concerning the demographic and geographic incidence of carcinoma of the lip and buccal cavity. *Cancer*, **40**, 343-348

Tabar, L., Fagerberg, C.J., Gad, A., Baldetrop, L., Holmberg, L.H., Grontoft, O., Ljungquist, U., Lundstrom, B., Manson, J.C., Ecklund, G. *et al.* (1985) Reduction in mortality from breast cancer after mass screening with mammography. Randomised trial from the Breast Cancer Screening Working Group of the Swedish National Board of Health and Welfare. *Lancet*, **i**, 829-832

Teppo, L., Pakkanen, M. & Hakulinen, T. (1978) Sunlight as a risk factor of malignant melanoma of the skin. *Cancer*, **41**, 2018-2027

Tivy, J., Mitchell, J. & Jansey, P. (1973) In: Tivy, J., ed., *The Organic Resources of Scotland*, Edinburgh, Oliver & Boyd

Torgersen, O. & Petersen, M. (1956) The epidemiology of gastric cancer in Oslo: Cartographic analysis of census tracts and mortality rates of sub-standard housing areas. *Br. J. Cancer*, **10**, 299-306

Trichopoulos, D., MacMahon, B. & Cole, P. (1972) Menopause and breast cancer risk. *J. natl Cancer Inst.*, **48**, 605-613

Truax, H., Barnett, R.N., Hukill, P.B., Campbell, P.C. & Eisenberg, H. (1966) Effect of inaccurate pathological diagnosis in survival statistics for melanoma: Survey of cases in the Connecticut tumor registry. *Cancer*, **19**, 1543-1547

Tulinius, H., Day, N.E., Johannesson, G., Bjarnason, O. & Gonzales, M. (1978) Reproductive factors and risk for breast cancer in Iceland. *Int. J. Cancer*, **21**, 724-730

Tuyns, A.J. (1970) Cancer of the esophagus: Further evidence of the relation to drinking habits in France. *Int. J. Cancer*, **5**, 152-156

Urbach, F. (1978) Evidence and epidemiology of ultraviolet-induced cancers in man. *Natl Cancer Inst. Monogr.*, **50**, 5-10

US Surgeon General (1982) *The Health Consequences of Smoking*, Washington DC, US Department of Health and Environmental Welfare

Vessey, M.P. (1978) Epidemiological studies of the occupational hazards of anaesthesia: A review. *Anaesthesia*, **33**, 430-438

Vianna, N.J. & Polan, A.K. (1973) Epidemiologic evidence for transmission of Hodgkin's disease. *New Engl. J. Med.*, **10**, 499-502

Villani, U. (1967) Tumore concordante del testicolo in una coppia di gemelli monozigoti. *Acta genet. med. gemellol.*, **16**, 172

de Waard, F. (1973) *Nurture and nature in cancer of the breast and endometrium.* In: Doll, R. & Vodopija, J., eds, *Host Environment Interactions in the Etiology of Cancer in Man* (*IARC Scientific Publications No. 7*), Lyon, International Agency for Research on Cancer, pp. 121-130

Wainwright, J.M. (1931) A comparison of conditions associated with breast cancers in Great-Britain and America. *Am. J. Cancer*, **15**, 2610-2645

Waterhouse, J.A.H., Muir, C.S., Correa, P. & Powell, J., eds (1976) *Cancer Incidence in Five Continents, Vol. III (IARC Scientific Publications No. 15)*, Lyon, International Agency for Research on Cancer

Waterhouse, J.A.H., Muir, C.S., Shanmugaratnam, K. & Powell, J., eds (1982) *Cancer Incidence in Five Continents, Vol. IV (IARC Scientific Publications No. 42)*, Lyon, International Agency for Research on Cancer

WHO (1967) *Manual of the International Statistical Classification of Diseases, Injuries and Causes of Death*, 8th Rev., Vol. 1, Geneva

Wigle, D.T. (1978) Malignant melanoma of skin and sunspot activity. *Lancet*, **ii**, 38

Williams, R.R. & Horm, J.W. (1977) Association of cancer sites with tobacco and alcohol consumption and socio-economic status of patients: Interview studies from the Third National Cancer Survey. *J. natl Cancer Inst.*, **58**, 525-547

Williams, E.D., Doniach, I., Bjarnason, O. & Michie, W. (1977) Thyroid cancer in an iodine-rich area: A histopathological study. *Cancer*, **39**, 215-222

Winship, T. & Rosvoli, R.V. (1970) Thyroid carcinoma in childhood: Final report on a 20 year study. *Clin. Proc. Child. Hosp.*, **26**, 327-348

Wynder, E.L. & Goldsmith, R. (1977) The epidemiology of bladder cancer. A second look. *Cancer*, **40**, 1246-1268

Wynder, E.L. & Gori, G.B. (1977) Contribution of the environment to cancer incidence: An epidemiological exercise. *J. natl Cancer Inst.*, **58**, 825-832

Wynder, E.L. & Stellman, S.D. (1977) Comparative epidemiology of tobacco-related cancers. *Cancer Res.*, **37**, 4608-4622

Wynder, E.L., Cornfield, J., Schroff, P.D. & Doraiswami, K.R. (1954) A study of environmental factors in carcinoma of the cervix. *Am. J. Obstet. Gynecol.*, **68**, 1016-1047

Wynder, E.L., Hultberg, S., Jacobsson, F. & Bross, T.J. (1957) Environmental factors in cancer of the upper alimentary tract. *Cancer*, **10**, 470-487

Wynder, E.L., Onderdonk, J. & Mantel, N. (1963) An epidemiological investigation of cancer of the bladder. *Cancer*, **16**, 1388-1407

Wynder, E.L., Escher, G.C. & Mantel, N. (1966) An epidemiological investigation of cancer of the endometrium. *Cancer*, **19**, 489-520

Young, M. & Russell, W.T. (1926) *An Investigation into the Statistics of Cancer in Different Trades and Professions (Medical Research Council, Special Report 99)*, London, HMSO

Zaridze, D.G. (1983) Environmental etiology of large bowel cancer. *J. natl Cancer Inst.*, **70**, 389-399

Zaridze, D.G., Boyle, P. & Smans, M. (1984) The epidemiology of cancer of the prostate. *Int. J. Cancer*, **33**, 213-230

Ziegler, J.L. and 27 co-authors (1984) Non-Hodgkin's lymphoma in 90 homosexual men. Relation to generalized lymphadenopathy and acquired immunodeficiency syndrome. *New Engl. J. Med.*, **311**, 565-570

APPENDIX I

INCIDENCE RATES

These tables present cumulative rates and standardized rates and their corresponding rank order for the principal sites and other sites. The headings are short site names which correspond to the following:

Principle sites		**Other sites**	
LIPC	Lip	TONG	Tongue
ORPH	Oral Cavity and Pharynx	SALG	Salivary Gland
OESO	Oesophagus	OMOU	Gum, Floor of Mouth, etc.
STOM	Stomach	PHA1	Oropharynx, Hypopharynx, etc.
LBWL	Large Bowel	PHA2	Nasopharynx
PANC	Pancreas	SINT	Small Intestine
LARY	Larynx	COLO	Colon
LUNG	Trachea, Bronchus and Lung	RECT	Rectum
MELA	Malignant Melanoma of Skin	LIVE	Liver
BREA	Breast	GALL	Gallbladder, etc.
CERV	Uterine cervix	NASA	Nasal Cavities, etc.
CORP	Uterine Corpus	BONE	Bone
OVAR	Ovary, Broad Ligament and	CONN	Connective and Other Soft Tissue
	Fallopian Tube	OUFG	Other and Unspecified Female
PROS	Prostate		Genital Organs
TEST	Testis	KIDN	Kidney
BLAD	Bladder	EYEZ	Eye
THYR	Thyroid	BRNS	Brain and Nervous System
HODG	Hodgkin's Disease	BRNS	Brain and Nervous System
NHOL	Non-Hodgkin's Lymphoma	LYLE	Lymphatic Leukaemia
LEUK	Leukaemia	MYLE	Myeloid Leukaemia

Table A1.1 Age-standardized incidence rates by principal cancer site, in each local government district and Islands area, Scotland, 1975-1980, by sex (male)

	LIPC	ORPH	OESO	STOM	LBWL	PANC	LARY	LUNG	MELA	PROS	TEST	BLAD	THYR	HODG	NHOL	LEUK
1 CAITHNESS	9.5	0.8	7.0	11.7	31.4	4.3	0.8	42.8	2.0	12.9	2.8	12.7	2.3	2.0	6.1	6.5
2 SUTHERLAND	9.6	5.4	6.8	13.9	38.7	6.6	2.1	64.9	7.7	22.0	3.4	3.2	0.0	2.1	3.7	6.2
3 ROSS-CROMARTY	9.2	3.5	6.9	18.3	47.3	6.6	4.8	52.3	1.7	22.2	3.3	12.6	1.5	1.5	3.6	8.5
4 SKYE-LOCHALSH	17.0	8.0	7.9	14.8	50.1	3.8	5.6	40.7	7.4	16.7	3.9	20.3	0.0	1.4	1.4	2.5
5 LOCHABER	8.5	1.2	9.6	17.1	36.5	6.0	2.4	56.6	1.5	26.5	6.5	15.3	1.7	1.7	3.7	5.6
6 INVERNESS	4.8	2.7	10.4	15.8	44.6	6.2	3.0	63.9	3.8	23.0	2.7	13.0	1.1	0.6	6.5	8.2
7 BADENOCH	5.1	11.2	7.6	5.2	53.9	9.4	3.1	55.0	3.3	38.4	7.2	12.8	0.0	0.0	6.5	0.0
8 NAIRN	7.3	0.0	5.7	19.2	51.9	5.6	0.0	68.2	3.6	20.4	7.0	10.1	3.3	0.0	12.5	11.3
9 MORAY	8.4	5.6	8.5	16.9	36.4	5.5	5.5	67.0	3.6	29.6	3.2	16.8	1.0	1.4	5.7	6.8
10 BANFF-BUCHAN	11.3	2.9	8.1	16.1	39.2	5.0	3.4	69.3	1.1	26.6	5.3	19.6	0.9	5.0	6.6	7.4
11 GORDON	7.5	4.3	8.2	11.5	39.2	5.0	2.1	52.8	5.0	24.4	4.8	16.2	0.2	4.4	6.2	8.5
12 ABERDEEN	3.5	2.2	8.7	20.7	44.1	3.1	5.7	98.5	2.3	26.4	4.8	23.2	1.3	2.9	9.6	7.3
13 KINCARDINE	5.5	2.4	4.7	15.1	40.9	9.6	3.3	53.4	0.8	26.4	6.9	22.9	0.8	2.7	17.2	6.2
14 ANGUS	3.9	5.8	10.3	23.6	40.9	10.9	5.0	73.8	2.1	23.7	4.1	12.6	0.0	1.5	5.8	12.6
15 DUNDEE	0.9	4.2	8.5	29.1	36.0	9.5	5.0	100.0	3.7	23.0	2.6	19.9	0.8	2.8	5.5	6.9
16 PERTH-KINROSS	2.8	4.0	5.6	19.4	38.4	8.2	3.0	69.7	3.8	24.0	3.4	15.0	0.6	3.1	6.9	7.5
17 KIRKCALDY	3.3	3.0	5.2	20.5	31.8	10.1	2.7	94.2	3.4	23.7	4.8	22.1	0.8	5.1	6.9	6.7
18 N-E FIFE	5.1	3.3	6.4	12.2	32.0	8.3	4.1	78.7	5.4	26.8	4.4	25.3	1.4	1.4	6.9	5.9
19 DUNFERMLINE	3.5	1.7	5.1	21.5	25.0	10.4	2.5	86.8	3.0	23.3	5.2	17.3	0.7	3.3	7.4	8.3
20 WEST LOTHIAN	2.8	4.8	6.8	21.5	30.5	9.2	3.8	97.0	1.5	27.7	3.8	22.0	0.4	4.2	5.9	7.5
21 EDINBURGH	1.0	5.3	6.6	20.8	33.4	9.7	4.6	103.2	3.2	25.4	4.5	19.2	0.5	2.8	10.2	7.3
22 MIDLOTHIAN	2.8	3.7	5.2	21.8	35.3	9.6	3.9	83.3	3.2	25.4	3.7	15.0	0.3	3.1	8.5	6.4
23 EAST LOTHIAN	2.8	3.4	5.7	21.8	31.5	11.4	3.1	84.0	2.4	35.6	7.4	20.7	0.4	4.9	4.1	5.6
24 TWEEDDALE	0.0	7.4	6.2	22.7	27.6	10.9	7.1	73.6	0.0	21.2	8.0	21.0	0.0	1.9	11.9	13.6
25 ETTRICK	5.5	4.2	5.9	27.6	28.4	10.4	4.1	67.9	0.0	36.1	4.1	7.2	0.0	6.9	4.4	7.8
26 ROXBURGH	3.5	5.0	3.8	22.4	27.9	7.8	3.9	67.9	2.9	20.9	4.6	10.2	1.0	1.2	4.7	7.5
27 BERWICKSHIRE	8.7	2.4	4.2	11.6	28.1	8.4	8.6	54.7	2.0	23.3	4.8	22.7	0.0	1.8	3.9	11.8
28 CLACKMANNAN	1.3	5.5	4.1	21.2	30.9	13.0	3.4	83.7	0.8	20.2	14.4	16.1	0.0	13.0	7.0	8.9
29 STIRLING	2.4	2.9	10.3	19.9	29.8	8.5	4.1	77.0	2.6	22.0	4.7	18.5	0.4	3.8	5.7	8.3
30 FALKIRK	1.3	4.1	6.3	18.0	31.9	5.8	3.1	81.7	1.9	28.7	1.3	18.0	0.7	3.2	5.4	4.5
31 ARGYLL-BUTE	4.4	6.8	8.3	15.7	34.7	5.8	1.4	61.0	1.5	22.7	2.7	14.9	0.4	4.0	7.3	6.7
32 DUMBARTON	2.5	3.6	8.3	23.0	36.3	8.7	3.1	97.5	3.3	14.7	3.4	15.5	0.5	1.9	8.6	6.1
33 GLASGOW	0.8	4.3	8.3	22.6	31.7	6.9	7.7	130.6	1.4	15.0	2.5	13.8	1.3	2.7	5.5	8.2
34 CLYDEBANK	1.9	4.5	9.0	22.8	29.5	7.5	5.3	108.0	2.0	23.1	2.5	12.3	2.3	3.7	11.0	6.0
35 BEARSDEN	0.7	3.5	5.9	19.2	37.3	7.3	4.4	58.5	1.4	20.5	4.9	24.2	0.4	2.5	4.8	11.3
36 STRATHKELVIN	0.4	3.5	6.8	22.3	32.1	9.5	6.7	81.2	4.4	15.9	2.0	18.8	0.0	0.8	3.8	4.3
37 CUMBERNAULD	0.8	2.6	7.7	28.0	30.0	6.8	4.9	109.1	1.3	16.8	2.4	14.3	0.0	3.6	5.4	6.1
38 MONKLANDS	2.3	7.7	8.2	26.3	33.3	7.8	4.7	103.9	3.1	15.6	2.6	16.8	0.2	1.8	6.0	5.1
39 MOTHERWELL	0.9	3.5	6.2	17.3	28.6	8.7	3.4	83.1	2.5	21.1	3.5	16.4	1.0	1.2	7.3	6.8
40 HAMILTON	1.2	2.8	6.7	13.9	34.8	6.9	3.4	93.3	2.0	18.9	3.8	16.8	0.0	5.4	8.5	5.7
41 EAST KILBRIDE	0.9	4.4	9.5	25.3	32.9	7.3	3.2	104.3	2.2	21.1	3.8	12.1	0.0	0.0	4.0	8.3
42 EASTWOOD	0.8	3.3	8.8	19.8	38.5	5.4	0.9	89.3	1.8	18.9	0.0	14.3	0.9	1.2	2.9	6.3
43 CLYDESDALE	3.7	1.7	7.2	17.3	27.5	10.8	2.2	81.6	1.1	20.3	6.7	15.1	1.4	5.5	5.5	6.3
44 RENFREW	1.6	3.5	8.2	20.0	30.3	8.8	3.3	93.2	1.5	16.5	2.7	16.2	0.0	3.7	4.4	6.0
45 INVERCLYDE	2.9	2.9	8.9	19.7	28.0	12.0	4.7	87.6	1.9	20.5	2.6	16.8	1.1	2.7	7.4	5.4
46 CUNNINGHAME	2.3	3.8	9.0	16.2	33.8	9.3	3.1	109.9	1.3	19.0	3.0	26.5	1.2	1.2	9.1	8.3
47 KILMARNOCK	0.9	2.7	8.8	22.3	32.9	7.5	3.2	80.6	2.8	20.4	4.1	24.3	0.0	3.6	6.1	8.0
48 KYLE-CARRICK	2.4	4.4	7.2	22.8	32.1	7.7	4.2	77.1	2.7	20.6	2.5	13.6	0.4	2.3	5.5	8.3
49 CUMNOCK-DOON	2.5	1.5	7.1	16.8	31.1	8.8	5.1	73.5	2.6	21.5	3.4	15.3	1.1	5.5	4.4	10.1
50 WIGTOWN	6.1	5.7	4.7	19.7	27.8	8.8	4.4	87.6	2.8	15.9	2.2	19.5	0.0	0.0	2.9	3.9
51 STEWARTRY	2.5	4.9	6.2	16.2	38.5	6.9	2.1	73.2	1.3	19.0	7.8	13.8	0.0	0.0	5.5	11.6
52 NITHSDALE	3.1	3.7	10.3	22.8	24.8	8.3	4.6	83.9	2.9	30.4	5.8	18.2	0.0	3.9	6.0	8.3
53 ANNANDALE	0.0	3.5	9.1	16.8	34.9	8.8	3.2	80.3	3.8	22.0	3.8	18.2	0.0	1.1	4.4	5.1
54 ORKNEY	8.0	1.3	5.9	18.7	21.2	8.8	0.0	40.2	4.1	27.2	7.8	18.2	0.0	5.5	3.0	5.1
55 SHETLAND	7.4	5.4	10.1	11.1	34.9	5.0	2.3	48.1	0.0	14.9	1.5	28.3	1.0	3.1	13.9	4.1
56 WESTERN ISLES	7.2	4.6	5.9	13.8	24.1	7.9	2.7	48.3	0.4	24.1	4.3	9.6	1.1	3.4	3.2	12.9
ALL SCOTLAND	2.7	3.9	7.5	20.7	34.2	8.7	4.1	91.4	2.4	21.9	3.7	18.0	0.8	2.9	6.2	7.1

(female)

		LIPC	ORPH	OESO	STOM	LBWL	PANC	LARY	LUNG	MELA	BREA	CERV	CORP	OVAR	BLAD	THYR	HODG	NHOL	LEUK
1	CAITHNESS	0.0	2.1	6.4	5.1	34.8	3.9	0.0	13.4	3.9	56.7	17.2	9.5	9.7	3.1	0.0	2.2	2.0	5.9
2	SUTHERLAND	0.0	1.5	0.9	2.9	24.3	5.3	0.0	11.6	4.3	53.2	21.8	5.1	20.6	2.7	0.6	0.6	3.0	2.8
3	ROSS-CROMARTY	0.4	1.6	4.2	8.0	40.2	2.4	0.9	17.5	4.3	62.0	9.1	12.6	8.8	9.0	2.6	0.6	3.6	7.1
4	SKYE-LOCHALSH	0.0	1.5	1.6	5.3	38.6	2.0	0.0	10.3	8.5	53.9	3.6	7.3	11.3	2.4	2.5	0.0	3.2	10.1
5	LOCHABER	1.9	0.8	3.0	9.5	34.2	1.9	0.0	11.6	5.2	49.5	8.9	11.0	13.9	4.6	0.0	3.1	2.3	0.8
6	INVERNESS	1.2	2.8	3.4	8.9	26.8	3.9	0.0	14.3	6.8	67.8	10.2	12.6	11.3	5.4	0.6	1.5	5.8	2.7
7	BADENOCH	0.0	0.9	3.6	1.6	38.1	0.0	0.0	31.8	0.0	80.9	4.9	17.6	12.9	2.3	0.0	0.0	2.8	6.5
8	NAIRN	0.4	0.9	1.7	4.8	37.5	5.2	2.5	19.3	0.0	69.4	6.0	6.6	12.8	5.0	9.2	0.0	3.4	0.9
9	MORAY	0.4	2.5	5.5	7.5	36.0	6.4	1.6	16.9	4.6	55.0	12.7	8.7	11.6	3.0	0.7	5.7	5.2	8.3
10	BANFF-BUCHAN	0.2	1.7	3.2	6.1	29.7	6.8	0.6	13.2	6.3	48.0	12.7	8.7	7.5	4.4	2.2	3.3	7.2	5.0
11	GORDON	0.0	2.6	2.2	4.2	29.5	3.6	0.9	14.4	2.8	51.4	9.7	6.2	8.1	5.3	1.3	2.0	4.4	5.1
12	ABERDEEN	0.5	2.6	3.9	10.7	35.3	5.8	1.2	26.4	3.1	61.0	9.5	8.2	14.6	8.6	0.7	1.9	6.6	4.4
13	KINCARDINE	0.0	1.7	4.2	6.4	39.6	4.7	0.0	15.0	6.6	66.6	15.4	5.8	12.6	3.2	2.8	1.1	6.3	7.4
14	ANGUS	0.1	1.5	5.6	11.2	32.1	7.2	1.1	22.6	4.0	57.3	11.8	5.2	7.8	4.9	1.2	2.0	8.3	5.9
15	DUNDEE	0.0	1.4	4.4	11.0	28.0	5.4	1.5	24.9	3.8	51.2	12.0	5.4	10.0	6.4	1.3	2.6	3.5	4.4
16	PERTH-KINROSS	0.3	2.1	3.9	8.2	28.9	6.1	1.3	16.0	4.8	61.5	30.4	7.4	13.2	5.5	1.9	1.1	5.5	4.0
17	KIRKCALDY	0.3	1.3	4.5	10.0	28.4	6.1	1.3	22.0	3.6	51.0	15.4	7.1	12.3	6.2	2.5	2.0	5.0	4.9
18	N-E FIFE	0.3	1.5	2.4	9.1	29.5	6.3	1.2	17.1	5.4	66.6	10.9	9.4	9.7	4.6	1.4	2.6	5.8	3.3
19	DUNFERMLINE	0.2	0.7	5.5	10.1	23.1	5.8	0.5	20.8	3.4	59.0	13.7	7.8	11.7	7.1	1.4	1.1	4.8	3.9
20	WEST LOTHIAN	0.3	3.5	4.2	11.7	28.6	6.9	0.8	25.7	5.0	85.8	12.3	11.5	13.1	6.9	3.9	1.6	6.3	5.9
21	EDINBURGH	0.2	2.6	3.8	9.9	27.3	6.6	0.6	25.9	5.1	62.6	15.4	7.9	10.7	6.8	2.7	2.1	5.4	4.5
22	MIDLOTHIAN	0.2	2.2	4.2	11.4	27.2	3.9	0.3	23.7	6.6	56.2	11.6	5.5	10.0	4.7	3.3	1.4	4.2	5.1
23	EAST LOTHIAN	0.2	2.1	3.7	10.1	23.7	7.6	1.0	19.7	4.4	58.6	8.3	10.5	14.8	4.1	2.6	2.5	6.8	4.6
24	TWEEDDALE	0.7	2.1	2.2	14.3	28.7	8.0	1.7	18.3	4.3	49.5	10.4	8.5	4.8	3.7	3.0	1.9	5.6	6.6
25	ETTRICK	0.2	2.9	4.5	5.7	27.3	5.3	1.0	17.5	3.9	63.7	18.4	9.0	15.1	2.7	0.5	3.8	6.8	7.9
26	ROXBURGH	0.0	0.2	5.1	12.3	18.7	2.7	0.8	15.1	1.2	52.5	8.5	8.5	10.3	6.7	2.1	0.8	3.7	11.4
27	BERWICKSHIRE	0.4	2.3	4.4	4.9	32.0	4.8	1.7	26.2	4.1	70.5	9.4	5.2	7.1	1.4	0.6	1.1	5.0	5.1
28	CLACKMANNAN	0.3	1.9	3.8	8.8	29.4	5.9	0.8	23.3	6.0	88.5	14.2	3.4	11.9	4.5	2.4	0.0	5.3	3.9
29	STIRLING	0.3	1.2	4.1	11.4	27.3	6.8	1.7	18.1	3.0	68.6	12.8	7.0	11.4	4.8	1.7	0.7	3.9	5.7
30	FALKIRK	0.0	2.6	5.3	10.4	30.3	5.7	0.8	19.3	2.7	83.1	10.1	6.2	12.9	6.3	1.7	3.5	4.0	7.6
31	ARGYLL-BUTE	0.0	1.8	3.3	10.9	23.7	6.1	0.2	28.4	2.3	61.0	10.2	4.9	11.5	6.5	0.3	2.2	4.0	3.3
32	DUMBARTON	0.1	1.9	4.0	7.6	28.0	8.3	1.1	23.3	2.5	57.3	8.8	3.0	9.3	4.9	1.7	1.9	4.5	6.4
33	GLASGOW	0.0	2.1	3.2	11.2	31.2	3.0	1.1	33.3	3.5	54.1	11.7	5.2	10.2	6.7	3.0	1.7	4.0	5.6
34	CLYDEBANK	0.0	2.1	3.3	13.5	30.9	6.1	1.4	23.3	4.8	52.2	6.4	6.1	11.6	4.9	2.3	1.6	6.2	3.4
35	BEARSDEN	0.2	0.8	3.4	9.7	26.3	7.4	0.3	18.8	3.0	70.2	4.3	5.4	6.0	3.9	1.9	2.7	4.9	3.6
36	STRATHKELVIN	0.0	1.9	2.6	10.3	27.2	5.0	0.6	21.1	1.2	58.0	6.6	6.9	10.6	5.3	2.2	2.1	4.0	4.3
37	CUMBERNAULD	0.0	0.8	5.7	15.4	24.7	4.3	1.5	21.6	1.9	56.0	8.1	4.2	8.5	6.1	2.2	2.2	1.8	3.5
38	MONKLANDS	0.0	2.8	5.4	11.3	22.9	4.6	1.4	17.9	2.9	46.7	12.4	4.1	8.7	4.3	0.7	1.4	4.1	5.8
39	MOTHERWELL	0.5	2.0	3.7	9.7	23.5	7.3	1.4	16.8	3.5	58.2	10.7	3.7	11.8	3.6	1.5	1.5	3.6	3.8
40	HAMILTON	0.0	1.8	8.1	12.4	28.5	5.6	1.4	16.6	5.1	47.8	11.2	7.3	13.4	3.5	2.1	1.0	4.1	3.8
41	EAST KILBRIDE	0.0	2.6	3.3	9.1	22.4	6.5	0.8	23.9	6.9	56.1	6.0	5.6	8.8	5.4	1.2	2.8	5.3	5.2
42	EASTWOOD	0.2	1.8	4.1	12.3	19.1	5.1	0.8	19.3	4.7	60.0	6.5	5.9	11.2	3.7	0.4	2.7	3.8	4.9
43	CLYDESDALE	0.2	2.1	4.5	8.7	24.8	7.2	0.5	13.3	1.7	47.7	10.8	4.0	10.1	4.7	1.5	3.7	2.1	5.3
44	RENFREW	0.0	2.7	5.3	13.1	25.8	4.2	0.8	21.1	5.0	48.6	6.3	4.6	7.6	4.4	0.5	2.6	3.9	4.0
45	INVERCLYDE	0.2	1.0	3.1	8.7	25.8	7.6	0.8	27.5	4.2	61.4	10.1	3.9	10.3	3.9	0.7	1.6	3.8	3.8
46	CUNNINGHAME	0.1	2.3	4.1	9.9	23.8	6.5	0.7	21.5	4.1	51.3	11.8	3.3	9.6	4.0	1.2	1.2	4.9	5.2
47	KILMARNOCK	0.2	2.8	4.5	9.6	25.9	6.2	0.8	13.8	3.9	80.5	12.2	5.6	12.0	7.3	1.3	1.2	2.1	4.9
48	KYLE-CARRICK	0.2	1.9	4.4	8.2	28.9	3.3	0.9	18.7	6.5	58.1	11.6	5.7	7.1	5.9	0.8	2.3	1.1	5.3
49	CUMNOCK-DOON	0.0	2.1	4.4	10.5	27.1	5.9	0.5	13.9	5.3	52.8	14.2	6.0	13.6	3.0	0.8	0.5	6.1	4.0
50	WIGTOWN	0.8	2.7	3.7	9.0	22.1	6.0	0.8	21.8	7.4	63.4	8.0	6.9	6.8	0.9	0.0	0.5	4.9	3.8
51	STEWARTRY	0.0	0.8	5.7	9.0	21.8	6.2	0.4	20.2	4.6	71.0	15.5	9.1	8.8	8.1	0.0	0.8	5.1	6.5
52	NITHSDALE	0.0	1.8	6.8	9.6	35.1	5.9	0.0	18.3	3.1	66.8	11.8	12.3	9.2	3.8	0.2	1.5	3.7	2.2
53	ANNANDALE	0.0	1.0	0.8	8.0	27.7	6.0	0.0	17.8	0.4	58.0	3.7	4.6	8.2	4.7	2.0	0.8	4.2	5.4
54	ORKNEY	0.0	2.8	0.8	5.4	8.9	6.2	0.0	13.6	3.1	49.7	17.6	9.9	16.1	2.7	2.6	0.0	4.0	5.0
55	SHETLAND	0.0	1.0	8.2	10.0	29.3	3.0	1.4	5.6	0.4	70.7	6.0	5.0	9.6	3.1	3.6	3.5	4.2	5.1
56	WESTERN ISLES	0.4	3.5	8.2	5.5	25.5	2.8	0.0	7.0	2.3	56.2	7.7	11.6	9.6	1.9	2.6	0.5	4.0	7.6
	ALL SCOTLAND	0.2	1.9	4.2	10.0	27.5	5.8	1.0	23.1	3.8	57.3	11.3	6.4	10.7	5.8	1.9	1.9	4.8	4.7

Table AI.2 Rank of age-standardized incidence rates by principal cancer site, in each local government district and Islands area, Scotland, 1975-1980, by sex (male)

		LIPC	ORPH	OESO	STOM	LBWL	PANC	LARY	LUNG	MELA	PROS	TEST	BLAD	THYR	HODG	NHOL	LEUK
1	CAITHNESS	4	55	27	52	36	53	54	54	31	56	41	46	2	37	28	31
2	SUTHERLAND	3	10	29	47	12	56	49	40	1	28	34	56	38	35	50	36
3	ROSS-CROMARTY	5	30	28	33	5	44	12	51	38	27	38	47	5	43	52	10
4	SKYE-LOCHALSH	1	2	21	45	4	54	6	55	2	48	28	14	38	45	56	55
5	LOCHABER	7	54	6	37	15	45	46	46	40	12	11	34	4	41	50	45
6	INVERNESS	19	44	1	42	6	51	40	41	37	24	42	44	10	53	22	15
7	BADENOCH	17	1	24	56	2	17	38	47	7	1	7	45	38	54	24	56
8	NAIRN	12	56	46	30	3	48	55	35	15	39	8	53	1	54	3	6
9	MORAY	8	8	8	41	17	2	7	39	11	5	39	24	13	45	35	27
10	BANFF-BUCHAN	2	40	16	38	10	49	27	32	50	11	13	17	17	7	22	23
11	GORDON	10	20	19	54	10	55	49	50	4	16	16	29	36	9	26	10
12	ABERDEEN	23	49	14	22	7	13	5	9	28	13	16	7	7	24	6	24
13	KINCARDINE	15	47	51	44	8	41	20	49	52	13	9	8	19	27	1	36
14	ANGUS	21	6	2	6	8	5	31	27	30	19	25	47	38	43	33	3
15	DUNDEE	46	22	16	1	9	15	10	8	10	24	45	16	19	25	37	26
16	PERTH-KINROSS	29	25	48	29	13	28	40	31	7	18	34	37	24	22	20	19
17	KIRKCALDY	26	39	49	23	33	10	42	12	13	19	16	10	24	6	20	28
18	N-E FIFE	17	37	34	51	31	28	20	24	3	10	23	4	19	45	26	42
19	DUNFERMLINE	23	50	50	18	53	8	44	15	19	21	14	23	22	19	15	12
20	WEST LOTHIAN	29	15	29	18	40	20	28	11	40	7	29	11	29	10	32	19
21	EDINBURGH	45	12	33	21	26	12	15	7	17	15	22	19	27	19	13	24
22	MIDLOTHIAN	32	27	46	16	20	13	24	18	13	3	31	13	35	38	8	32
23	EAST LOTHIAN	29	36	38	16	35	4	36	16	27	33	6	12	29	8	24	45
24	TWEEDDALE	55	4	56	11	1	5	3	28	55	2	3	55	38	2	5	1
25	ETTRICK	15	22	42	50	51	11	20	38	49	35	25	52	38	25	13	18
26	ROXBURGH	23	13	54	53	46	8	24	36	20	21	21	9	13	48	46	19
27	BERWICKSHIRE	6	47	55	20	49	30	1	48	31	42	1	32	38	1	4	4
28	CLACKMANNAN	42	9	2	32	47	1	27	17	52	28	16	20	38	34	44	9
29	STIRLING	36	40	38	25	38	25	20	26	24	6	20	1	38	13	43	33
30	FALKIRK	42	24	36	34	43	46	52	20	35	26	56	22	29	21	48	51
31	ARGYLL-BUTE	20	5	34	43	32	34	36	44	40	55	42	38	38	11	19	28
32	DUMBARTON	33	29	18	8	23	32	2	10	15	8	34	33	22	38	35	38
33	GLASGOW	50	20	15	7	18	18	8	1	31	52	29	21	27	38	40	19
34	CLYDEBANK	40	17	10	12	34	34	17	4	29	34	48	41	14	27	17	15
35	BEARSDEN	52	30	38	46	44	37	27	45	45	45	51	49	7	30	9	40
36	STRATHKELVIN	54	30	42	30	14	15	44	21	31	23	15	15	2	32	37	7
37	CUMBERNAULD	50	46	29	13	29	41	4	3	45	43	54	6	29	52	33	52
38	MONKLANDS	38	3	23	2	42	46	11	6	5	37	51	24	38	16	42	38
39	MOTHERWELL	48	30	7	4	27	30	13	19	47	50	40	39	36	40	49	49
40	HAMILTON	44	43	38	15	25	24	27	13	18	47	45	28	13	35	40	30
41	EAST KILBRIDE	46	18	32	35	45	39	33	5	26	52	33	24	48	48	30	44
42	EASTWOOD	53	37	53	47	22	37	53	32	31	34	29	50	17	3	17	33
43	CLYDESDALE	22	50	22	5	52	50	48	43	29	45	2	39	29	14	10	33
44	RENFREW	41	30	19	35	41	7	31	14	39	41	10	36	6	30	47	40
45	INVERCLYDE	28	40	12	3	24	2	13	23	50	49	42	29	36	27	55	48
46	CUNNINGHAME	38	28	10	13	28	18	36	2	35	37	45	24	9	48	37	12
47	KILMARNOCK	48	44	13	9	38	34	33	25	11	36	25	2	24	16	44	17
48	KYLE-CARRICK	36	18	25	26	29	32	19	29	22	32	48	5	29	32	15	45
49	CUMNOCK-DOON	33	52	28	24	48	21	9	37	23	50	34	43	10	3	7	8
50	WIGTOWN	14	7	51	27	37	41	17	30	24	44	53	34	38	54	28	54
51	STEWARTRY	33	14	38	40	50	39	49	41	47	4	4	18	38	42	12	5
52	NITHSDALE	27	27	2	9	15	26	15	23	20	28	12	41	38	12	30	12
53	ANNANDALE	55	30	9	39	54	21	33	34	7	9	32	29	38	51	54	49
54	ORKNEY	8	53	42	27	56	21	55	56	6	28	4	3	13	3	11	42
55	SHETLAND	11	10	5	55	21	52	47	53	55	54	55	51	38	22	2	53
56	WESTERN ISLES	13	18	42	49	55	29	42	52	54	17	24	54	10	18	53	2

(female)

		LIPC	ORPH	OESO	STOM	LBWL	PANC	LARY	LUNG	MELA	BREA	CERV	CORP	OVAR	BLAD	THYR	HODG	NHOL	LEUK
1	CAITHNESS	31	17	2	50	9	43	42	49	30	33	5	10	36	45	54	16	53	13
2	SUTHERLAND	31	36	54	55	44	32	42	52	33	41	2	44	1	49	45	50	48	52
3	ROSS-CROMARTY	7	32	24	40	1	53	20	36	24	17	37	2	42	1	9	46	43	8
4	SKYE-LOCHALSH	31	36	53	49	3	54	42	54	1	40	56	21	25	52	14	50	47	2
5	LOCHABER	1	51	47	30	10	55	42	52	13	50	38	7	6	29	54	7	51	56
6	INVERNESS	2	4	39	34	35	43	42	45	5	8	30	2	25	17	45	32	14	53
7	BADENOCH	31	48	38	56	4	56	1	2	55	22	53	1	11	53	1	50	49	10
8	NAIRN	31	48	52	52	5	34	14	25	55	5	50	27	13	21	41	50	46	55
9	MORAY	7	13	9	42	6	14	30	39	21	38	13	27	20	47	18	44	17	3
10	BANFF-BUCHAN	17	30	44	44	16	9	30	51	9	53	13	15	51	32	33	1	2	27
11	GORDON	31	8	50	53	17	47	20	44	46	45	34	29	48	19	41	6	25	23
12	ABERDEEN	5	8	31	16	7	26	12	6	39	20	35	17	5	19	7	21	4	35
13	KINCARDINE	31	30	24	43	2	38	42	43	6	10	7	34	14	44	36	11	1	7
14	ANGUS	28	36	7	12	11	6	14	15	29	31	19	41	49	22	33	23	6	13
15	DUNDEE	31	40	19	14	27	30	5	10	34	47	18	38	34	11	23	40	45	35
16	PERTH-KINROSS	12	17	19	38	21	30	10	33	18	18	1	20	9	16	14	21	12	39
17	KIRKCALDY	7	42	31	22	26	20	10	16	34	48	7	23	15	13	30	11	19	29
18	N-E FIFE	12	36	16	31	17	15	12	38	11	10	26	11	36	29	30	40	9	50
19	DUNFERMLINE	17	54	49	20	48	26	34	22	38	25	16	19	19	5	30	29	23	41
20	WEST LOTHIAN	17	1	9	8	25	8	28	9	16	12	23	6	10	6	2	19	5	13
21	EDINBURGH	12	23	24	24	29	10	23	8	14	16	41	18	28	7	8	35	12	33
22	MIDLOTHIAN	17	16	33	9	32	43	38	12	6	34	29	37	34	26	4	14	27	23
23	EAST LOTHIAN	17	17	24	20	23	43	17	24	23	26	9	8	4	35	9	23	3	32
24	TWEEDDALE	4	51	35	54	55	17	42	3	2	50	3	16	56	40	48	2	11	9
25	ETTRICK	17	3	50	46	23	1	2	30	24	13	40	14	3	49	20	44	55	29
26	ROXBURGH	31	56	16	6	54	32	17	36	30	43	36	13	30	8	52	40	41	4
27	BERWICKSHIRE	31	14	14	51	54	52	42	42	52	3	42	41	52	55	45	50	19	1
28	CLACKMANNAN	7	40	19	35	12	37	23	7	27	7	9	54	17	31	16	27	14	23
29	STIRLING	12	23	33	9	19	24	2	13	10	6	12	24	23	25	30	4	37	43
30	FALKIRK	31	8	28	18	29	10	23	32	41	15	30	29	11	12	25	16	31	41
31	ARGYLL-BUTE	31	32	12	15	46	28	17	25	44	20	39	46	22	10	51	23	31	17
32	DUMBARTON	31	29	41	41	15	20	40	4	44	31	22	56	40	22	25	38	24	5
33	GLASGOW	31	23	30	12	38	15	2	1	48	39	48	41	32	8	25	27	31	33
34	CLYDEBANK	31	44	44	3	13	49	14	13	47	44	54	31	23	22	17	29	6	50
35	BEARSDEN	31	17	41	26	14	4	38	28	18	4	46	38	20	37	23	6	21	12
36	STRATHKELVIN	17	44	39	19	37	36	42	20	41	29	42	25	55	19	23	9	31	20
37	CUMBERNAULD	31	51	48	1	32	40	30	18	52	37	15	49	29	14	18	23	31	18
38	MONKLANDS	31	23	5	11	43	40	5	34	50	56	28	50	46	34	41	19	54	49
39	MOTHERWELL	17	8	11	2	49	40	36	40	43	27	25	53	45	42	28	16	30	46
40	HAMILTON	5	22	35	26	47	39	7	41	36	54	50	21	18	43	20	35	43	38
41	EAST KILBRIDE	31	32	4	5	38	5	7	11	14	36	47	38	8	17	9	32	14	48
42	EASTWOOD	31	8	41	31	50	29	40	25	4	24	27	36	42	40	36	43	39	16
43	CLYDESDALE	31	32	7	6	53	12	23	50	20	55	49	33	27	26	50	8	37	43
44	RENFREW	31	17	15	36	42	35	34	20	51	52	27	51	33	32	28	2	9	46
45	INVERCLYDE	17	6	12	4	40	6	28	5	16	19	49	47	50	37	48	9	37	43
46	CUNNINGHAME	28	44	46	36	45	42	28	19	26	46	32	52	30	36	41	11	39	22
47	KILMARNOCK	17	14	28	24	39	2	30	47	44	23	19	55	38	4	36	29	21	29
48	KYLE-CARRICK	17	23	16	38	21	12	20	29	27	28	17	35	16	15	33	35	52	20
49	CUMNOCK-DOON	31	55	19	17	34	17	42	46	30	42	23	32	52	47	39	15	55	39
50	WIGTOWN	3	6	19	28	51	48	30	17	8	14	9	25	7	56	39	48	8	10
51	STEWARTRY	31	48	35	33	52	46	42	23	12	1	44	12	54	3	54	50	50	54
52	NITHSDALE	31	23	5	28	8	24	36	30	3	9	6	4	42	39	52	32	18	35
53	ANNANDALE	31	44	1	45	28	23	42	35	21	29	19	47	41	26	22	46	41	19
54	ORKNEY	31	4	56	48	56	17	42	48	39	49	55	9	47	49	9	50	27	27
55	SHETLAND	31	44	55	22	20	49	7	56	54	2	4	45	2	45	3	4	27	23
56	WESTERN ISLES	7	1	3	47	41	51	42	55	48	34	45	5	38	54	9	48	31	5

Table AI.3 District codes ordered by rank of age-standardized incidence rates, Scotland, 1975-1980, by principal cancer site, by sex (male)

	LIPC	ORPH	OESO	STOM	LBWL	PANC	LARY	LUNG	MELA	PROS	TEST	BLAD	THYR	HODG	NHOL	LEUK
1	4	7	8	15	24	28	27	33	2	7	27	29	8	27	13	24
2	10	4	14	38	7	9	32	45	4	24	42	47	35	24	55	56
3	2	38	52	45	8	45	24	37	18	22	24	54	1	49	8	14
4	1	24	28	39	4	23	37	34	11	51	54	18	5	42	27	27
5	3	31	55	43	3	14	12	41	37	8	51	48	3	54	24	51
6	27	14	5	14	6	24	4	38	54	28	23	37	44	17	12	8
7	5	50	39	33	12	44	9	21	18	20	7	12	12	10	49	38
8	9	9	9	32	14	18	33	15	53	32	8	13	34	23	22	49
9	54	28	53	52	13	28	49	12	7	53	13	28	46	11	35	28
10	11	55	48	47	10	17	15	32	15	18	43	17	8	20	43	3
11	55	2	34	24	11	25	38	20	47	10	5	20	49	31	54	11
12	8	21	45	34	2	21	3	17	9	5	52	23	56	52	51	46
13	56	26	47	46	18	22	39	40	22	12	10	22	54	29	21	19
14	50	51	12	37	36	12	45	44	17	13	19	4	28	43	25	52
15	13	20	33	40	52	38	52	19	32	21	36	36	40	34	48	6
16	25	56	15	23	5	15	21	23	8	11	17	15	9	38	19	34
17	7	34	10	22	9	7	50	28	21	56	28	10	10	47	42	47
18	18	48	32	19	33	48	34	22	39	18	12	51	42	56	34	25
19	8	41	44	20	15	33	48	39	19	17	11	21	18	21	31	16
20	31	11	11	27	22	20	13	30	52	14	29	28	15	19	18	20
21	14	33	4	21	55	54	18	38	28	28	26	33	13	30	17	33
22	43	25	43	12	42	49	29	48	48	19	21	30	19	18	8	28
23	28	15	38	17	32	53	25	52	49	35	18	19	32	55	10	10
24	12	30	7	49	45	40	26	18	50	8	56	38	17	12	7	21
25	19	16	48	28	40	28	22	47	28	15	25	46	47	25	23	12
26	17	46	49	48	21	18	20	29	40	30	47	9	16	15	18	15
27	52	52	1	54	39	52	28	14	23	3	14	41	21	13	11	9
28	45	22	3	50	48	18	10	24	12	2	4	40	33	45	50	31
29	23	32	37	18	48	56	35	48	42	52	41	11	20	33	1	17
30	20	44	20	8	37	39	40	50	14	28	20	53	30	44	52	40
31	16	35	2	38	18	27	14	18	27	54	22	45	48	35	41	1
32	22	39	41	28	31	48	44	42	35	48	53	27	38	48	20	22
33	49	3	21	3	17	32	47	10	41	23	40	32	43	36	37	43
34	51	53	31	30	34	47	53	53	1	41	2	5	23	28	14	42
35	32	38	18	41	23	31	41	8	30	25	49	50	22	40	32	29
36	28	23	30	44	1	34	7	28	46	47	18	44	11	2	2	2
37	48	18	29	5	50	42	31	49	8	45	32	16	39	1	36	13
38	38	42	40	10	29	35	23	25	3	37	3	31	51	32	46	32
39	38	17	35	53	47	51	46	9	43	39	9	43	2	22	15	38
40	34	28	51	51	20	41	8	2	5	46	46	39	55	39	33	44
41	44	45	23	9	44	37	18	51	45	43	1	34	4	5	40	35
42	28	10	25	8	38	13	17	8	20	27	8	52	53	51	38	18
43	30	40	56	31	30	50	56	43	31	36	31	49	52	29	29	54
44	40	47	54	13	35	3	19	31	33	50	44	6	7	14	28	41
45	21	6	36	4	41	5	36	35	34	42	39	7	14	9	47	48
46	38	37	8	35	28	38	5	5	36	33	15	1	41	18	28	23
47	47	13	22	42	28	30	43	7	51	39	45	3	38	4	44	5
48	41	27	18	2	49	8	11	27	38	4	34	14	37	28	30	45
49	44	45	17	56	27	10	51	13	25	44	33	35	31	46	39	39
50	15	12	19	25	51	43	51	11	10	38	48	42	29	41	2	53
51	37	19	50	18	25	8	2	3	44	49	35	55	50	53	5	30
52	35	48	13	1	43	55	30	56	13	40	38	25	45	37	3	37
53	42	54	42	28	19	1	42	55	28	34	50	8	24	8	56	55
54	38	5	26	11	53	4	1	4	56	55	37	56	25	6	53	50
55	53	1	27	55	56	11	8	4	24	31	55	24	27	7	45	4
56	24	8	24	7	54	2	54	54	55	1	30	2	28	50	4	7

(female)

	LIPC	ORPH	OESO	STOM	LBWL	PANC	LARY	LUNG	MELA	BREA	CERV	CORP	OVAR	BLAD	THYR	HODG	NHOL	LEUK
1	5	56	53	37	3	25	8	33	4	51	16	7	2	3	7	10	13	27
2	6	20	1	39	13	23	25	7	24	55	2	3	55	12	19	24	10	4
3	50	25	56	34	4	47	29	24	52	27	24	6	25	51	55	43	23	9
4	24	6	41	45	7	36	33	32	42	35	54	52	23	47	21	29	12	26
5	40	54	52	41	8	41	15	45	6	8	1	56	12	19	34	55	20	56
6	12	45	38	26	9	45	38	12	13	29	51	20	5	20	23	11	14	32
7	9	50	14	43	12	14	41	28	22	28	17	5	50	21	12	5	34	13
8	3	39	43	20	52	20	40	21	50	6	13	23	41	26	20	42	50	3
9	28	31	9	22	1	10	55	20	10	52	28	54	16	33	3	35	18	24
10	17	12	20	29	5	30	17	15	29	13	49	1	20	31	56	44	43	50
11	56	42	39	38	14	21	16	41	18	18	19	18	7	15	54	13	24	35
12	21	11	31	33	28	48	18	22	51	20	29	51	30	30	22	17	16	20
13	29	9	45	14	34	43	12	34	5	25	10	26	8	17	41	45	21	14
14	18	27	27	15	35	9	9	29	41	50	9	25	13	37	16	22	41	1
15	16	47	44	31	10	34	14	14	21	30	38	10	17	48	4	49	6	42
16	30	22	28	12	18	18	34	17	20	21	20	24	48	16	28	30	28	31
17	19	16	18	49	11	54	26	50	45	3	47	12	28	6	35	38	9	37
18	45	35	48	30	29	24	23	37	18	18	15	21	40	41	9	1	52	53
19	36	1	28	36	55	49	31	48	35	45	14	19	19	11	37	37	17	48
20	23	44	15	23	48	33	11	44	43	31	52	16	35	36	40	20	27	36
21	48	23	49	19	18	17	48	36	9	12	48	4	9	8	25	12	48	46
22	10	40	50	55	23	32	3	19	53	7	33	40	31	14	53	18	35	55
23	22	38	16	17	25	53	21	51	23	47	48	17	29	32	15	31	19	11
24	47	21	3	47	20	29	30	23	3	42	21	29	34	34	36	23	32	28
25	25	48	23	21	17	52	28	8	25	19	40	50	4	28	33	36	48	22
26	20	29	21	40	15	12	43	42	48	23	18	38	8	53	30	14	11	54
27	39	52	13	35	53	19	45	31	28	39	43	8	43	22	31	33	55	10
28	46	34	47	52	21	31	46	35	48	48	39	9	21	43	44	28	22	17
29	14	33	30	50	28	42	20	48	14	53	23	11	37	5	39	46	54	25
30	33	13	33	5	30	15	37	52	1	38	6	30	28	18	29	34	39	47
31	44	10	17	42	22	18	10	25	49	32	31	34	48	28	17	19	33	23
32	7	41	12	18	37	28	50	30	26	14	45	49	33	44	18	6	31	21
33	13	32	22	51	49	22	47	18	2	1	30	43	44	10	10	52	37	33
34	51	43	29	8	8	8	19	16	15	56	11	13	15	38	14	40	56	52
35	11	3	51	28	41	44	44	38	17	22	28	48	22	23	48	21	30	12
36	2	18	40	46	38	37	52	53	34	41	3	42	1	46	47	47	38	15
37	55	14	24	44	33	28	39	28	40	37	5	22	18	35	13	39	29	40
38	52	4	7	48	47	13	35	3	19	9	32	15	56	45	42	32	44	18
39	4	2	36	18	45	40	22	18	12	33	25	35	47	42	50	48	45	49
40	54	28	35	3	56	39	32	9	54	4	22	41	32	42	11	18	42	19
41	53	15	42	32	19	38	42	39	36	2	37	27	53	39	46	26	53	30
42	8	17	32	9	39	46	4	40	30	49	27	14	3	40	38	15	28	45
43	37	30	34	13	42	8	5	27	39	28	50	33	52	40	8	41	3	43
44	15	46	10	10	2	22	7	13	47	34	56	2	42	13	14	9	40	29
45	35	36	48	53	46	1	2	11	31	11	38	55	39	1	6	25	15	39
46	34	55	5	25	31	51	8	6	11	48	42	31	38	55	27	53	8	44
47	32	53	37	56	40	11	13	49	33	15	34	53	54	49	2	3	4	41
48	31	8	19	54	19	50	54	47	32	17	44	45	11	9	24	56	2	38
49	38	7	11	4	39	35	53	54	56	54	55	37	14	54	45	50	7	18
50	27	51	25	1	42	55	51	1	38	5	41	38	45	25	43	2	51	34
51	28	5	4	27	50	56	49	43	44	24	8	44	10	2	32	4	5	2
52	49	37	2	8	51	27	38	10	37	44	7	46	49	4	52	54	47	6
53	43	24	55	11	43	3	24	5	27	10	35	39	27	7	28	51	1	51
54	42	19	54	24	27	4	27	2	55	40	53	28	51	56	51	8	38	8
55	41	49	55	2	24	5	56	4	8	43	4	47	38	27	5	7	25	5
56	1	26	54	7	54	7	1	56	7	38		32	24	50	1	27	49	

Table A1.4 Cumulative rate for ages 0-75 years, by principal cancer site, in each local government district and Islands area, Scotland, 1975-1980, by sex (male)

		LIPC	ORPH	OESO	STOM	LBWL	PANC	LARY	LUNG	MELA	PROS	TEST	BLAD	THYR	HODG	NHOL	LEUK
1	CAITHNESS	1.23	0.13	1.14	1.85	3.82	0.54	0.13	8.00	0.20	1.84	0.23	1.08	0.24	0.27	0.70	0.37
2	SUTHERLAND	0.97	0.50	0.52	1.78	4.67	0.25	0.26	8.12	0.77	3.01	0.52	0.59	0.00	0.28	0.24	0.47
3	ROSS-CROMARTY	0.86	0.43	0.82	2.79	5.01	0.85	0.43	8.99	0.18	2.95	0.23	1.49	0.17	0.09	0.34	0.72
4	SKYE-LOCHALSH	1.64	1.00	0.86	1.89	6.24	0.33	0.84	4.99	0.67	1.28	0.24	1.87	0.00	0.35	0.35	0.35
5	LOCHABER	0.77	0.29	1.27	2.72	4.31	0.98	0.50	7.12	0.19	2.89	0.41	1.39	0.14	0.11	0.19	0.44
6	INVERNESS	0.52	0.30	1.38	1.81	5.09	0.60	0.43	7.85	0.17	2.22	0.24	1.84	0.05	0.03	0.60	0.90
7	BADENOCH	0.89	1.57	1.23	0.83	8.31	1.19	0.31	5.53	0.32	3.82	0.57	1.81	0.00	0.00	1.43	0.00
8	NAIRN	0.82	0.00	0.82	1.18	6.01	0.72	0.00	7.70	0.33	2.11	0.58	1.30	0.33	0.00	1.12	0.62
9	MORAY	0.95	0.58	1.11	1.72	4.70	1.11	0.56	9.15	0.33	3.57	0.28	2.28	0.13	0.15	0.75	0.88
10	BANFF-BUCHAN	1.09	0.37	0.98	1.16	4.61	0.86	0.52	9.17	0.07	3.02	0.42	2.11	0.09	0.42	0.57	0.59
11	GORDON	0.83	0.49	1.00	1.16	4.20	0.38	0.23	7.21	0.46	2.90	0.38	2.00	0.00	0.33	0.72	1.08
12	ABERDEEN	0.39	0.24	1.08	2.41	4.95	1.24	0.71	12.38	0.21	2.88	0.36	2.76	0.11	0.26	0.92	0.84
13	KINCARDINE	0.61	0.41	0.43	1.69	4.84	0.66	0.38	8.50	0.08	2.83	0.57	2.74	0.06	0.27	1.80	0.80
14	ANGUS	0.50	0.60	1.27	2.97	4.70	1.55	0.33	9.70	0.13	2.22	0.28	1.58	0.00	0.12	0.88	1.24
15	DUNDEE	0.10	0.49	1.01	3.51	3.93	1.09	0.89	12.92	0.31	2.49	0.22	2.29	0.07	0.20	0.66	0.46
16	PERTH-KINROSS	0.29	0.40	0.71	2.10	4.23	0.79	0.41	9.19	0.31	2.24	0.28	1.75	0.06	0.31	0.73	0.65
17	KIRKCALDY	0.35	0.41	0.72	2.40	3.50	1.04	0.39	11.81	0.34	2.38	0.39	2.58	0.07	0.40	0.63	0.79
18	N-E FIFE	0.78	0.28	0.80	1.29	3.73	0.90	0.58	10.08	0.58	2.90	0.39	3.11	0.09	0.08	0.82	0.56
19	DUNFERMLINE	0.36	0.21	0.59	2.74	2.89	1.17	0.27	11.79	0.37	2.88	0.44	1.86	0.08	0.27	0.92	0.82
20	WEST LOTHIAN	0.27	0.63	0.81	3.00	3.45	1.09	0.48	12.95	0.18	2.70	0.32	2.78	0.06	0.32	0.88	0.51
21	EDINBURGH	0.12	0.53	0.79	2.21	3.78	1.14	0.57	13.15	0.32	2.59	0.35	2.21	0.05	0.28	0.82	0.67
22	MIDLOTHIAN	0.10	0.35	0.85	3.13	3.90	1.24	0.56	11.10	0.28	3.51	0.29	2.12	0.08	0.17	1.13	0.53
23	EAST LOTHIAN	0.12	0.20	0.83	2.32	3.32	1.30	0.38	11.32	0.19	2.35	0.55	2.34	0.03	0.40	0.74	0.38
24	TWEEDDALE	0.00	0.72	0.50	2.49	6.75	1.85	0.97	9.45	0.00	4.37	0.87	0.75	0.00	0.62	1.53	1.44
25	ETTRICK	0.32	0.48	0.72	1.28	3.50	1.27	0.84	8.32	0.07	2.11	0.30	1.30	0.00	0.26	0.89	0.34
26	ROXBURGH	0.40	0.54	0.28	1.17	3.48	1.78	0.86	9.60	0.24	3.38	0.38	2.82	0.08	0.18	0.73	0.94
27	BERWICKSHIRE	1.29	0.17	0.34	2.28	3.02	0.88	1.29	7.19	0.16	2.84	1.07	2.48	0.00	1.11	1.12	1.09
28	CLACKMANNAN	0.13	0.48	1.18	1.78	3.48	1.51	0.49	11.67	0.08	2.30	0.37	2.34	0.00	0.28	0.37	0.87
29	STIRLING	0.23	0.39	0.75	2.14	3.33	0.89	0.58	10.14	0.34	3.13	0.35	2.91	0.00	0.30	0.55	0.57
30	FALKIRK	0.12	0.52	0.87	2.04	3.45	0.79	0.19	10.57	0.17	2.76	0.10	2.13	0.04	0.25	0.43	0.44
31	ARGYLL-BUTE	0.53	0.74	0.74	1.81	3.83	0.91	0.42	8.35	0.15	1.83	0.20	1.67	0.00	0.42	0.93	0.55
32	DUMBARTON	0.19	0.33	1.17	2.58	3.56	0.70	0.96	13.17	0.28	2.76	0.27	1.81	0.05	0.21	0.74	0.58
33	GLASGOW	0.08	0.47	0.97	2.72	3.97	1.08	0.82	16.28	0.17	1.67	0.20	1.94	0.05	0.23	0.58	0.66
34	CLYDEBANK	0.23	0.57	1.37	2.94	3.86	0.83	0.81	13.19	0.19	1.67	0.23	1.49	0.11	0.40	0.85	0.74
35	BEARSDEN	0.12	0.47	0.57	1.87	3.38	0.70	0.53	8.98	0.34	2.30	0.19	1.89	0.20	0.19	1.07	0.41
36	STRATHKELVIN	0.05	0.50	0.51	2.45	3.95	1.23	0.32	9.91	0.14	1.96	0.37	2.80	0.05	0.18	0.59	0.81
37	CUMBERNAULD	0.13	0.31	0.94	3.27	3.47	0.96	0.57	14.55	0.19	0.83	0.14	2.23	0.00	0.07	0.75	0.23
38	MONKLANDS	0.25	0.55	1.04	3.12	3.04	0.70	0.56	13.98	0.19	1.17	0.17	1.79	0.00	0.35	0.45	0.63
39	MOTHERWELL	0.12	0.38	1.13	3.78	3.78	0.98	0.55	10.82	0.33	1.86	0.22	1.75	0.04	0.19	0.38	0.55
40	HAMILTON	0.10	0.32	0.78	2.47	4.08	1.13	0.45	12.09	0.22	1.21	0.25	2.02	0.22	0.12	0.69	0.49
41	EAST KILBRIDE	0.18	0.45	0.88	1.99	3.72	0.63	0.39	13.11	0.19	2.07	0.28	2.23	0.00	0.11	0.66	0.59
42	EASTWOOD	0.05	0.24	0.59	1.60	3.70	0.93	0.14	8.71	0.20	2.24	0.84	1.30	0.05	0.32	0.91	0.84
43	CLYDESDALE	0.30	0.12	1.01	2.83	3.07	0.75	0.42	8.09	0.20	2.42	0.46	1.41	0.10	0.35	0.81	0.55
44	RENFREW	0.12	0.35	1.08	1.85	3.30	1.20	0.39	11.83	0.09	1.74	0.25	1.85	0.15	0.18	0.45	0.44
45	INVERCLYDE	0.28	0.27	1.00	3.17	3.85	1.32	0.80	14.11	0.15	2.35	0.21	2.04	0.00	0.21	0.35	0.61
46	CUNNINGHAME	0.28	0.44	1.11	2.84	3.48	1.12	0.32	10.52	0.17	1.81	0.24	2.04	0.15	0.12	0.57	0.82
47	KILMARNOCK	0.04	0.34	0.84	2.74	3.35	0.89	0.43	10.19	0.26	2.00	0.37	2.49	0.07	0.35	0.39	0.83
48	KYLE-CARRICK	0.31	0.52	0.94	2.13	3.30	1.08	0.84	9.59	0.25	2.40	0.20	2.75	0.07	0.28	0.98	0.56
49	CUMNOCK-DOON	0.30	0.18	1.01	1.85	3.49	1.04	0.49	8.07	0.25	1.63	0.27	1.55	0.00	0.67	0.78	1.03
50	WIGTOWN	0.53	0.70	0.42	1.58	3.81	0.74	0.53	10.44	0.24	2.85	0.85	1.47	0.00	0.00	0.84	0.38
51	STEWARTRY	0.33	0.47	0.56	1.88	3.71	0.58	0.52	8.43	0.13	3.08	0.19	2.08	0.00	0.13	0.99	0.77
52	NITHSDALE	0.38	0.35	1.29	2.75	4.48	0.83	0.66	11.00	0.28	2.19	0.49	1.45	0.10	0.35	0.53	0.55
53	ANNANDALE	0.00	0.36	0.78	2.08	2.35	0.78	0.50	8.27	0.33	2.48	0.33	2.06	0.00	0.10	0.38	0.55
54	ORKNEY	1.20	0.18	0.56	1.86	2.38	0.84	0.00	4.90	0.34	2.83	0.58	2.50	0.17	0.44	0.88	0.58
55	SHETLAND	0.85	0.58	0.98	1.34	3.45	0.81	0.20	5.81	0.00	1.46	0.12	1.88	0.00	0.30	1.80	0.38
58	WESTERN ISLES	0.88	0.41	0.52	1.73	2.94	1.07	0.32	5.91	0.00	2.63	0.41	1.11	0.22	0.32	0.30	1.23
	ALL SCOTLAND	0.28	0.43	0.89	2.39	3.83	1.03	0.51	11.82	0.23	2.27	0.30	2.07	0.08	0.25	0.68	0.84

(female)

		LIPC	ORPH	OESO	STOM	LBWL	PANC	LARY	LUNG	MELA	BREA	CERV	CORP	OVAR	BLAD	THYR	HODG	NHOL	LEUK
1	CAITHNESS	0.00	0.15	0.87	0.62	4.04	0.47	0.00	1.85	0.46	6.27	1.87	1.24	1.05	0.51	0.00	0.16	0.23	0.42
2	SUTHERLAND	0.00	0.00	0.21	0.63	3.00	0.44	0.00	1.70	0.46	5.80	1.90	0.43	2.44	0.22	0.00	0.00	0.43	0.22
3	ROSS-CROMARTY	0.10	0.17	0.60	1.04	5.42	0.26	0.05	2.28	0.34	6.91	0.92	1.43	1.06	1.16	0.22	0.08	0.32	0.81
4	SKYE-LOCHALSH	0.00	0.25	0.25	0.49	4.69	0.25	0.00	1.19	0.90	6.05	0.30	0.85	1.28	0.25	0.31	0.00	0.31	0.93
5	LOCHABER	0.38	0.00	0.58	0.83	4.21	0.21	0.00	1.79	0.31	5.20	0.71	0.99	1.84	0.72	0.00	0.31	0.14	0.00
6	INVERNESS	0.13	0.30	0.37	1.15	2.71	0.60	0.00	1.62	0.69	7.67	1.03	1.33	1.12	0.84	0.08	0.00	0.60	0.37
7	BADENOCH	0.00	0.00	0.31	0.00	4.62	0.00	0.00	3.54	0.00	8.43	0.61	1.96	1.29	0.36	1.22	0.00	0.31	0.53
8	NAIRN	0.00	0.00	0.28	0.92	4.16	0.67	0.31	2.47	0.00	8.10	0.53	0.69	1.45	0.62	0.00	0.00	0.31	0.00
9	MORAY	0.04	0.35	0.56	0.88	4.33	0.76	0.07	2.22	0.48	5.99	1.48	0.79	1.26	0.29	0.24	0.08	0.51	0.69
10	BANFF-BUCHAN	0.05	0.12	0.24	0.54	3.52	0.79	0.04	1.55	0.25	4.85	1.43	1.02	0.79	0.49	0.16	0.50	0.70	0.43
11	GORDON	0.00	0.26	0.28	0.48	2.79	0.39	0.12	2.09	0.37	5.52	0.91	0.75	0.93	0.54	0.04	0.23	0.43	0.49
12	ABERDEEN	0.09	0.27	0.41	1.17	3.88	0.52	0.16	1.81	0.53	6.83	0.97	1.06	1.69	1.06	0.23	0.16	0.64	0.42
13	KINCARDINE	0.00	0.17	0.51	0.87	4.88	0.52	0.00	1.81	0.53	6.88	1.50	0.67	0.85	0.43	0.08	0.34	0.99	0.60
14	ANGUS	0.00	0.10	0.44	1.24	3.40	0.85	0.13	2.88	0.39	6.42	1.30	0.59	0.85	0.62	0.16	0.17	0.69	0.53
15	DUNDEE	0.00	0.15	0.53	1.19	3.06	0.61	0.18	3.16	0.36	5.50	1.23	0.72	1.13	0.67	0.18	0.11	0.36	0.37
16	PERTH-KINROSS	0.02	0.24	0.45	0.85	3.33	0.62	0.14	2.25	0.50	6.79	2.76	0.87	1.48	0.71	0.26	0.17	0.63	0.31
17	KIRKCALDY	0.02	0.15	0.49	1.12	3.29	0.78	0.14	2.90	0.35	5.68	1.53	0.84	1.44	0.77	0.13	0.22	0.59	0.40
18	N-E FIFE	0.04	0.26	0.51	1.12	3.20	0.89	0.13	2.09	0.56	7.30	1.17	1.17	1.25	0.61	0.12	0.18	0.58	0.43
19	DUNFERMLINE	0.03	0.05	0.33	1.19	2.51	0.64	0.06	2.62	0.33	6.17	1.42	0.96	1.21	0.77	0.42	0.15	0.50	0.29
20	WEST LOTHIAN	0.04	0.37	0.58	1.11	3.48	0.85	0.05	3.33	0.43	7.32	1.37	1.38	1.44	0.76	0.29	0.20	0.77	0.48
21	EDINBURGH	0.04	0.22	0.46	1.13	3.07	0.70	0.09	3.28	0.51	6.73	1.23	0.94	1.26	0.76	0.30	0.12	0.83	0.37
22	MIDLOTHIAN	0.00	0.28	0.46	1.47	3.11	0.41	0.04	2.85	0.87	8.09	0.85	0.78	1.25	0.58	0.23	0.18	0.47	0.55
23	EAST LOTHIAN	0.04	0.21	0.41	1.40	3.52	0.92	0.14	2.35	0.43	6.31	1.06	1.22	1.61	0.60	0.30	0.23	0.65	0.26
24	TWEEDDALE	0.18	0.00	0.35	0.36	1.41	0.74	0.00	3.58	0.73	5.73	0.98	1.32	0.59	0.36	0.00	0.40	0.52	0.38
25	ETTRICK	0.00	0.40	0.15	0.59	3.28	1.12	0.23	2.04	0.62	7.26	0.88	0.95	1.89	0.26	0.17	0.08	0.17	0.67
26	ROXBURGH	0.00	0.00	0.38	1.15	2.67	0.45	0.15	2.34	0.38	8.15	1.22	0.85	1.29	0.76	0.00	0.06	0.30	0.53
27	BERWICKSHIRE	0.00	0.44	0.79	0.58	2.55	0.44	0.07	1.91	0.15	7.82	0.80	0.59	0.75	0.18	0.00	0.00	0.77	1.54
28	CLACKMANNAN	0.00	0.10	0.47	0.94	3.60	0.59	0.00	3.16	0.36	7.17	1.55	0.40	1.09	0.55	0.16	0.20	0.53	0.59
29	STIRLING	0.04	0.24	0.38	1.30	3.28	0.58	0.16	3.05	0.50	7.08	1.45	0.82	1.30	0.62	0.16	0.36	0.40	0.43
30	FALKIRK	0.05	0.13	0.48	1.20	3.01	0.87	0.09	2.27	0.33	6.98	1.03	0.88	1.51	0.79	0.17	0.19	0.46	0.31
31	ARGYLL-BUTE	0.23	0.23	0.54	1.33	2.66	0.58	0.13	2.38	0.22	6.35	1.04	0.58	1.18	0.82	0.13	0.22	0.36	0.56
32	DUMBARTON	0.00	0.26	0.45	0.86	3.61	0.70	0.05	3.78	0.18	5.98	1.04	0.34	1.18	0.51	0.04	0.15	0.49	0.76
33	GLASGOW	0.01	0.22	0.48	1.27	2.80	0.71	0.21	4.05	0.28	5.87	1.23	0.64	1.15	0.77	0.16	0.17	0.43	0.36
34	CLYDEBANK	0.00	0.25	0.49	1.60	3.76	0.72	0.19	3.07	0.39	5.83	0.75	0.50	1.21	0.61	0.24	0.13	0.87	0.38
35	BEARSDEN	0.06	0.23	0.36	0.99	3.31	0.88	0.00	2.11	0.47	8.11	0.64	0.61	1.20	0.50	0.22	0.34	0.49	0.47
36	STRATHKELVIN	0.08	0.13	0.37	1.06	2.99	0.88	0.00	2.48	0.31	6.53	0.72	0.81	0.70	0.50	0.19	0.16	0.58	0.35
37	CUMBERNAULD	0.00	0.10	0.25	1.58	2.88	0.80	0.06	2.59	0.15	8.12	0.77	0.48	1.14	0.62	0.09	0.22	0.51	0.47
38	MONKLANDS	0.02	0.25	0.57	1.10	2.66	0.54	0.17	2.28	0.18	4.87	1.20	0.50	0.91	0.42	0.06	0.23	0.20	0.22
39	MOTHERWELL	0.02	0.31	0.83	1.57	2.73	0.50	0.07	2.08	0.28	8.37	1.08	0.50	1.01	0.43	0.06	0.10	0.39	0.34
40	HAMILTON	0.06	0.25	0.56	1.03	2.78	0.59	0.14	2.04	0.38	5.40	1.23	0.96	1.34	0.34	0.17	0.15	0.35	0.38
41	EAST KILBRIDE	0.00	0.21	0.71	1.27	3.20	0.88	0.17	2.92	0.53	8.09	0.80	0.66	1.56	0.74	0.22	0.11	0.75	0.45
42	EASTWOOD	0.00	0.28	0.27	0.87	2.34	0.51	0.00	2.59	0.58	6.85	0.76	0.72	1.00	0.55	0.10	0.19	0.46	0.49
43	CLYDESDALE	0.00	0.25	0.71	1.31	2.40	0.56	0.12	1.88	0.56	5.72	1.26	0.70	1.21	0.85	0.33	0.33	0.60	0.45
44	RENFREW	0.00	0.23	0.53	0.91	2.81	0.61	0.04	2.61	0.15	5.26	0.71	0.47	1.15	0.53	0.15	0.29	0.43	0.34
45	INVERCLYDE	0.04	0.29	0.85	1.26	3.02	0.91	0.09	3.24	0.45	8.67	1.09	0.81	0.96	0.44	0.03	0.23	0.39	0.39
46	CUNNINGHAME	0.02	0.12	0.35	1.09	2.77	0.49	0.07	2.71	0.37	5.48	1.23	0.46	1.20	0.44	0.04	0.15	0.54	0.44
47	KILMARNOCK	0.05	0.21	0.43	1.11	2.84	0.91	0.08	1.88	0.24	8.40	1.27	0.40	1.00	1.03	0.15	0.12	0.31	0.37
48	KYLE-CARRICK	0.02	0.15	0.59	0.85	3.34	0.83	0.13	2.37	0.34	6.38	1.17	0.71	1.32	0.85	0.15	0.12	0.56	0.48
49	CUMNOCK-DOON	0.00	0.07	0.63	1.00	2.70	0.70	0.00	1.86	0.37	5.78	1.35	0.85	0.84	0.39	0.05	0.17	0.13	0.37
50	WIGTOWN	0.10	0.39	0.30	1.18	2.57	0.29	0.10	2.43	0.84	7.01	0.87	0.80	1.45	0.23	0.05	0.11	0.79	0.67
51	STEWARTRY	0.00	0.00	0.49	1.00	2.38	0.36	0.00	2.56	0.39	8.12	1.82	0.98	0.98	0.73	0.00	0.00	0.28	0.37
52	NITHSDALE	0.00	0.15	0.76	0.91	3.77	0.54	0.05	2.32	0.85	7.43	1.24	1.58	1.03	0.45	0.06	0.15	0.84	0.47
53	ANNANDALE	0.00	0.08	0.75	0.61	3.58	0.99	0.00	2.10	0.47	6.51	0.42	0.88	0.92	0.61	0.16	0.10	0.32	0.60
54	ORKNEY	0.17	0.16	0.00	0.67	1.37	0.83	0.00	1.74	0.31	5.75	1.80	0.83	0.99	0.15	0.50	0.00	0.47	0.29
55	SHETLAND	0.00	0.16	0.00	1.14	3.33	0.33	0.17	0.65	0.00	8.40	0.66	0.65	1.72	0.48	0.28	0.48	0.85	0.65
56	WESTERN ISLES	0.00	0.47	0.84	0.63	3.30	0.20	0.00	0.80	0.28	6.68	0.77	1.35	1.38	0.19	0.37	0.08	0.58	0.62
	ALL SCOTLAND	0.02	0.21	0.48	1.12	3.09	0.67	0.12	2.88	0.37	8.25	1.18	0.77	1.22	0.65	0.18	0.17	0.50	0.41

Table AI.5 Age-standardized incidence rates by other cancer site, in each local government district and islands area, Scotland, 1975-1980, by sex (male)

		TONG	SALG	OMQU	PHA1	PHA2	SINT	COLO	RECT	LIVE	GALL	NASA	BONE	CONN	KIDN	EYEZ	BRNS	LYLE	MYLE
1	CAITHNESS	0.8	2.9	0.0	0.8	0.0	2.1	25.2	6.1	0.0	0.0	1.0	0.0	3.6	2.2	1.0	3.0	4.3	2.3
2	SUTHERLAND	0.0	0.0	3.5	5.1	4.6	1.8	18.7	20.0	0.0	0.0	0.0	0.0	3.5	5.1	0.8	7.9	0.9	2.1
3	ROSS-CROMARTY	1.5	3.1	0.0	2.0	0.4	0.5	30.7	16.7	0.5	0.7	0.0	2.0	4.0	4.8	0.0	4.3	4.5	3.1
4	SKYE-LOCHALSH	4.0	1.9	4.0	0.0	0.0	0.0	30.9	19.2	8.2	1.1	0.0	0.0	0.0	5.2	0.0	6.3	1.1	1.4
5	LOCHABER	0.0	0.0	0.0	2.7	0.0	0.0	25.7	10.8	0.0	1.1	0.0	1.5	1.5	4.5	1.1	4.4	3.9	0.0
6	INVERNESS	0.4	0.5	1.7	1.4	1.0	1.5	29.3	15.3	1.4	3.0	0.5	0.7	0.9	7.4	1.0	2.6	5.0	3.2
7	BADENOCH	0.0	0.0	8.0	3.1	0.0	0.0	34.0	19.8	0.0	1.7	0.0	3.1	0.0	4.5	2.4	0.0	0.0	0.0
8	NAIRN	0.0	0.0	0.0	1.6	0.0	0.0	30.7	21.2	2.4	0.0	0.3	0.0	3.4	4.9	0.6	4.8	11.3	0.0
9	MORAY	1.6	2.1	3.0	1.2	1.1	1.2	20.8	15.6	2.1	0.7	0.3	0.4	3.2	3.4	0.6	5.3	2.6	3.2
10	BANFF-BUCHAN	0.0	2.4	1.5	1.4	0.4	1.2	24.3	14.9	2.8	1.6	0.0	1.2	2.8	4.0	0.7	7.2	4.2	2.6
11	GORDON	2.5	0.6	1.1	0.7	0.9	0.0	25.1	14.1	1.9	1.1	2.3	1.1	1.1	7.2	0.4	6.7	4.8	2.7
12	ABERDEEN	0.5	1.3	1.1	0.8	0.1	1.3	26.7	17.4	2.8	1.1	0.5	1.4	1.6	7.4	0.9	6.2	3.6	2.8
13	KINCARDINE	0.6	1.8	1.3	1.1	0.0	0.9	29.0	11.9	1.5	1.3	0.7	0.0	1.1	4.5	0.0	7.7	3.2	2.4
14	ANGUS	1.0	0.5	3.1	1.6	0.3	0.5	26.4	14.6	0.5	1.3	0.3	0.8	1.5	5.5	0.2	7.2	3.9	6.6
15	DUNDEE	1.1	0.6	1.7	1.6	0.9	0.7	20.7	15.3	3.2	0.6	0.8	1.3	2.2	5.9	0.9	6.5	3.0	3.0
16	PERTH-KINROSS	0.6	1.8	2.3	1.4	0.5	0.1	24.5	13.9	0.7	1.6	0.4	1.0	3.2	6.5	1.4	3.7	3.0	2.1
17	KIRKCALDY	0.8	0.7	1.1	1.4	0.6	0.0	19.3	12.5	1.1	1.6	0.4	0.4	1.8	6.6	0.5	6.3	3.7	1.7
18	N-E FIFE	0.5	0.7	0.9	2.3	0.8	1.5	17.4	14.7	0.6	1.9	0.3	1.4	1.9	6.8	0.5	5.0	2.7	2.6
19	DUNFERMLINE	1.0	2.0	1.1	0.8	0.2	0.2	17.4	7.8	1.4	0.5	0.8	1.1	1.4	3.0	1.0	7.5	4.6	2.8
20	WEST LOTHIAN	1.0	2.5	2.2	1.6	0.3	0.5	17.2	13.4	1.8	0.2	0.8	0.8	1.5	7.0	0.0	8.3	3.7	2.9
21	EDINBURGH	1.4	1.9	2.1	1.9	0.9	1.5	20.2	13.3	3.8	1.8	0.3	1.4	1.7	6.2	0.5	8.0	3.5	3.0
22	MIDLOTHIAN	1.0	1.3	1.6	1.1	0.3	0.8	22.0	13.4	1.1	0.7	0.8	0.7	1.6	5.1	1.4	6.8	2.9	2.9
23	EAST LOTHIAN	2.6	1.8	2.2	0.8	0.3	0.2	17.8	13.7	1.8	0.5	0.3	0.6	1.8	5.0	0.3	7.2	3.5	1.4
24	TWEEDDALE	2.0	0.0	4.0	1.4	0.0	2.5	26.9	28.3	3.7	2.7	1.7	0.0	3.4	4.7	0.0	0.0	0.0	4.3
25	ETTRICK	1.3	2.6	1.4	1.5	1.0	0.6	14.6	13.0	2.8	0.5	0.0	0.8	2.5	6.8	1.0	1.7	4.8	3.1
26	ROXBURGH	0.8	5.9	3.0	1.1	0.0	0.0	16.1	12.3	1.5	0.6	0.0	0.0	1.7	7.0	0.8	8.6	3.4	4.1
27	BERWICKSHIRE	1.7	0.0	0.7	0.0	0.0	2.1	13.7	14.2	0.7	4.1	0.0	0.0	1.5	8.2	0.0	3.9	2.7	5.3
28	CLACKMANNAN	1.1	1.1	1.7	2.8	0.0	0.0	14.7	13.4	3.2	1.2	0.6	2.3	0.0	6.7	0.0	5.3	7.3	1.1
29	STIRLING	0.5	1.7	2.0	0.4	0.7	0.2	17.5	13.4	1.5	1.3	0.7	0.4	3.3	4.3	1.1	5.7	3.0	2.1
30	FALKIRK	0.8	0.8	1.9	1.5	0.4	0.1	16.7	12.9	1.5	1.8	0.7	0.7	1.9	4.8	0.5	3.8	3.1	4.6
31	ARGYLL-BUTE	3.2	1.1	1.9	2.0	1.3	0.3	19.8	12.3	0.3	1.1	0.0	0.2	2.3	3.8	0.8	5.0	3.1	4.7
32	DUMBARTON	1.2	0.6	2.0	1.0	0.4	0.6	22.7	12.0	0.8	3.8	0.3	1.9	1.1	6.4	0.4	5.7	3.5	2.6
33	GLASGOW	1.0	0.5	1.8	1.8	0.6	0.5	21.9	14.4	2.1	2.0	0.6	0.7	1.4	6.2	0.8	5.8	3.5	3.3
34	CLYDEBANK	0.4	0.9	2.3	2.3	0.0	1.2	21.4	10.3	0.4	1.5	0.7	3.4	0.4	8.0	1.2	7.9	2.6	4.5
35	BEARSDEN	2.0	0.9	1.7	0.7	0.0	0.0	20.6	8.9	2.0	0.0	1.8	0.8	1.3	4.7	0.8	6.1	3.3	2.6
36	STRATHKELVIN	0.4	0.3	1.4	1.4	0.4	0.8	23.7	13.6	1.8	1.8	0.5	1.1	0.7	7.9	0.4	5.3	5.9	5.0
37	CUMBERNAULD	0.7	0.0	1.6	0.5	0.0	0.8	21.5	10.6	2.9	1.7	0.0	0.6	2.7	4.5	1.6	4.9	1.3	1.2
38	MONKLANDS	3.3	0.7	1.6	3.6	0.0	0.0	20.1	9.8	2.5	1.1	1.8	0.8	0.8	6.3	1.2	5.3	1.3	3.9
39	MOTHERWELL	0.9	0.4	1.4	1.4	0.2	0.3	17.9	15.4	1.9	1.1	0.7	0.8	0.5	4.1	0.3	6.0	2.1	2.7
40	HAMILTON	0.3	0.6	1.3	1.5	0.3	0.8	16.3	17.3	1.6	2.5	0.5	1.2	1.3	4.3	1.1	5.5	4.5	1.5
41	EAST KILBRIDE	1.5	0.7	2.3	0.8	0.5	0.0	16.4	12.2	0.3	1.3	0.9	2.4	0.7	4.9	1.8	5.5	1.0	4.2
42	EASTWOOD	1.2	0.0	0.6	0.8	0.0	0.0	25.4	9.5	1.9	1.3	1.3	1.7	0.8	7.9	0.8	5.4	3.6	2.1
43	CLYDESDALE	1.3	0.8	0.4	0.5	0.0	0.8	18.7	8.8	1.9	0.5	0.0	0.0	2.1	3.7	0.7	4.7	2.6	2.1
44	RENFREW	0.8	0.3	1.9	1.0	0.1	0.7	19.0	11.3	1.2	1.1	0.9	0.5	0.9	4.6	0.8	5.5	4.0	1.7
45	INVERCLYDE	0.8	0.9	1.6	1.3	0.4	0.4	19.1	14.7	0.4	0.9	0.6	0.5	0.3	7.1	2.5	4.6	2.8	1.6
46	CUNNINGHAME	1.8	0.7	1.1	1.2	0.0	1.4	18.5	14.4	1.4	2.7	0.2	1.3	2.1	4.5	0.7	4.8	3.0	3.0
47	KILMARNOCK	0.6	0.4	1.5	1.0	0.3	0.0	21.7	9.2	1.4	0.7	0.0	1.5	1.2	5.2	1.0	4.8	4.4	2.7
48	KYLE-CARRICK	1.8	1.1	1.3	1.6	0.2	0.4	22.7	9.4	1.1	1.7	0.9	1.2	1.7	4.3	0.9	6.7	2.6	2.5
49	CUMNOCK-DOON	0.0	0.6	0.0	1.5	0.3	0.6	17.5	10.5	1.6	0.5	0.0	0.7	0.6	4.4	0.6	4.4	4.5	4.0
50	WIGTOWN	1.8	0.7	1.4	2.5	0.0	1.2	23.8	7.3	1.8	2.9	0.0	0.0	3.1	4.5	0.6	9.2	1.0	4.0
51	STEWARTRY	0.0	0.9	2.6	2.4	1.3	0.0	18.2	9.6	1.6	0.0	0.0	0.0	2.6	1.3	0.4	6.1	2.9	5.3
52	NITHSDALE	0.5	0.8	1.1	2.6	0.0	1.6	22.3	14.2	0.9	2.7	0.2	0.0	2.0	3.8	0.0	6.9	5.0	2.8
53	ANNANDALE	0.6	0.6	2.6	0.9	0.0	0.0	16.3	8.5	1.4	3.2	0.0	1.3	2.0	1.3	0.4	8.9	4.5	2.1
54	ORKNEY	1.3	0.0	0.0	0.0	2.6	3.0	10.8	10.4	0.9	4.9	0.9	1.3	0.0	4.0	1.9	1.3	5.9	0.0
55	SHETLAND	2.8	0.0	2.8	0.0	0.0	0.0	22.6	12.3	2.3	0.4	0.0	0.0	1.6	2.3	0.9	5.2	0.7	3.4
56	WESTERN ISLES	0.8	1.4	1.5	2.5	0.0	1.0	15.4	8.7	1.2	0.4	0.6	0.0	0.8	4.3	2.4	6.8	6.9	3.3
	ALL SCOTLAND	1.0	1.0	1.7	1.5	0.4	0.6	20.9	13.3	1.9	1.5	0.5	0.9	1.6	5.5	0.8	6.0	3.5	2.8

(female)

	MYLE	LYLE	BRNS	EYEZ	KIDN	OUFG	CONN	BONE	NASA	GALL	LIVE	RECT	COLO	SINT	PHA2	PHA1	OMOU	SALG	TONG
1 CAITHNESS	1.5	3.6	4.7	2.2	2.2	1.0	2.2	0.6	0.0	0.0	0.0	11.0	23.8	0.0	0.0	0.0	0.8	1.6	1.2
2 SUTHERLAND	0.6	0.0	9.1	0.6	0.0	1.5	1.2	0.0	0.0	3.0	3.0	4.3	20.0	0.0	0.0	0.0	0.6	1.8	0.9
3 ROSS-CROMARTY	3.4	2.2	6.6	0.4	1.1	4.6	4.2	0.7	0.0	0.7	0.0	11.5	28.8	0.9	0.0	1.5	0.5	5.2	1.1
4 SKYE-LOCHALSH	6.3	3.2	4.1	0.0	0.0	0.6	3.6	0.0	0.8	3.2	1.0	11.0	27.6	0.0	0.0	1.5	0.0	3.8	0.8
5 LOCHABER	0.0	0.6	5.1	1.9	5.5	0.5	2.1	0.0	0.6	0.6	0.0	7.8	26.4	0.0	0.0	0.0	0.0	0.0	0.0
6 INVERNESS	1.6	0.6	4.8	0.0	5.1	3.5	0.7	0.9	0.0	0.3	0.0	8.5	18.3	0.8	0.0	1.7	2.7	0.0	0.0
7 BADENOCH	0.0	6.5	3.5	0.7	0.0	3.2	2.4	0.0	0.0	0.9	2.3	22.1	16.0	0.0	3.5	0.7	0.9	2.4	0.9
8 NAIRN	0.0	0.9	6.7	0.4	5.0	2.5	1.4	0.0	0.4	2.6	1.6	11.6	25.9	0.3	0.0	0.9	0.6	1.7	0.0
9 MORAY	4.4	3.0	5.1	1.7	1.8	2.1	1.0	0.9	0.4	3.2	0.9	12.1	23.9	0.4	0.0	0.9	0.6	2.2	1.0
10 BANFF-BUCHAN	2.1	2.2	3.3	1.9	3.3	2.5	0.2	1.0	0.5	1.6	0.0	8.9	20.8	0.4	0.0	0.6	1.2	2.6	0.0
11 GORDON	2.0	2.2	4.0	0.8	4.4	2.0	3.0	0.2	0.6	3.2	1.0	9.1	20.4	0.4	0.0	0.5	1.5	1.3	0.6
12 ABERDEEN	1.6	2.3	6.2	0.6	1.8	1.7	2.8	1.2	0.6	1.6	0.4	10.6	24.7	0.7	0.0	0.8	0.9	2.2	0.8
13 KINCARDINE	3.2	3.7	4.0	1.4	4.4	0.9	1.7	1.0	0.6	1.1	1.0	13.3	26.3	0.9	0.0	0.0	1.3	3.0	0.5
14 ANGUS	2.5	1.5	6.2	1.1	1.8	3.8	1.4	0.8	0.3	2.2	0.4	8.5	23.6	0.5	0.3	0.4	0.8	0.5	0.3
15 DUNDEE	2.0	1.3	2.5	1.1	1.2	2.0	1.5	0.7	0.0	1.6	1.0	6.9	22.0	0.4	0.5	0.7	0.5	0.7	0.2
16 PERTH-KINROSS	1.8	1.6	4.9	0.3	3.0	2.2	1.3	1.2	0.1	1.2	0.8	7.3	20.7	0.3	0.3	1.2	0.4	2.6	0.5
17 KIRKCALDY	2.1	2.5	5.7	1.9	3.6	2.8	1.9	1.5	0.7	2.3	1.2	7.9	22.0	0.5	0.5	0.6	0.5	0.8	0.1
18 N-E FIFE	1.7	0.6	6.2	0.2	3.2	2.2	2.2	1.7	0.5	0.9	0.8	11.5	20.6	0.2	0.1	0.8	0.4	2.6	0.4
19 DUNFERMLINE	1.5	1.8	3.7	0.7	2.6	2.1	1.5	0.3	0.3	2.4	1.2	7.5	17.9	0.5	0.2	0.6	0.5	0.8	0.5
20 WEST LOTHIAN	2.6	2.1	7.2	0.8	4.2	2.9	1.9	1.5	0.2	1.4	0.8	9.7	18.9	1.0	0.1	2.2	1.0	2.9	0.3
21 EDINBURGH	1.6	2.4	6.5	0.0	3.5	3.3	2.1	0.8	0.3	2.4	2.1	8.4	16.9	0.3	0.5	0.6	0.8	2.2	0.6
22 MIDLOTHIAN	1.6	1.0	3.7	1.3	4.2	2.0	1.5	1.1	0.8	1.4	1.0	8.3	18.9	0.7	0.0	1.6	0.6	1.8	0.3
23 EAST LOTHIAN	5.7	1.1	7.1	0.6	3.0	3.7	0.7	0.8	0.5	0.3	0.6	3.3	11.0	1.1	0.0	0.6	0.4	3.3	0.6
24 TWEEDDALE	2.0	0.9	7.0	1.0	3.1	2.1	2.1	1.4	0.0	1.3	1.1	8.0	20.7	0.5	0.0	2.4	0.5	3.1	0.0
25 ETTRICK	1.0	2.5	7.0	0.6	4.0	0.8	1.4	0.0	0.4	0.4	0.0	5.9	21.4	0.4	0.0	0.9	0.0	1.6	0.2
26 ROXBURGH	1.6	3.6	6.2	1.0	3.6	4.5	2.5	0.8	0.0	1.7	1.1	6.0	16.6	2.2	0.0	1.4	0.0	0.9	1.5
27 BERWICKSHIRE	2.8	2.6	3.1	0.6	2.5	0.6	1.8	0.8	0.0	1.3	1.2	10.4	21.6	0.4	0.0	0.9	1.0	1.1	0.0
28 CLACKMANNAN	2.0	1.6	3.3	0.7	2.3	2.1	1.8	0.8	0.5	1.3	0.4	9.5	19.9	0.3	0.4	1.4	0.0	0.9	0.0
29 STIRLING	3.1	1.6	3.1	1.3	3.2	2.0	0.4	0.8	0.6	2.8	0.7	8.7	18.6	0.2	0.3	0.9	1.0	1.1	0.2
30 FALKIRK	5.6	2.4	4.3	0.5	3.4	0.9	1.5	0.8	1.1	1.3	1.4	7.9	15.9	0.2	0.4	0.1	2.0	0.0	0.5
31 ARGYLL-BUTE	3.1	1.9	4.2	0.4	3.6	1.2	0.8	0.6	0.4	1.7	1.0	8.4	21.9	0.7	0.3	0.7	0.6	0.5	0.7
32 DUMBARTON	1.4	2.4	2.3	0.9	2.7	1.6	1.2	0.2	0.9	1.7	0.7	7.6	18.5	0.4	0.2	0.9	0.6	0.3	0.5
33 GLASGOW	2.3	1.9	2.9	1.2	2.0	1.3	0.7	0.6	1.0	1.2	1.7	14.0	16.9	0.3	0.6	1.2	1.1	1.4	0.7
34 CLYDEBANK	1.3	0.6	3.1	0.3	3.4	2.5	1.5	0.9	0.8	3.5	0.3	9.5	16.8	0.2	0.5	0.6	0.7	0.4	0.0
35 BEARSDEN	3.7	4.0	4.0	2.1	4.0	1.7	0.4	1.1	1.0	1.4	0.0	9.6	18.7	1.4	0.5	1.1	0.8	0.4	0.6
36 STRATHKELVIN	2.4	2.5	5.1	0.5	2.5	2.0	1.0	0.2	0.6	1.7	0.8	6.0	15.8	0.2	0.1	1.3	0.7	0.6	1.1
37 CUMBERNAULD	1.8	1.9	6.1	0.2	2.9	1.5	0.6	0.3	0.3	3.6	0.2	7.1	16.3	0.9	0.7	0.3	0.8	0.3	0.9
38 MONKLANDS	1.7	1.6	3.3	1.1	3.8	1.7	0.6	1.1	0.6	2.2	1.8	7.2	20.2	0.3	0.7	1.1	0.6	0.3	1.3
39 MOTHERWELL	3.0	1.9	4.4	0.2	2.7	0.8	0.1	1.5	1.1	2.2	0.9	6.1	16.3	1.2	0.3	0.9	0.8	0.5	0.8
40 HAMILTON	0.8	1.7	3.2	0.8	4.2	1.2	1.0	0.4	0.2	4.3	1.0	4.3	14.8	0.5	0.4	0.8	0.6	0.2	0.7
41 EAST KILBRIDE	2.4	1.3	4.4	1.2	3.0	2.3	0.8	0.7	0.5	1.8	0.6	7.2	17.6	0.6	0.3	1.1	0.1	0.0	0.3
42 EASTWOOD	1.6	1.9	3.2	0.7	3.4	1.6	0.8	0.7	0.3	1.7	0.5	6.8	17.0	0.4	0.3	0.8	0.6	0.5	0.6
43 CLYDESDALE	3.8	1.7	5.9	1.2	2.0	1.6	1.0	0.4	0.7	0.9	1.6	8.4	17.5	0.1	0.0	1.4	0.2	0.2	0.4
44 RENFREW	2.5	1.9	3.5	0.7	2.3	2.4	0.8	1.6	0.6	0.9	1.2	7.3	21.6	0.4	0.8	0.8	1.0	1.0	0.0
45 INVERCLYDE	3.2	1.8	1.4	1.4	2.3	1.5	1.0	0.3	0.3	0.9	0.7	8.4	16.8	0.1	0.1	0.5	0.6	0.2	0.7
46 CUNNINGHAME	4.1	1.9	5.1	0.8	2.0	2.1	1.0	0.5	0.0	2.9	0.0	5.3	16.8	0.0	1.2	0.9	1.0	0.5	0.0
47 KILMARNOCK	0.9	3.7	6.1	0.0	7.1	3.2	1.7	0.0	0.3	1.1	0.5	5.2	16.6	0.3	0.0	0.6	0.1	0.7	1.1
48 KYLE-CARRICK	1.7	2.6	3.3	0.0	3.8	3.6	1.0	0.5	0.7	3.1	1.2	8.0	27.1	1.0	0.0	0.9	0.6	1.3	0.6
49 CUMNOCK-DOON	2.4	2.7	1.4	1.4	2.3	1.5	1.0	0.0	0.3	0.9	0.7	8.4	18.8	0.0	0.0	0.5	1.0	0.0	0.0
50 WIGTOWN	1.6	0.5	1.4	0.4	2.0	1.5	1.0	0.5	0.3	0.9	1.6	5.3	16.8	0.3	0.0	0.5	1.0	0.5	0.7
51 STEWARTRY	3.8	0.5	5.9	0.0	7.1	2.1	1.0	0.0	0.0	1.1	1.2	3.6	16.6	2.9	1.2	0.9	1.0	0.7	0.0
52 NITHSDALE	2.5	1.5	3.5	1.2	3.8	3.2	1.7	1.9	0.0	3.1	0.0	8.0	27.1	1.0	0.0	0.6	0.2	1.3	1.1
53 ANNANDALE	3.2	1.8	1.3	0.0	4.1	3.6	2.7	1.6	0.0	2.2	0.5	5.8	5.3	1.8	0.0	0.3	0.7	0.5	0.0
54 ORKNEY	4.1	1.8	2.0	2.0	6.8	2.8	0.4	0.0	0.0	0.9	1.2	3.6	5.3	0.0	0.0	0.0	1.7	3.8	1.2
55 SHETLAND	0.9	2.9	2.7	0.5	2.7	0.9	4.8	0.0	0.0	0.9	0.9	5.8	23.5	1.8	0.0	0.0	0.7	3.9	0.4
56 WESTERN ISLES	1.7	4.4	5.3	0.0	6.1	2.1	2.4	0.0	0.4	0.5	0.2	9.7	15.8	1.0	0.4	1.5	1.0	3.1	1.0
ALL SCOTLAND	2.1	2.0	4.5	0.8	3.0	1.9	1.3	0.8	0.4	1.7	0.8	8.1	19.3	0.5	0.2	0.8	0.7	1.1	0.5

Table AI.6 Rank of age-standardized incidence rates by other cancer site, in each local government district and Islands area, Scotland, 1975-1980, by sex (male)

#		TONG	SALG	OMOU	PHA1	PHA2	SINT	COLO	RECT	LIVE	GALL	NASA	BONE	CONN	KIDN	EYEZ	BRNS	LYLE	MYLE
1	CAITHNESS	29	3	51	43	32	3	12	56	53	51	5	43	2	54	14	51	18	37
2	SUTHERLAND	48	46	4	1	1	5	36	3	53	51	35	43	3	24	23	6	54	38
3	ROSS-CROMARTY	13	2	51	12	16	28	3	8	46	38	35	5	1	29	47	47	14	19
4	SKYE-LOCHALSH	1	9	2	53	32	41	2	5	53	31	35	43	53	22	47	19	51	47
5	LOCHABER	48	46	51	5	32	7	10	40	1	51	23	9	30	34	11	45	22	52
6	INVERNESS	44	39	23	26	5	7	5	11	30	5	35	30	42	5	14	52	7	17
7	BADENOCH	48	46	1	3	32	41	1	4	53	17	23	2	53	34	2	55	56	52
8	NAIRN	48	46	51	17	32	41	3	2	12	51	35	43	7	27	47	41	1	52
9	MORAY	11	7	6	33	4	11	27	9	14	38	28	39	7	51	33	33	42	17
10	BANFF-BUCHAN	48	6	30	26	16	11	15	13	7	21	35	16	10	46	30	10	19	31
11	GORDON	5	34	42	47	7	41	13	21	17	31	1	20	39	7	40	15	10	28
12	ABERDEEN	40	17	42	43	30	10	8	6	7	17	23	11	27	5	19	21	26	26
13	KINCARDINE	36	11	37	35	32	16	6	38	26	25	13	43	39	34	47	8	33	36
14	ANGUS	23	39	5	17	22	28	9	16	46	25	28	25	30	21	46	10	22	1
15	DUNDEE	21	34	23	17	7	22	28	11	4	42	11	13	15	20	19	18	35	21
16	PERTH-KINROSS	36	14	11	26	14	39	14	22	43	42	13	23	7	15	7	50	9	38
17	KIRKCALDY	29	28	42	35	11	41	33	32	35	21	27	39	22	14	36	19	24	43
18	N-E FIFE	40	28	47	10	11	32	44	14	45	13	28	11	20	11	36	38	40	31
19	DUNFERMLINE	48	8	42	43	27	7	44	54	30	45	11	20	34	52	14	4	12	14
20	WEST LOTHIAN	23	5	14	17	22	37	46	25	20	50	33	25	30	9	47	9	24	24
21	EDINBURGH	15	9	16	14	7	28	30	28	2	14	28	16	24	18	36	5	27	21
22	MIDLOTHIAN	23	17	27	35	32	18	22	28	35	38	35	30	27	24	7	13	37	24
23	EAST LOTHIAN	36	11	14	49	22	18	41	23	3	11	3	35	22	26	44	10	27	47
24	TWEEDDALE	6	46	2	26	32	2	7	1	7	8	6	43	4	31	47	55	2	7
25	ETTRICK	16	4	34	22	32	24	54	30	7	45	35	43	13	11	14	53	10	19
26	ROXBURGH	29	1	6	35	5	41	51	33	26	42	35	25	24	9	23	3	31	9
27	BERWICKSHIRE	10	46	48	53	32	3	55	19	43	2	35	43	30	1	47	48	40	2
28	CLACKMANNAN	21	19	23	4	32	41	53	25	4	29	19	4	53	13	47	33	3	50
29	STIRLING	40	13	17	52	16	37	42	31	26	25	13	39	6	40	11	27	35	38
30	FALKIRK	29	24	20	22	10	39	47	33	26	14	13	30	20	29	36	49	34	51
31	ARGYLL-BUTE	3	14	17	12	18	35	32	37	51	31	35	42	14	48	23	38	46	5
32	DUMBARTON	19	34	17	39	16	24	18	17	41	3	28	6	39	16	40	27	27	31
33	GLASGOW	23	39	22	16	11	28	23	17	14	11	19	30	34	18	23	26	42	14
34	CLYDEBANK	44	22	11	10	32	11	26	44	49	24	6	1	51	18	9	6	42	6
35	BEARSDEN	6	24	30	47	32	41	29	50	16	51	13	25	36	31	23	22	32	31
36	STRATHKELVIN	44	44	23	26	32	41	17	24	20	14	2	20	47	3	40	33	5	4
37	CUMBERNAULD	35	46	34	50	32	41	25	41	6	17	35	35	11	34	6	40	48	49
38	MONKLANDS	2	28	27	32	2	24	31	45	11	36	23	7	44	17	9	33	48	11
39	MOTHERWELL	27	42	40	26	27	35	40	10	17	31	13	25	50	45	44	24	46	28
40	HAMILTON	47	34	37	22	22	18	49	7	24	10	23	16	36	40	11	29	14	46
41	EAST KILBRIDE	13	28	11	43	14	41	48	36	51	25	6	3	47	27	5	28	53	8
42	EASTWOOD	19	46	49	14	32	41	11	47	20	21	4	8	44	27	23	32	51	12
43	CLYDESDALE	16	24	50	50	32	18	36	51	17	45	35	43	16	50	30	43	42	38
44	RENFREW	29	44	27	39	30	32	35	39	33	37	6	37	42	33	23	29	21	43
45	INVERCLYDE	29	46	20	32	16	22	34	14	41	29	19	13	52	8	1	24	39	21
46	CUNNINGHAME	8	28	40	33	32	9	38	17	30	38	33	13	16	34	30	44	19	21
47	KILMARNOCK	38	42	30	39	22	41	24	49	41	38	35	9	38	22	14	41	17	28
48	KYLE-CARRICK	11	19	37	17	27	32	18	48	30	17	6	16	24	40	19	15	42	35
49	CUMNOCK-DOON	48	21	51	22	32	24	42	42	35	45	35	30	49	40	33	45	12	10
50	WIGTOWN	8	28	34	7	32	11	16	55	20	7	35	43	9	34	33	1	37	52
51	STEWARTRY	48	22	9	9	3	41	39	46	24	51	35	43	12	55	47	22	7	2
52	NITHSDALE	40	24	42	6	16	16	21	19	39	8	35	43	18	48	40	17	14	26
53	ANNANDALE	48	34	9	42	32	6	49	53	46	5	35	23	18	55	47	2	48	38
54	ORKNEY	18	48	51	53	2	1	56	43	39	4	35	13	53	46	4	54	5	52
55	SHETLAND	4	48	8	53	32	41	20	33	13	1	35	43	27	53	19	37	55	13
56	WESTERN ISLES	27	18	30	7	32	15	52	52	33	49	19	43	44	40	2	13	4	14

(female)

	TONG	SALG	OMOU	PHA1	PHA2	SINT	COLO	RECT	LIVE	GALL	NASA	BONE	CONN	OUFG	KIDN	EYEZ	BRNS	LYLE	MYLE
1 CAITHNESS	4	20	18	46	25	44	9	8	44	56	40	32	10	48	38	1	22	7	43
2 SUTHERLAND	11	17	27	46	25	44	25	52	1	8	40	43	29	38	53	27	1	49	53
3 ROSS-CROMARTY	8	1	35	5	25	9	1	8	44	49	40	27	28	1	49	36	8	25	9
4 SKYE-LOCHALSH	44	3	49	5	25	44	2	8	15	5	40	43	3	53	53	48	28	8	1
5 LOCHABER	14	33	49	46	25	44	4	33	44	50	8	43	12	56	4	4	15	48	53
6 INVERNESS	44	51	1	3	25	13	38	22	44	54	12	17	41	8	5	48	21	49	36
7 BADENOCH	44	13	14	30	25	44	49	1	2	43	40	43	8	8	53	48	36	1	53
8 NAIRN	11	19	49	16	1	44	8	5	44	11	40	43	25	14	8	7	56	46	4
9 MORAY	9	14	27	16	25	33	8	4	6	29	23	17	31	20	51	36	5	10	24
10 BANFF-BUCHAN	44	11	8	33	25	28	19	20	22	5	23	15	51	14	45	4	15	25	26
11 GORDON	23	24	4	38	25	17	23	19	44	28	18	15	4	28	22	19	39	25	38
12 ABERDEEN	14	14	14	25	25	28	7	10	15	18	28	40	5	32	7	27	29	24	10
13 KINCARDINE	27	8	5	46	25	14	5	3	37	41	12	9	18	48	45	8	8	5	10
14 ANGUS	36	35	18	41	25	9	10	22	15	15	40	15	25	3	48	14	48	40	18
15 DUNDEE	40	43	35	30	8	21	20	36	25	28	28	20	20	28	28	40	20	42	28
16 PERTH-KINROSS	27	31	35	11	21	26	13	42	25	39	39	9	28	19	16	27	36	37	28
17 KIRKCALDY	43	11	42	25	8	33	22	31	9	14	10	27	15	11	23	4	13	18	31
18 N-E FIFE	33	30	35	33	25	39	37	8	25	43	18	5	10	20	41	42	8	49	24
19 DUNFERMLINE	27	20	45	46	25	21	53	35	40	12	36	2	20	10	33	42	33	34	20
20 WEST LOTHIAN	36	10	9	2	19	7	27	14	3	20	28	37	15	7	8	24	23	9	33
21 EDINBURGH	23	14	18	33	21	33	27	24	15	32	38	5	12	28	18	19	2	25	43
22 MIDLOTHIAN	2	17	35	38	8	21	27	28	32	12	28	20	20	45	8	19	7	21	17
23 EAST LOTHIAN	44	8	27	4	25	14	27	13	13	32	8	11	38	42	26	48	33	45	45
24 TWEEDDALE	36	5	42	46	25	6	55	55	44	54	40	20	41	4	25	19	35	2	36
25 ETTRICK	44	6	35	1	25	21	20	28	44	36	18	8	12	20	12	10	3	48	36
26 ROXBURGH	40	20	49	48	25	26	17	47	13	32	40	43	25	51	18	27	4	18	2
27 BERWICKSHIRE	1	28	49	18	25	2	45	56	44	53	40	43	7	2	53	16	55	10	28
28 CLACKMANNAN	44	28	49	8	10	44	15	11	9	23	40	20	17	53	34	17	8	15	48
29 STIRLING	44	28	9	18	25	44	28	17	37	36	18	20	52	20	37	27	43	15	36
30 FALKIRK	40	47	47	18	15	28	33	21	29	20	12	20	48	28	23	24	39	37	18
31 ARGYLL-BUTE	27	51	2	45	10	39	50	31	8	10	1	20	20	48	47	10	50	12	4
32 DUMBARTON	17	35	27	30	15	14	14	24	15	36	40	43	38	42	16	32	25	21	28
33 GLASGOW	27	43	27	16	14	26	35	34	28	23	23	32	28	38	30	36	49	28	13
34 CLYDEBANK	27	51	8	33	35	9	18	12	37	23	5	40	41	41	41	17	28	56	3
35 BEARSDEN	17	23	22	11	5	33	42	2	5	39	4	37	20	14	20	12	46	49	48
36 STRATHKELVIN	44	41	22	41	39	39	43	17	40	4	12	11	48	32	12	46	4	4	13
37 CUMBERNAULD	44	51	18	46	10	4	38	18	44	32	8	11	31	26	34	40	43	18	45
38 MONKLANDS	23	41	42	13	8	17	32	48	15	23	28	40	52	53	41	2	29	28	38
39 MOTHERWELL	8	33	22	10	25	9	51	41	25	23	12	37	44	38	29	32	43	42	23
40 HAMILTON	11	43	18	43	21	39	47	38	42	3	14	11	44	32	14	32	15	37	47
41 EAST KILBRIDE	17	43	27	13	39	33	24	44	4	15	28	5	44	51	30	42	11	28	8
42 EASTWOOD	3	51	35	25	33	5	47	45	22	18	1	35	55	32	49	14	15	38	20
43 CLYDESDALE	14	35	18	48	5	21	54	52	15	1	18	27	13	13	8	48	39	49	31
44 RENFREW	17	47	27	18	21	17	38	38	32	22	38	27	52	42	26	42	24	42	33
45 INVERCLYDE	17	47	8	15	17	44	33	38	35	23	28	27	42	18	34	19	42	28	15
46 CUNNINGHAME	38	27	47	25	44	44	41	43	8	50	10	35	31	31	41	12	31	5	52
47 KILMARNOCK	23	35	9	8	44	28	40	24	9	43	8	3	38	38	37	24	36	15	20
48 KYLE-CARRICK	33	31	49	25	43	43	15	38	28	43	40	37	31	17	40	46	52	21	38
49 CUMNOCK-DOON	44	24	9	38	44	44	31	24	44	48	12	43	31	38	40	8	15	14	7
50 WIGTOWN	17	35	9	18	44	44	43	49	44	9	23	34	31	20	1	38	54	54	48
51 STEWARTRY	44	31	49	16	33	33	45	50	35	41	28	43	55	8	14	19	31	40	18
52 NITHSDALE	8	24	45	33	7	7	3	28	44	7	40	1	18	5	11	48	53	55	10
53 ANNANDALE	44	35	22	43	17	17	12	50	9	2	28	3	8	11	52	48	50	12	8
54 ORKNEY	4	3	3	48	1	1	58	54	22	15	40	43	48	46	2	3	47	3	51
55 SHETLAND	33	2	22	48	3	3	11	48	42	43	40	43	1	48	30	32	14	55	33
56 WESTERN ISLES	9	8	9	5	10	44	51	14	42	52	23	43	8	20	3	48	14	3	33

Table AI.7 District codes ordered by rank of age-standardized incidence rates, Scotland, 1975-1980, for other cancer sites, by sex (male)

	TONG	SALG	OMOU	PHA1	PHA2	SINT	COLO	RECT	LIVE	GALL	NASA	BONE	CONN	KIDN	EYEZ	BRNS	LYLE	MYLE
1	4	28	7	2	2	54	7	24	5	55	11	34	3	27	45	50	8	14
2	38	3	4	38	54	24	4	8	21	27	38	7	1	34	56	53	24	27
3	31	1	24	7	51	27	8	2	23	32	23	41	2	42	7	28	28	51
4	55	25	2	28	9	1	3	7	28	54	42	28	24	36	54	18	58	38
5	11	20	14	5	26	2	6	4	15	53	1	3	8	12	41	21	38	31
6	35	10	28	52	8	53	13	12	37	8	48	32	29	8	37	2	54	34
7	24	9	9	56	6	6	24	40	24	50	44	38	9	6	18	34	6	24
8	50	18	55	50	11	19	12	3	12	52	24	42	18	11	22	13	51	41
9	46	21	53	51	21	48	14	9	10	24	41	5	50	45	38	20	18	28
10	27	4	51	18	15	12	5	39	25	40	34	47	10	28	34	10	11	49
11	9	23	41	34	30	9	1	8	38	23	15	18	37	20	29	14	25	38
12	48	13	34	3	18	10	11	15	8	33	19	12	51	18	40	23	49	42
13	41	29	16	31	17	34	16	10	55	18	13	46	25	25	5	58	19	55
14	3	31	23	21	33	50	10	45	9	21	30	54	31	28	8	22	52	56
15	21	18	20	42	16	56	18	18	33	36	18	15	15	17	19	11	40	33
16	54	56	21	33	41	13	50	14	35	30	35	10	43	32	1	48	3	19
17	25	22	28	14	3	52	38	33	11	12	39	40	48	38	47	52	47	9
18	43	12	31	15	29	43	48	48	39	7	29	48	53	33	25	15	1	8
19	32	39	32	20	32	40	32	52	28	37	28	21	52	21	55	17	10	3
20	42	28	28	38	10	23	55	27	29	48	33	38	18	15	12	4	46	25
21	15	48	44	8	45	22	52	11	30	42	56	11	30	14	15	12	44	21
22	28	49	30	48	40	45	22	16	19	17	45	19	17	47	48	35	33	15
23	14	51	33	25	5	15	33	23	18	23	38	18	23	4	42	51	28	46
24	33	52	36	49	7	32	47	36	36	10	40	53	28	2	26	45	35	20
25	20	30	15	30	23	25	37	28	40	34	12	28	48	22	35	39	17	22
26	39	35	8	40	47	38	34	20	51	14	8	39	21	23	2	33	12	52
27	56	56	28	18	20	38	9	28	13	41	17	35	22	41	44	28	32	12
28	30	43	38	8	18	49	15	22	28	13	14	20	12	8	33	32	23	47
29	44	46	22	39	48	3	35	21	30	29	9	14	55	30	31	44	21	11
30	26	17	45	24	39	14	21	25	8	46	32	33	27	3	43	40	33	39
31	1	50	56	36	12	33	38	30	19	28	18	38	5	24	46	41	28	35
32	17	18	47	10	44	21	31	17	47	11	21	8	20	35	10	42	35	18
33	45	38	10	45	13	18	17	28	44	44	20	49	14	44	50	28	13	32
34	37	41	35	46	4	48	45	55	58	39	48	22	18	13	9	38	30	10
35	23	40	50	25	5	44	44	31	22	31	2	23	33	37	49	38	29	48
36	47	37	37	22	7	39	43	41	49	38	54	37	40	48	17	38	31	13
37	16	32	40	17	8	31	17	32	17	45	55	44	35	50	18	55	37	32
38	13	53	13	13	56	20	2	13	48	9	4	45	47	7	21	18	53	1
39	12	11	48	44	22	29	48	44	52	22	10	9	13	5	30	31	15	18
40	29	15	13	32	43	30	51	5	54	47	8	17	11	48	11	37	50	29
41	52	8	48	47	46	16	39	37	32	3	3	28	32	56	52	47	22	43
42	18	33	39	13	49	5	23	49	46	16	5	31	44	40	32	8	45	2
43	6	39	17	44	50	8	49	54	27	28	7	8	6	49	36	43	27	53
44	34	47	52	32	52	51	29	34	18	15	53	4	56	48	48	46	18	17
45	36	36	11	47	53	7	19	38	18	19	43	2	38	28	39	46	43	44
46	40	44	19	29	55	55	18	51	53	25	31	13	42	39	23	5	48	45
47	53	42	12	55	42	11	20	42	3	49	47	55	38	10	14	49	34	40
48	5	2	18	35	38	37	30	48	14	43	27	52	41	54	4	3	39	4
49	2	55	27	11	37	38	41	47	34	56	28	51	49	52	53	27	31	23
50	7	54	42	23	24	30	53	35	45	20	25	50	39	31	3	30	37	37
51	49	24	43	43	25	16	28	43	41	8	7	43	34	43	8	18	53	28
52	19	7	54	37	27	42	58	34	31	2	53	58	45	9	27	1	38	30
53	51	19	3	29	28	28	58	56	7	51	43	24	54	19	24	6	42	54
54	8	27	5	55	34	17	28	53	2	35	22	25	4	55	20	25	4	50
55	10	45	8	27	35	47	25	19	4	5	50	27	7	1	51	54	41	8
56	22	8	49	4	1	28	54	50	1	1	51	1	28	53	13	7	2	7

(female)

#	TONG	SALG	OMOU	PHA1	PHA2	SINT	COLO	RECT	LIVE	GALL	NASA	BONE	CONN	OUFG	KIDN	EYEZ	BRNS	LYLE	MYLE
1	27	3	6	25	8	54	3	7	2	43	31	52	55	3	50	1	2	7	4
2	22	55	31	20	51	27	4	35	7	53	38	18	3	27	54	37	21	24	27
3	42	54	54	6	41	55	52	13	20	40	41	47	4	14	56	54	25	56	35
4		4	11	23	34	37	5	9	41	36	35	53	11	24	5	5	26	36	9
5	54	24	13	3	35	42	13	8	35	4	34	20	12	52	6	17	9	47	32
6	39	56	10	56	22	24	8	18	9	10	5	17	53	6	8	10	3	13	54
7	3	25	45	4	15	20	12	3	47	52	46	41	27	20	12	8	22	1	50
8	52	23	34	28	37	52	9	4	31	2	36	25	7	7	43	13	13	4	42
9	9	13	50	47	17	3	1	1	54		23	12	56	51	20	48	18	20	3
10	56	20	48	39	28	39	14	12	17	50	17	15	18	19	22	24	28	9	14
11	40	17	20	35	47	14	55	28	28	31	45	22	1	53	52	30	41	27	53
12	8	10	56	16	31	34	53	34	48	8	13	37	25	17	25	34	47	55	13
13	2	7	29	38	36	6	16	23	26	19	6	40	5	43	36	45	17	31	37
14	5	21	12	41	56	32	32	20	23	22	37	36	21	35	40	14	56	50	34
15	12	9	7	45	32	23	48	56	32	17	48	13	17	8	51	41	50	28	46
16	43	12	1	33	30	13	28	37	12	14	30	10	20	10	32	26	10	48	31
17	50	50	43	29	18	53	26	29	21	54	40	35	28	48	26	33	42	29	23
18	45	22	21	8	45	11	34	36	43	41	29	9	52	45	16	27	40	26	15
19	32	8	37	50	33	38	10	11	4	12	42	6	13	16	21	44	5	17	52
20	35	19	14	9	20	44	15	10	38	42	11	29	35	29	46	50	15	37	43
21	44	1	40	44	39	43	25	30	14	30	25	28	31	18	35	23	6	32	20
22	41	26	53	51	18	22	17	6	42	20	18	31	19	56	11	11	1	49	48
23	21	35	35	30	21	19	11	14	55	44	10	14	15	25	17	6	20	22	40
24	47	11	39	27	48	15	41	21	10	45	9	21	22	50	30	21	44	12	19
25	38	52	55	42	5	25	2	32	16	38	33	23	8	30	24	46	32	10	10
26	11	29	36	48	53	16	29	2	39	34	49	30	26	15	23	20	35	3	11
27	16	49	32	17	2	47	23	47	18	39	56	3	14	37	44	29	33	21	17
28	33	27	33	12	3	33	21	49	15	33	12	45	16	11	15	12	4	11	16
29	19	18	44	46	54	28	22	22	49	28	22	44	2	21	39	25	38	45	28
30	13	16	2	32	14	12	20	25	33	11	15	16	33	46	33	2	12	46	33
31	31	51	47	7	13	10	49	52	30	15	39	43	46	42	41	28	52	41	44
32	34		41	15	12	30	38	17	45	9	52	33	48	36	55	16	46	33	18
33	48	5	9	18	6	41	30	31	44	23	44	1	50	40	19	31	19	38	56
34	55	39	23	52	10	35	45	5	22	21	50	50	37	12	45	39	23	19	45
35	18	14	18	21	9	21	33	33	52	26	20	46	9	33	37	38	24	53	21
36	14	32	15	34	55	9	6	19	46	37	43	42	45	47	28	55	7	42	30
37	24	43	3	10	4	17	18	15	29	29	21	19	49	39	29	9	16	30	6
38	20	53	22	49	7	51	44	48	13	32	19	39	47	49	48	49	48	16	39
39	46	47	42	11	11	31	37	45	34	25	16	48	32	2	1	32	43	40	26
40	30	50	25	22	43	36	47	40	19	35	2	38	23	34	49	3	11	52	25
41	15	38	16	36	40	18	46	44	36	16	7	34	24	23	38	36	30	14	49
42	26	36	24	14	38	40	35	39	56	51	4	11	6	44	18	15	45	39	12
43	17	33	17	53	44	48	36	16	40	13	3	5	34	32	47	43	37	15	1
44	4	41	38	40	46	2	50	46	24	47	8	54	40	1	34	19	29	44	22
45	7	15	52	2	49	4	51	47	53	7	14	24	42	54	13	40	39	23	24
46	6	40	19	5	29	5	27	42	3	55	55	4	39	31	10	18	36	25	38
47	53	45	46	55	50	7	42	38	51	18	54	55	41	55	31	4	55	8	41
48	10	48	30	54	27	56	40	26	5	48	53	2	36	13	14	53	14	5	29
49	37	44	51	13	42	8	7	55	6	49	51	7	30	41	42	6	34	6	51
50	36	30	27	43	23	28	31	50	50	3	47	8	54	26	3	7	31	43	36
51	49	42	4	37	24	50	56	53	8	46	32	26	10	38	9	52	54	2	55
52	25	34	5	26	26	49	39	51	25	5	24	27	29	4	53	35	53	18	47
53	51	37	28	24	25	46	19	2	27	56	26	32	38	28	2	22	49	35	2
54	29	48	8	19	1	45	43	43	11	27	27	49	44	5	4	42	51	51	5
55	23	6	28	1		29	24	54	37	24	28	51	51		7	47	27	54	8
56	28	31				1	54	24	1	6	1	56	43		27	56	8	34	7

Table AI.8 Cumulative rates for ages 0-75 years, by other cancer site, in each local government district and Islands area, Scotland, 1975-1980, by sex (male)

		TONG	SALG	OMOU	PHA1	PHA2	SINT	COLO	RECT	LIVE	GALL	NASA	BONE	CONN	KIDN	EYEZ	BRNS	LYLE	MYLE
1	CAITHNESS	0.13	0.37	0.00	0.13	0.00	0.23	2.88	0.74	0.00	0.00	0.10	0.00	0.33	0.23	0.10	0.38	0.18	0.19
2	SUTHERLAND	0.00	0.00	0.50	0.56	0.47	0.00	2.01	2.85	0.00	0.00	0.00	0.00	0.50	0.51	0.00	0.58	0.00	0.26
3	ROSS-CROMARTY	0.15	0.21	0.00	0.28	0.00	0.13	3.11	1.90	0.13	0.08	0.00	0.09	0.37	0.64	0.00	0.59	0.36	0.30
4	SKYE-LOCHALSH	0.67	0.31	0.33	0.00	0.00	0.00	3.99	2.25	0.46	0.00	0.00	0.19	0.00	0.65	0.00	0.56	0.00	0.35
5	LOCHABER	0.00	0.00	0.00	0.49	0.00	0.00	2.78	1.54	0.22	0.00	0.00	0.04	0.15	0.52	0.13	0.49	0.30	0.00
6	INVERNESS	0.11	0.04	0.20	0.22	0.12	0.24	2.87	2.22	0.00	0.41	0.08	0.38	0.15	0.94	0.40	0.32	0.57	0.33
7	BADENOCH	0.00	0.00	1.26	0.31	0.00	0.00	4.50	1.81	0.00	0.00	0.00	0.04	0.00	0.92	0.00	0.00	0.00	0.00
8	NAIRN	0.00	0.00	0.00	0.00	0.00	0.00	3.74	2.27	0.40	0.00	0.05	0.38	0.42	0.42	0.00	0.79	0.62	0.00
9	MORAY	0.25	0.23	0.30	0.00	0.14	0.13	2.51	2.19	0.27	0.09	0.00	0.05	0.30	0.47	0.05	0.66	0.18	0.41
10	BANFF-BUCHAN	0.00	0.24	0.18	0.04	0.00	0.11	2.91	1.70	0.26	0.09	0.05	0.14	0.18	0.68	0.08	0.86	0.33	0.20
11	GORDON	0.18	0.04	0.24	0.19	0.05	0.14	2.60	1.60	0.33	0.05	0.31	0.13	0.12	0.87	0.07	0.73	0.51	0.41
12	ABERDEEN	0.06	0.11	0.09	0.07	0.02	0.04	3.06	1.81	0.47	0.16	0.07	0.14	0.13	0.82	0.15	0.61	0.31	0.24
13	KINCARDINE	0.10	0.15	0.22	0.21	0.00	0.24	3.53	1.31	0.19	0.22	0.09	0.00	0.06	0.47	0.00	0.69	0.30	0.20
14	ANGUS	0.09	0.03	0.24	0.28	0.04	0.03	2.94	1.76	0.09	0.21	0.03	0.06	0.15	0.80	0.04	0.66	0.46	0.50
15	DUNDEE	0.14	0.04	0.18	0.19	0.09	0.08	2.20	1.73	0.34	0.07	0.06	0.08	0.18	0.64	0.05	0.69	0.11	0.25
16	PERTH-KINROSS	0.08	0.21	0.21	0.17	0.05	0.00	2.70	1.53	0.10	0.09	0.09	0.10	0.31	0.79	0.17	0.39	0.37	0.21
17	KIRKCALDY	0.09	0.10	0.17	0.15	0.04	0.00	1.91	1.59	0.11	0.17	0.06	0.04	0.18	0.76	0.04	0.65	0.32	0.29
18	N-E FIFE	0.05	0.07	0.05	0.21	0.05	0.05	1.85	1.88	0.12	0.18	0.07	0.13	0.20	0.67	0.05	0.41	0.13	0.31
19	DUNFERMLINE	0.00	0.18	0.12	0.09	0.03	0.16	1.78	0.91	0.20	0.07	0.07	0.05	0.13	0.41	0.11	0.78	0.36	0.40
20	WEST LOTHIAN	0.09	0.27	0.37	0.18	0.00	0.06	2.01	1.43	0.29	0.04	0.04	0.08	0.21	0.79	0.00	0.72	0.32	0.19
21	EDINBURGH	0.16	0.19	0.21	0.16	0.10	0.09	2.29	1.49	0.46	0.22	0.03	0.10	0.16	0.70	0.05	0.84	0.30	0.31
22	MIDLOTHIAN	0.04	0.17	0.12	0.19	0.00	0.05	2.47	1.44	0.17	0.13	0.00	0.06	0.14	0.64	0.10	0.61	0.14	0.35
23	EAST LOTHIAN	0.00	0.16	0.16	0.04	0.05	0.17	1.90	1.42	0.42	0.26	0.23	0.08	0.17	0.76	0.04	0.65	0.21	0.13
24	TWEEDDALE	0.25	0.00	0.23	0.23	0.00	0.00	3.25	3.50	0.47	0.22	0.00	0.00	0.51	0.50	0.00	0.00	0.96	0.48
25	ETTRICK	0.09	0.25	0.18	0.23	0.00	0.10	1.74	1.76	0.18	0.13	0.00	0.00	0.19	0.77	0.13	0.18	0.18	0.16
26	ROXBURGH	0.08	0.60	0.37	0.09	0.07	0.00	2.02	1.46	0.34	0.09	0.00	0.00	0.18	0.84	0.08	0.72	0.29	0.64
27	BERWICKSHIRE	0.17	0.00	0.00	0.00	0.00	0.17	1.55	1.48	0.00	0.78	0.00	0.00	0.21	0.91	0.00	0.32	0.34	0.46
28	CLACKMANNAN	0.07	0.19	0.08	0.33	0.00	0.00	1.89	1.59	0.33	0.21	0.08	0.30	0.00	0.85	0.00	0.50	0.45	0.09
29	STIRLING	0.05	0.11	0.30	0.04	0.03	0.00	1.71	1.62	0.17	0.09	0.10	0.04	0.30	0.47	0.08	0.51	0.30	0.21
30	FALKIRK	0.08	0.06	0.25	0.21	0.11	0.03	1.83	1.62	0.17	0.25	0.07	0.05	0.14	0.49	0.05	0.34	0.28	0.08
31	ARGYLL-BUTE	0.26	0.13	0.27	0.21	0.06	0.05	2.32	2.32	0.05	0.24	0.00	0.06	0.21	0.46	0.10	0.43	0.21	0.34
32	DUMBARTON	0.06	0.06	0.23	0.04	0.05	0.04	2.09	1.48	0.10	0.36	0.00	0.11	0.03	0.80	0.02	0.66	0.20	0.38
33	GLASGOW	0.08	0.06	0.22	0.20	0.06	0.04	2.43	1.55	0.24	0.22	0.07	0.09	0.11	0.68	0.07	0.57	0.29	0.30
34	CLYDEBANK	0.00	0.14	0.24	0.39	0.00	0.06	2.59	1.27	0.07	0.12	0.16	0.29	0.07	0.77	0.12	0.79	0.30	0.35
35	BEARSDEN	0.26	0.10	0.22	0.17	0.00	0.00	2.33	1.05	0.18	0.00	0.12	0.06	0.17	0.38	0.10	0.71	0.12	0.29
36	STRATHKELVIN	0.03	0.04	0.29	0.18	0.00	0.00	2.53	1.42	0.27	0.26	0.00	0.14	0.29	0.73	0.00	0.48	0.34	0.47
37	CUMBERNAULD	0.00	0.00	0.27	0.04	0.00	0.00	2.03	1.44	0.37	0.13	0.06	0.06	0.06	0.62	0.06	0.57	0.06	0.16
38	MONKLANDS	0.20	0.05	0.11	0.37	0.03	0.03	1.96	1.08	0.26	0.04	0.00	0.14	0.10	0.61	0.10	0.54	0.22	0.38
39	MOTHERWELL	0.10	0.05	0.14	0.14	0.05	0.05	2.04	1.73	0.30	0.13	0.06	0.05	0.07	0.44	0.05	0.51	0.28	0.22
40	HAMILTON	0.04	0.06	0.17	0.15	0.03	0.11	1.86	2.20	0.11	0.39	0.07	0.09	0.19	0.57	0.17	0.52	0.35	0.11
41	EAST KILBRIDE	0.14	0.07	0.23	0.12	0.00	0.00	1.97	1.75	0.04	0.20	0.11	0.26	0.06	0.71	0.11	0.47	0.07	0.44
42	EASTWOOD	0.12	0.00	0.05	0.17	0.00	0.00	2.57	1.13	0.29	0.16	0.20	0.21	0.10	1.00	0.17	0.61	0.13	0.40
43	CLYDESDALE	0.05	0.10	0.07	0.07	0.00	0.20	2.00	1.06	0.32	0.06	0.00	0.00	0.25	0.61	0.00	0.40	0.22	0.24
44	RENFREW	0.05	0.03	0.18	0.16	0.02	0.07	1.93	1.37	0.14	0.13	0.12	0.06	0.07	0.54	0.06	0.56	0.22	0.19
45	INVERCLYDE	0.07	0.00	0.14	0.12	0.03	0.12	2.10	1.55	0.00	0.14	0.06	0.00	0.03	0.82	0.21	0.53	0.31	0.26
46	CUNNINGHAME	0.16	0.07	0.16	0.15	0.00	0.19	1.89	1.59	0.13	0.12	0.03	0.18	0.20	0.52	0.11	0.42	0.33	0.35
47	KILMARNOCK	0.11	0.04	0.12	0.15	0.07	0.00	2.25	1.10	0.16	0.09	0.00	0.10	0.11	0.68	0.07	0.45	0.37	0.33
48	KYLE-CARRICK	0.19	0.09	0.17	0.16	0.00	0.07	2.34	0.97	0.08	0.12	0.09	0.09	0.18	0.66	0.05	0.61	0.26	0.27
49	CUMNOCK-DOON	0.00	0.14	0.00	0.23	0.00	0.07	2.09	1.39	0.15	0.12	0.00	0.04	0.07	0.61	0.07	0.48	0.41	0.47
50	WIGTOWN	0.23	0.12	0.24	0.12	0.03	0.10	2.63	0.97	0.16	0.12	0.12	0.00	0.21	0.56	0.16	0.97	0.30	0.00
51	STEWARTRY	0.00	0.15	0.17	0.30	0.13	0.00	2.46	1.25	0.13	0.21	0.00	0.00	0.27	0.17	0.00	0.53	0.27	0.37
52	NITHSDALE	0.06	0.13	0.07	0.27	0.00	0.12	2.74	1.75	0.12	0.00	0.00	0.06	0.21	0.48	0.07	0.59	0.21	0.23
53	ANNANDALE	0.00	0.10	0.28	0.08	0.00	0.18	1.56	0.79	0.14	0.29	0.00	0.00	0.18	0.22	0.00	0.96	0.22	0.16
54	ORKNEY	0.16	0.00	0.00	0.00	0.33	0.39	1.27	1.11	0.22	0.38	0.00	0.16	0.00	0.82	0.16	0.16	0.56	0.16
55	SHETLAND	0.19	0.00	0.39	0.00	0.00	0.00	1.97	1.47	0.20	0.85	0.10	0.00	0.16	0.16	0.23	0.65	0.00	0.38
56	WESTERN ISLES	0.11	0.21	0.10	0.20	0.00	0.10	1.73	1.21	0.11	0.00	0.10	0.00	0.12	0.52	0.18	0.82	0.79	0.31
	ALL SCOTLAND	0.10	0.11	0.19	0.17	0.04	0.07	2.29	1.53	0.23	0.17	0.06	0.09	0.15	0.65	0.08	0.59	0.29	0.28

(female)

#	District	MYLE	LYLE	BRNS	EYEZ	KIDN	OUFG	CONN	BONE	NASA	GALL	LIVE	RECT	COLO	SINT	PHA2	PHA1	OMOU	SALG	TONG
1	CAITHNESS	0.08	0.22	0.44	0.22	0.38	0.13	0.06	0.15	0.00	0.00	0.00	1.28	2.79	0.00	0.00	0.00	0.00	0.27	0.15
2	SUTHERLAND	0.00	0.00	0.93	0.00	0.00	0.00	0.20	0.00	0.00	0.43	0.43	0.64	2.38	0.00	0.00	0.00	0.00	0.22	0.00
3	ROSS-CROMARTY	0.43	0.25	0.64	0.10	0.08	0.40	0.58	0.09	0.00	0.08	0.00	1.76	3.66	0.18	0.00	0.25	0.08	0.54	0.09
4	SKYE-LOCHALSH	0.83	0.30	0.23	0.00	0.00	0.00	0.30	0.00	0.00	0.31	0.25	1.55	3.14	0.00	0.00	0.25	0.00	0.31	0.00
5	LOCHABER	0.00	0.00	0.62	0.17	0.49	0.00	0.35	0.00	0.21	0.00	0.00	0.83	3.38	0.00	0.00	0.00	0.00	0.00	0.21
6	INVERNESS	0.24	0.13	0.30	0.00	0.59	0.35	0.07	0.10	0.13	0.07	0.00	0.84	1.87	0.05	0.00	0.17	0.30	0.00	0.00
7	BADENOCH	0.00	0.53	0.32	0.28	0.82	0.32	0.30	0.00	0.00	0.28	0.38	2.80	2.02	0.00	0.28	0.00	0.00	0.30	0.00
8	NAIRN	0.00	0.00	0.00	0.08	0.14	0.19	0.35	0.00	0.04	0.22	0.00	1.85	2.51	0.00	0.00	0.00	0.04	0.28	0.00
9	MORAY	0.38	0.23	0.70	0.09	0.20	0.24	0.13	0.06	0.05	0.33	0.10	1.44	2.89	0.00	0.00	0.13	0.08	0.22	0.18
10	BANFF-BUCHAN	0.20	0.14	0.54	0.20	0.20	0.13	0.05	0.07	0.05	0.14	0.08	0.99	2.53	0.04	0.00	0.04	0.08	0.21	0.00
11	GORDON	0.25	0.11	0.36	0.13	0.44	0.18	0.28	0.10	0.07	0.27	0.08	1.05	1.74	0.08	0.00	0.08	0.19	0.12	0.00
12	ABERDEEN	0.18	0.17	0.40	0.05	0.47	0.18	0.28	0.10	0.03	0.14	0.11	1.19	2.89	0.04	0.00	0.08	0.09	0.21	0.09
13	KINCARDINE	0.33	0.18	0.60	0.18	0.18	0.08	0.16	0.06	0.08	0.00	0.09	1.66	3.22	0.08	0.00	0.00	0.09	0.33	0.08
14	ANGUS	0.38	0.09	0.28	0.13	0.24	0.35	0.10	0.06	0.05	0.23	0.09	0.83	2.57	0.13	0.00	0.03	0.03	0.07	0.03
15	DUNDEE	0.25	0.08	0.48	0.01	0.31	0.23	0.15	0.09	0.05	0.17	0.07	0.74	2.32	0.04	0.04	0.08	0.08	0.05	0.02
16	PERTH-KINROSS	0.16	0.12	0.35	0.04	0.35	0.23	0.14	0.04	0.00	0.12	0.10	0.77	2.56	0.05	0.03	0.10	0.05	0.10	0.10
17	KIRKCALDY	0.20	0.13	0.46	0.18	0.42	0.32	0.18	0.10	0.09	0.26	0.09	0.94	2.38	0.04	0.04	0.08	0.07	0.22	0.00
18	N-E FIFE	0.27	0.04	0.57	0.04	0.25	0.30	0.23	0.15	0.05	0.04	0.13	1.29	1.91	0.04	0.00	0.09	0.08	0.09	0.09
19	DUNFERMLINE	0.18	0.11	0.42	0.03	0.30	0.33	0.13	0.02	0.03	0.29	0.08	0.91	1.60	0.05	0.00	0.00	0.00	0.18	0.05
20	WEST LOTHIAN	0.27	0.18	0.48	0.06	0.57	0.32	0.22	0.13	0.04	0.20	0.33	1.08	2.40	0.18	0.04	0.28	0.11	0.33	0.00
21	EDINBURGH	0.15	0.16	0.74	0.08	0.42	0.20	0.18	0.05	0.02	0.16	0.10	0.97	2.10	0.03	0.02	0.07	0.08	0.21	0.07
22	MIDLOTHIAN	0.14	0.22	0.69	0.08	0.44	0.13	0.16	0.11	0.04	0.20	0.08	1.05	2.06	0.04	0.10	0.05	0.05	0.20	0.16
23	EAST LOTHIAN	0.11	0.07	0.39	0.03	0.36	0.07	0.07	0.07	0.07	0.20	0.19	1.21	2.31	0.07	0.00	0.14	0.07	0.29	0.00
24	TWEEDDALE	0.17	0.21	0.38	0.17	0.37	0.38	0.18	0.00	0.00	0.00	0.00	0.33	1.07	0.18	0.00	0.00	0.00	0.40	0.00
25	ETTRICK	0.26	0.09	0.89	0.08	0.43	0.18	0.25	0.08	0.08	0.16	0.08	0.88	2.31	0.08	0.00	0.32	0.08	0.25	0.00
26	ROXBURGH	0.15	0.22	0.57	0.08	0.43	0.00	0.08	0.08	0.00	0.15	0.08	0.81	2.06	0.07	0.00	0.14	0.08	0.18	0.00
27	BERWICKSHIRE	0.76	0.45	0.16	0.14	0.00	0.61	0.18	0.00	0.00	0.00	0.00	0.31	2.24	0.16	0.00	0.10	0.00	0.14	0.30
28	CLACKMANNAN	0.29	0.30	0.51	0.06	0.34	0.09	0.11	0.04	0.00	0.23	0.05	1.20	2.40	0.00	0.07	0.12	0.00	0.06	0.00
29	STIRLING	0.09	0.30	0.39	0.07	0.24	0.22	0.01	0.05	0.07	0.13	0.04	1.08	2.22	0.02	0.04	0.08	0.12	0.12	0.02
30	FALKIRK	0.32	0.14	0.32	0.07	0.39	0.19	0.22	0.03	0.07	0.21	0.09	0.98	2.03	0.04	0.05	0.12	0.02	0.02	0.04
31	ARGYLL-BUTE	0.32	0.25	0.16	0.09	0.22	0.08	0.22	0.00	0.13	0.26	0.18	0.84	1.82	0.07	0.04	0.00	0.19	0.00	0.04
32	DUMBARTON	0.48	0.20	0.43	0.03	0.42	0.09	0.12	0.05	0.00	0.15	0.18	0.92	2.69	0.04	0.05	0.08	0.08	0.09	0.14
33	GLASGOW	0.17	0.14	0.38	0.07	0.28	0.18	0.11	0.00	0.05	0.16	0.10	0.83	1.96	0.07	0.03	0.11	0.08	0.04	0.06
34	CLYDEBANK	0.36	0.00	0.23	0.13	0.21	0.12	0.05	0.00	0.11	0.22	0.08	1.28	2.50	0.04	0.12	0.04	0.16	0.00	0.05
35	BEARSDEN	0.37	0.09	0.57	0.00	0.36	0.38	0.15	0.05	0.18	0.00	0.12	1.45	1.87	0.10	0.06	0.18	0.09	0.14	0.09
36	STRATHKELVIN	0.12	0.23	0.37	0.03	0.32	0.20	0.03	0.06	0.04	0.36	0.05	0.82	2.06	0.00	0.04	0.03	0.09	0.08	0.00
37	CUMBERNAULD	0.28	0.19	0.36	0.10	0.26	0.16	0.10	0.08	0.08	0.20	0.12	1.09	1.79	0.10	0.04	0.11	0.10	0.00	0.05
38	MONKLANDS	0.13	0.09	0.38	0.03	0.17	0.05	0.00	0.03	0.12	0.30	0.07	0.66	2.00	0.11	0.00	0.17	0.07	0.07	0.05
39	MOTHERWELL	0.18	0.13	0.22	0.08	0.29	0.14	0.03	0.04	0.04	0.21	0.05	0.80	1.94	0.12	0.02	0.16	0.05	0.08	0.13
40	HAMILTON	0.19	0.18	0.56	0.03	0.32	0.21	0.05	0.10	0.05	0.42	0.03	0.82	1.96	0.04	0.00	0.03	0.08	0.02	0.14
41	EAST KILBRIDE	0.25	0.13	0.66	0.13	0.26	0.08	0.03	0.09	0.22	0.40	0.12	0.84	2.37	0.00	0.15	0.14	0.03	0.04	0.11
42	EASTWOOD	0.13	0.13	0.60	0.00	0.19	0.15	0.05	0.05	0.05	0.30	0.00	0.82	1.72	0.10	0.00	0.05	0.06	0.00	0.16
43	CLYDESDALE	0.32	0.04	0.31	0.00	0.52	0.42	0.05	0.04	0.00	0.57	0.12	0.59	1.81	0.08	0.00	0.00	0.11	0.05	0.14
44	RENFREW	0.19	0.11	0.43	0.07	0.39	0.11	0.00	0.08	0.03	0.18	0.05	0.83	1.97	0.05	0.00	0.10	0.06	0.02	0.07
45	INVERCLYDE	0.21	0.15	0.29	0.11	0.28	0.22	0.11	0.08	0.05	0.19	0.03	0.97	2.06	0.00	0.07	0.11	0.15	0.03	0.05
46	CUNNINGHAME	0.28	0.14	0.43	0.06	0.37	0.17	0.11	0.01	0.10	0.08	0.05	0.88	1.88	0.00	0.00	0.07	0.00	0.00	0.03
47	KILMARNOCK	0.04	0.34	0.49	0.00	0.14	0.22	0.06	0.18	0.00	0.08	0.18	0.93	1.91	0.07	0.08	0.14	0.07	0.03	0.02
48	KYLE-CARRICK	0.27	0.18	0.33	0.05	0.22	0.25	0.13	0.03	0.08	0.10	0.15	0.80	2.44	0.02	0.00	0.07	0.08	0.02	0.00
49	CUMNOCK-DOON	0.22	0.15	0.12	0.07	0.27	0.17	0.09	0.00	0.09	0.08	0.07	0.82	1.86	0.00	0.00	0.07	0.00	0.14	0.00
50	WIGTOWN	0.50	0.18	0.59	0.10	0.76	0.28	0.09	0.11	0.00	0.20	0.00	0.62	1.86	0.00	0.12	0.09	0.21	0.11	0.09
51	STEWARTRY	0.12	0.13	0.12	0.00	0.36	0.40	0.00	0.14	0.03	0.00	0.03	0.47	1.91	0.00	0.00	0.00	0.00	0.11	0.00
52	NITHSDALE	0.25	0.18	0.35	0.11	0.59	0.44	0.22	0.08	0.05	0.30	0.05	0.84	2.92	0.00	0.12	0.05	0.15	0.11	0.10
53	ANNANDALE	0.36	0.14	0.09	0.06	0.18	0.46	0.16	0.00	0.00	0.52	0.00	0.70	2.88	0.10	0.00	0.00	0.00	0.08	0.00
54	ORKNEY	0.00	0.00	0.11	0.17	1.02	0.18	0.64	0.00	0.00	0.32	0.15	0.87	0.70	0.32	0.00	0.00	0.17	0.32	0.00
55	SHETLAND	0.00	0.49	0.30	0.00	0.33	0.00	0.00	0.00	0.00	0.00	0.00	0.81	2.52	0.17	0.00	0.00	0.18	0.49	0.00
56	WESTERN ISLES	0.11	0.44	0.49	0.00	0.89	0.27	0.21	0.00	0.08	0.09	0.00	1.13	2.17	0.00	0.09	0.28	0.10	0.29	0.10
	ALL SCOTLAND	0.21	0.15	0.44	0.07	0.34	0.20	0.12	0.08	0.05	0.19	0.09	0.83	2.16	0.05	0.02	0.09	0.07	0.11	0.08

Figure AII.1. Colour disc

Showing hue on a circular axis; chroma or saturation is deepest at the periphery, becoming progressively less deep towards the centre. The colour scales used in the cancer mortality atlas of England and Wales (A), Japan (B) and for the present atlas (C) are shown in relation to the order of colour and location on the disc.

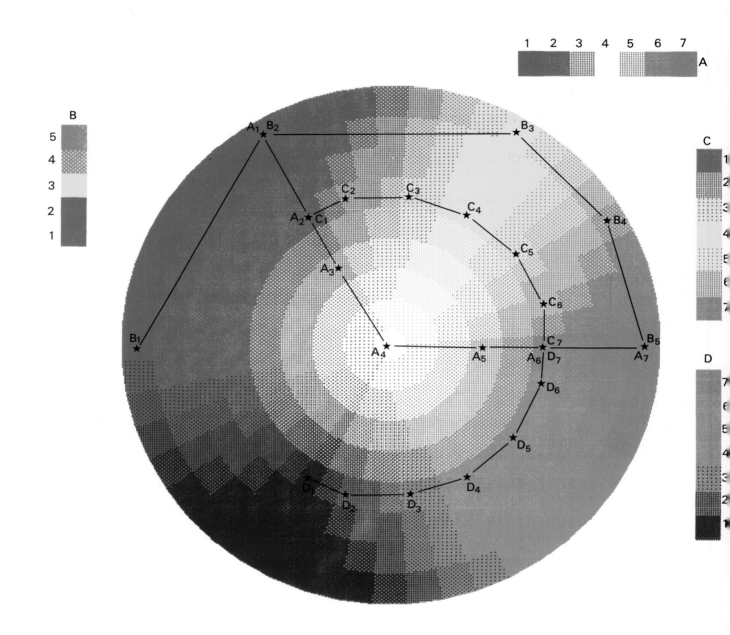

APPENDIX II

CANCER MAPS — TECHNICAL CONSIDERATIONS

A. *Use of an absolute and a relative scale for incidence*

The overall aim in constructing the maps has been to aid understanding of the underlying distribution of cancer incidence throughout Scotland. The choice of scale was a primary consideration, since this greatly influences the final appearance and interpretation. Two options are available, an absolute and a relative scale.

An *absolute scale* is one in which class limits are assigned after taking into account the overall range of values. The number of areas falling into a given class is greatly influenced by the shape of the underlying distribution. When a grey scale is chosen (from light grey for low rates to dark grey for high rates), if the distribution contains a relatively large number of high values, the map will appear generally dark. In a scale of this kind, colour is not required, as intensity of shading indicates the distribution of values. This type of scale is used on the sixth and seventh pages for the maps presented in Chapter 7.

A *relative scale* is one in which the rates are shown according to their position in the overall rate distribution. An example of this is a percentile scale. The top class contains the areas with rates lying in, say, the top 5% of the distribution: the rates in the next class contain perhaps the next 10%, and so on. The limitations of this scale are, first, that the shape of the distribution of incidence is lost, and second, the percentages for the cut-off points for the intervals have to be selected.

When colour is available, a relative scale has great advantages. For example, it is possible to choose a neutral colour, like pale yellow, for the central class, and to add progressively more green for each of the classes towards the lower end of the distribution, and more red for classes towards the upper end. This gives an even, visually satisfying, colour scale, very close to human feelings on colour (red, high; green, low) (Appendix IIB). In many respects, use of a relative scale with colour combines the advantages of an absolute scale with those associated with the mapping of statistical significance, thereby drawing attention to areas both of high and low incidence.

Because both methods have different advantages and disadvantages, it was decided to present both in this atlas — a black-and-white map with an absolute scale and a colour map with a relative scale — for each cancer site.

In selecting parameters for the absolute scale, the obvious outliers were first removed and the range of values divided into a number of classes to provide suitable cut-off points. For example, the range of lung cancer rates in men is 40.2-130.7; 130.7 is an outlier, the second highest rate being less than 110. The classes are then chosen as (1) <50, (2) 50<60, (3) 60<70, (4) 70<80, (5) 80<90, (6) 90<100, (7) 100, (8) 110 and over.

For the relative scale, it was decided to use seven classes: lowest 5%, next 10%, next 20%, middle 30%, next 20%, next 10%, and highest 5%. These are more useful than classes of equal size because, on average, extreme values are less frequent. For a normal distribution, this choice gives classes of almost equal length, in terms of the incidence values observed. The colours used are those described above and below.

The scheme described was used only as a guideline, so that one or more areas could be moved from one class to another if the percentage rule created an unsuitable cut-off point. To facilitate the choice of cut-off point, the conventional histogram was not used, since it does not allow obwervation of a natural grouping of the values of the age-standardized rates, which is 'lost' by the aggregation into histogram classes. Smoothed histograms, the peaks and valleys of which bring out groups (peaks) and 'natural' cut-off points (valleys), are hence presented at the top left-hand corner of the colour maps. The levels of age-standardized incidence presented on the vertical scale thus do not fall into regular groupings, such as 0-4, 5-9, etc, but rather provide the range of values observed for a given aggregation of rates.

B. *Colour notation*

There are a number of methods of colour notation. The one chosen is based on the Munsell system (Nickerson, 1940; Newhall *et al.*, 1943), which uses three axes or parameters:

(1) colour or hue, e.g., red, blue, orange;

(2) value or luminosity (degree of lightness), e.g., orange and brown;

(3) chroma or saturation (purity of colour), e.g., pink and red.

Perception of colour can be expressed simply in terms of these axes; in practical terms, only two need be considered here — hue and chroma.

Most people, except perhaps those with severe colour blindness, can appreciate, without explanation, the meaning of hue. Chroma (saturation) may be a little more difficult to understand; however, a simple way is to consider red as a fully saturated colour, and pink, which is red mixed with white, as being less saturated.

Study of the colour disc (Fig. A II.1) shows hue as a circular axis. With a starting point at red, the hue moves to orange, yellow, green, cyan (sky blue), blue, purple and back to red.

Different degrees of chroma may be incorporated with hue, so that the visual steps between colours become less abrupt and a smooth scale is obtained. The different degrees of chroma, also, are displayed on the colour disc. At the centre, there is no chroma, giving a state of complete whiteness; moving towards the circumference in a straight line, the degree of chroma increases progressively as whiteness is lost. This system is used by many national bureaux.

For the purpose of this atlas, the most effective result is obtained by using only two of the three primary hues (red, green, blue). If all three are used, the hue axis, being circular, fails to give a sense of the direction of gradation.

The advantages and disadvantages of the various combinations of two hues are :

(1) *green to blue (through cyan)*: It is difficult to decide instinctively which hue represents a high value and which a low value.

(2) *blue to red*: This combination gives a very satisfactory subjective scale where temperature is involved, blue representing cold and red heat.

(3) *green to red*: This combination was chosen for the atlas because it gives an intuitive depiction of cancer, red representing high rates and green low rates.

The colours on a scale may be represented as a series of points on the colour disc (Fig. A II.1). Example of scales used previously are discussed below.

The scale shown at A, that chosen for the cancer mortality atlas of England and Wales (Gardner *et al.*, 1984), begins at a point on green on the circumference, and passes in a straight line with decreasing chroma to the centre (white), and then by another straight line with increasing chroma through red, again to a point on the circumference. Interposing white has led to an extreme visual leap (green-white-red). In the atlas of England and Wales, this feature was accentuated by

use of the statistical significance of standardized mortality ratios, giving a preponderance of white areas.

The scale shown at B, that chosen for the cancer mortality atlas of Japan (Segi, 1977) passes from cyan (sky blue) to green, yellow, orange and red, keeping close to the circumference and using maximum chroma throughout. The visual steps in this scale are large, and use of the three primary colours negates the concept of natural order associated with scales confined to red/green.

The scale chosen for the Scottish cancer atlas is shown by line C. This passes from a point in the green region by short, direct steps through yellow to orange-red. Each step has been kept short by employing partial saturation, avoiding angles in the line as far as possible. It is obvious that the ideal scale should be represented by a line that is not too long and not too bent.

The scale shown at D is the complement of C, and is the result of applying the same principles to the blue-red axis. As discussed earlier, this was not used.

Other colour schemes were used in the atlas of cancer mortality for US counties (Mason *et al.*, 1975) and in the Chinese cancer mortality atlas (Editorial Committee for the Atlas of Cancer Mortality in the People's Republic of China, 1979). In the US atlas, a red, orange, yellow, green scale has been used, which conveys gradient well; a red, blue scale was also used. Both scales are used to indicate levels of significance, again raising the relative/absolute question: should a significantly high rate be placed in the scale at a point above a rate that is higher in absolute terms, but not significantly different from the national average?

The Chinese atlas, using a variety of scales of various shades of brown and yellow, successfully conveys gradient. Much less successful is the scale that was used to indicate differences in the significance of rates, where the colours move erratically from red to purple, orange, green and yellow. Thus, orange is of higher value than purple and yellow, of lower value than green. Intuitively, this seems to be the wrong order — an impression confirmed when the natural ordering of hues on the colour disc is studied.

C.   *Randomness of spatial pattern of incidence*

In assessing the randomness of the areal distribution of cancer incidence (Table A.II.1), the ranks of rates, rather than the rates themselves are used.

Table AII.1 Summary of indices of non-randomness (D) and their statistical significance

Site	Males		Females	
	D	p-value	D	p-value
Lip	10.88	0.000	16.50	0.013
Oral cavity and pharynx	20.03	0.833	19.56	0.694
Oesophagus	17.35	0.067	17.63	0.102
Stomach	15.98	0.004	15.38	0.001
Large bowel	14.28	0.000	14.93	0.000
Pancreas	16.43	0.011	17.41	0.073
Larynx	18.83	0.426	17.93	0.158
Trachea, bronchus, lung	13.00	0.000	15.73	0.002
Malignant melanoma	19.46	0.658	18.30	0.253
Breast	NA	NA	19.02	0.499
Cervix	NA	NA	17.13	0.045
Corpus	NA	NA	15.80	0.003
Ovary	NA	NA	18.11	0.200
Prostate	15.77	0.002	NA	NA
Testis	17.14	0.046	NA	NA
Bladder	17.30	0.060	18.34	0.264
Thyroid	17.39	0.070	17.65	0.105
Hodgkin's disease	17.96	0.164	17.03	0.036
Non-Hodgkin's lymphoma	16.28	0.008	18.07	0.191
Leukaemia	20.05	0.838	19.03	0.502

[a]0.000 should be taken to be less than 0.0005, i.e. less than 5 chances in 10 000 of being random.

NA, not applicable

Consider the difference in rank between two adjacent regions with a common border: if, for all possible pairs of such adjacent regions, a low value is frequently observed (i.e., the ranks are close), it may be concluded that there is clustering.

Suppose the map has N regions with K pairs of adjacent regions ($[R_i, R_j]$, $i < j$). It is clear that K must be less than or equal to $N \cdot (N-1)/2$. The term d_k is defined as the absolute value of the difference in rank between the regions making the kth adjacency, for $k = 1, 2, ..., K$.

Then consider the random variable

$$D = \frac{1}{K} \sum_1^K d_k$$

which is the average absolute difference in rank between adjacent regions.

The probability distribution of D under the null hypothesis would give an appropriate test of randomness. Unfortunately, it is difficult to determine this distribution theoretically.

For any k, the probability function of d_k is simply

$$P[d_k = V] = \frac{(N-V) \, 2}{N \, (N-1)}$$

Its expectation can be computed as:

$$\sum_{V=1}^{N-1} \frac{V(N-V)2}{N(N-1)} = \cdots = \frac{N+1}{3}.$$

Hence, $[ED] = (N+1)/3$. $Var[D]$, and higher moments can also be derived, although the calculation becomes very tedious.

The computer provides an easier solution: with a simulation, random ranks are assigned to the regions (under the null hypothesis), D is computed iteratively to define the null distribution of D (the VAX 11/780 at IARC, running over a weekend, generated over 10 million random maps to provide a kernel estimate of the density function).

With the Scottish map (56 regions, 132 adjacencies), the result shown in Figure A II.2 for (E[D] = 19) was obtained.

For example, the map of lip cancer incidence in males gives a value of about 10 for D — a value never obtained in the 10 million trials. By this criterion, the map showed a very high degree of clustering. Inspection of the map for lung cancer in males, with a D value of about 13, clearly showed a pattern. This value occurred with a frequency of less than 0.0005 in the simulation.

Calculation of the D statistic could yield non-unique results in the case of equal rates for two or several regions on the map. For example, what ranks should be assigned to two regions with the same rate, and what does the D value become if these ranks are assigned? This problem is particularly pertinent when dealing with a cancer site for which there are a number of zero rates.

It was thus decided to use classical tied ranks (the mean of ranks involved in a tie). Individual d_k occurring between a region in the tie and a region outside will be, on average, consistent with situations in which no tie exists. A d_k corresponding to two regions within a tie will, of course, be 0, a situation that does not arise under the null hypothesis. This would tend to provide an overestimate of the degree of clustering in such a group.

The problem arose, fortunately, in only a few maps, e.g., thyroid in males, on which a relatively large proportion of zero rates occur. Care should always be taken when investigating clusters of these rare cancer types.

These D values are presented in Table A II.1 for the sites mapped, and give an insight into the geographical pattern of cancer incidence in Scotland. For the examples discussed previously, it is interesting that the pattern of lung cancer is much more strongly aggregated for males than for females (surprisingly, there is no statistically significant pattern for laryngeal or mouth cancer, which are also strongly linked to tobacco use). The extreme nature of the aggregation of lip cancer in males is outstanding.

Stomach cancer appears to be distributed in a significantly non-random way for people of both sexes, and the clustering is slightly stronger for females. Large-bowel cancer incidence is strongly aggregated for people of both sexes. Interestingly, this aggregation appears stronger for colonic cancer in males but stronger for rectal cancer in females.

Pancreatic cancer incidence shows a weak aggregation in people of both sexes; unexpectedly, there is no evidence of clustering or of a gradient in the incidence of malignant melanoma.

Of the female cancers, neither that of the breast, ovary nor cervix has a strong pattern of aggregation; however, there appears to be a significant clustering of districts with similar rates of endometrial cancer.

Testicular cancer incidence shows only weak aggregation, but there is significant clustering of that of prostatic cancer. Bladder cancer incidence demonstrates no pattern for females, but a weak aggregation is seen for males. No definite pattern emerges for thyroid cancer.

The incidence of Hodgkin's disease demonstrates no pattern among males, but there is evidence of weak aggregation for females. In contrast, for non-Hodgkin's lymphoma, there is no evidence of spatial aggregation of female rates, but there is a pronounced aggregation of male rates. Finally, there appears to be no pattern to the rates of leukaemia for people of either sex.

D. *Technical aspects of the production of the maps*

Data handling: The numerators and denominators, received from Edinburgh, were stored in a VAX 11/780 computer at the International Agency for Research on Cancer. These data were reorganized into files with a structure compatible with the programs used for computing various rates and indices. The programs computed rate(s) as desired for a given sex, for any one or combination of sites and/or areas.

Graphic handling: Software for general mapping was developed and installed on the VAX and on a decentralized graphic station (Hewlett-Packard). The latter comprises a colour screen, an eight-pen plotter with paper advance feature, a digitizing tablet and a floppy disc unit. This station was connected to the VAX. Outline maps were drawn using the digitizing tablet, and completed maps were presented on the colour screen or on the plotter. A subset of the software running on the VAX creates monochromatic maps on a Tektronix screen or *via* the VAX dot matrix printer, and colour maps are produced on a smaller colour-dot matrix printer, also attached to the VAX.

Production of Scottish maps: After the desired sets of rates and indices had been computed, the files containing the rates for each region or combination of regions were processed by the VAX mapping software for production of black-and-white draft maps on the printer. If the results looked suitable, the files were transmitted to the decentralized graphic station for inspection on the colour screen. A paper copy in colour could then be produced on the plotter, or on the colour printer.

Two options were considered for the production and printing of the final maps:

(1) colour maps could be produced on the plotter and sent directly to a commercial printing firm, which would then use classical colour separation and reproduction techniques;

(2) for each map, separate plates could be produced for each colour, these being merged during the printing process by polychromatic techniques.

The second of these options was chosen.

Figure AII.2. Kernel estimate of D-distribution as a result of simulation of ten million maps

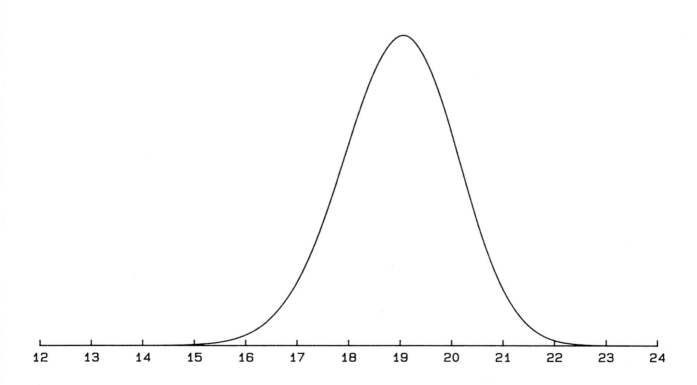

APPENDIX III

ANNOTATED DEFINITIONS OF SOME EPIDEMIOLOGICAL
TERMS AND CONCEPTS USED IN THE TEXT

The terms listed below have been selected from a much wider range of terms used in this atlas. For other terms and for further information, see *A Dictionary of Epidemiology* (Last, 1983)

Age-adjustment (age-standardization): see *Incidence*

Classification of cancer: see **International Classification of Diseases**

Cumulative risk: the chance of developing a particular cancer over a given age-span (usually, and in this atlas, from birth to 75 years of age)

While epidemiologists are used to dealing with incidence rates per 100 000 population (see below), to others this approach is unfamiliar. Nearly everybody can think in terms of chance. In horse racing, for example, one is trying to guess the likelihood that a given horse would come in first — the favourite is believed to have a good chance, an outsider is very unlikely to be placed, and the betting odds are fixed accordingly. The risk of developing cancer can be estimated, and with much greater precision, in the same way. For example, the Glasgow male has slightly more than 16 chances in 100 of developing lung cancer between birth and 75 years of age, compared with the Aberdeen male, who has around 12 chances in 100. While the epidemiologist cannot, without further information, e.g., on smoking habits, predict which individuals are likely to develop the disease, he can predict the *number of people* within the population as a whole that will develop it.

Incidence: the number of newly diagnosed cases of cancer occurring in a given population in a particular timespan.

Incidence rate: Although it is of interest to know that there were 4579 males with lung cancer in 1975-1980 in Glasgow, and 2087 in Edinburgh, unless the size of the population of these cities is also known, it is not possible to say whether lung cancer is commoner in one city than in the other. It is usual to calculate the number of cases of cancer at a given site that would occur in 100 000 people living in the area considered. To continue the example, in the 1975-1980 period there were 1186 male cases of lung cancer for every

100 000 males living in Glasgow and 973 for every 100 000 males of Edinburgh. As these cancers appeared over six years, it is the convention to divide the total by six to obtain an average annual rate of incidence — in our example, 197.7 and 162.1 per 100 000 *per annum* respectively. In Caithness, however, there were 48 male cases of lung cancer, but the male population of 13 865 is well below 100 000. The same technique is used; a calculation is made of the number of lung cancers that would have occurred if there had been 100 000 men in Caithness. The end result is an average annual incidence rate of 57.7 per 100 000. Such rates are often referred to as *crude rates*.

Cancer is more frequent in the old. This would be of no importance if the age composition of the populations of each of the areas to be compared was the same. This is clearly not so — the cities attract younger persons for work, leaving an older population in the region they come from. To allow for this, the incidence rates are usually adjusted so that differences in the age-structure of areas compared are taken into account. The age-standardization is done in a way that permits comparison, not only between areas of the same country, as in this atlas, but also between different countries, as in the publications of the International Agency for Research on Cancer (Waterhouse *et al.*, 1976, 1982).

As the World Standard Population used for international comparisons has a younger age structure than that of the Scottish population, the net effect is that the age-standardized rates are usually lower than the crude (non-standardized) rates. The age-standardized rate can be regarded as giving the magnitude of the risk in each of the districts or countries compared; the crude rate gives the size of the burden of cancer in terms of the number of cases of cancer per 100 000 population in each district or country.

International Classification of Diseases: a classification given in a publication of the World Health Organization (WHO, 1967), used by vital statistics offices, cancer registries and others, in which diseases are grouped or classified and assigned a code number.

Cancers are described by the organ or part of the body in which they arise, e.g., cancer of the stomach. In the 8th revision of the *International Classification of Diseases* (ICD8), 140 has been allotted to cancer of the lip, 141 to tongue, etc. (Chapter 7). These numbers are used in text, tables and maps and are sometimes referred to as ICD rubrics.

Mortality: the number of deaths occurring in a defined population in a particular timespan. As for *incidence*, it is usual to present mortality as rates per 100 000 or per million, and for comparison these are usually age-adjusted.

Survival: the number or proportion of persons with newly diagnosed cancer who are alive at a given time after diagnosis. Survival of cancer patients is a measure of the efficiency of treatment and is frequently presented as the proportion surviving five years after first diagnosis. For lung and stomach cancer, this proportion is around 5%, for breast cancer, 60%.

Statistical significance: A large branch of mathematics deals with probability. In the context of this atlas, it may be questioned whether the difference observed between the incidence of cancer in district A is likely to be truly (or significantly) different from that in the rest of Scotland. In other words, what is the likelihood that the difference seen could have occurred by chance alone? Usually, mathematics are used to test the degree of chance involved, e.g., less than one chance in 20 or less than one chance in 100, expressed as $p < 0.05$ or $p < 0.01$, respectively. To conserve space in the tables, the symbols $+$ and $++$ are used, respectively, for rates significantly higher than the average rate for the rest of Scotland; rates that are significantly lower are denoted $-$ and $--$. However, the tables given alongside the maps in this atlas do not show whether the incidence in district A is significantly different from that in district B. It is relatively easy to calculate this significance by computing

$$\frac{R_A - R_B}{\sqrt{(SE_A)^2 + (SE_B)^2}}$$

where R_A is the age-standardized rate in district A, and SE_A the standard error of the rate; R_B is the age-standardized rate in district B and SE_B the standard error of that rate. If the result obtained is greater than 1.96, the difference is statistically significant at the 5% level (i.e., there is less than one chance in 20 of the difference occurring by chance); if the result is greater than 2.58, the difference is statistically significant at the 1% level (i.e., less than one chance in 100).

APPENDIX IV

LIST OF CANCER REGISTRIES OUTSIDE SCOTLAND
MENTIONED IN THE TEXT

Cancer registry title	Text title
Senegal, Dakar	Dakar, Senegal
Canada, Alberta	Alberta
Canada, British Columbia	British Columbia
Canada, Manitoba	Manitoba
Canada, Newfoundland	Newfoundland
Canada, Northwest Territories and Yukon	Northwest Territories and Yukon
Canada, Ontario	Ontario
Canada, Saskatchewan	Saskatchewan
Netherlands Antilles (less Aruba)	Netherlands Antilles
USA, California, Alameda County	Alameda
USA, California, San Francisco Bay Area	San Francisco Bay Area
USA, California, Los Angeles County	Los Angeles
USA, Connecticut	Connecticut
USA, Georgia, Atlanta	Atlanta
USA, Iowa	Iowa
USA, Louisiana, New Orleans	New Orleans
USA, Michigan, Detroit	Detroit
USA, New York State (less New York City)	New York State
USA, Puerto Rico	Puerto Rico
USA, Utah	Utah
USA, Washington, Seattle	Seattle
China, Shanghai	Shanghai
Hong Kong	Hong Kong
India, Bombay	Bombay
India, Poona	Poona
Japan, Nagasaki City	Nagasaki
Japan, Osaka Prefecture	Osaka
Czechoslovakia, Western Slovakia	Czechoslovakia
Denmark	Denmark
Federal Republic of Germany, Hamburg	Hamburg
Finland	Finland
France, Bas-Rhin	Bas-Rhin
France, Doubs	Doubs
Italy, Varese	Varese
Norway	Norway
Poland, Cracow City	Cracow
Poland, Warsaw City	Warsaw
Romania, County Cluj	Romania, County Cluj
Spain, Navarra	Navarra
Spain, Zaragoza	Zaragoza
Sweden	Sweden

Cancer Registry Title

Switzerland, Geneva
Switzerland, Neuchatel
Switzerland, Vaud
UK, England, Birmingham and
    West Midlands Region
UK, England, North-western Regions
UK, England, Oxford Region
UK, England, South Thames Region
UK, England, Trent Region
UK, England and Wales, Mersey Region
Yugoslavia, Slovenia
Australia, South Australia
New Zealand
USA, Hawaii

Text Title

Geneva
Neuchatel
Vaud, Switzerland
Birmingham

North-western
Oxford
South Thames
Trent
Mersey
Yugoslavia
South Australia
New Zealand
Hawaii

PUBLICATIONS OF THE INTERNATIONAL AGENCY FOR RESEARCH ON CANCER

SCIENTIFIC PUBLICATIONS SERIES

(Available from Oxford University Press)

No. 1 LIVER CANCER (1971)
176 pages; £10-

No. 2 ONCOGENESIS AND HERPES VIRUSES (1972)
Edited by P.M. Biggs, G. de Thé & L.N. Payne, 515 pages; £30.-

No. 3 N-NITROSO COMPOUNDS - ANALYSIS AND FORMATION (1972)
Edited by P. Bogovski, R. Preussmann & E.A. Walker, 140 pages; £8.50

No. 4 TRANSPLACENTAL CARCINOGENESIS (1973)
Edited by L. Tomatis & U. Mohr, 181 pages; £11.95

No. 5 PATHOLOGY OF TUMOURS IN LABORATORY ANIMALS. VOLUME 1. TUMOURS OF THE RAT. PART 1 (1973)
Editor-in-Chief V.S. Turusov, 214 pages; £17.50

No. 6 PATHOLOGY OF TUMOURS IN LABORATORY ANIMALS. VOLUME 1. TUMOURS OF THE RAT. PART 2 (1976)
Editor-in-Chief V.S. Turusov 319 pages; £17.50

No. 7 HOST ENVIRONMENT INTER-ACTIONS IN THE ETIOLOGY OF CANCER IN MAN (1973)
Edited by R. Doll & I. Vodopija, 464 pages; £30.-

No. 8 BIOLOGICAL EFFECTS OF ASBESTOS (1973)
Edited by P. Bogovski, J.C. Gilson, V. Timbrell & J.C. Wagner, 346 pages; £25.-

No. 9 N-NITROSO COMPOUNDS IN THE ENVIRONMENT (1974)
Edited by P. Bogovski & E.A. Walker 243 pages; £15.-

No. 10 CHEMICAL CARCINOGENESIS ESSAYS (1974)
Edited by R. Montesano & L. Tomatis, 230 pages; £15.-

No. 11 ONCOGENESIS AND HERPES-VIRUSES II (1975)
Edited by G. de-Thé, M.A. Epstein & H. zur Hausen
Part 1, 511 pages; £30.-
Part 2, 403 pages; £30.-

No. 12 SCREENING TESTS IN CHEMICAL CARCINOGENESIS (1976)
Edited by R. Montesano, H. Bartsch & L. Tomatis, 666 pages; £30.-

No. 13 ENVIRONMENTAL POLLUTION AND CARCINOGENIC RISKS (1976)
Edited by C. Rosenfeld & W. Davis 454 pages; £17.50

No. 14 ENVIRONMENTAL N-NITROSO COMPOUNDS - ANALYSIS AND FORMATION (1976)
Edited by E.A. Walker, P. Bogovski & L. Griciute, 512 pages; £35.-

No. 15 CANCER INCIDENCE IN FIVE CONTINENTS. VOL. III (1976)
Edited by J. Waterhouse, C.S. Muir, P. Correa & J. Powell, 584 pages; £35.-

No. 16 AIR POLLUTION AND CANCER IN MAN (1977)
Edited by U. Mohr, D. Schmahl & L. Tomatis, 331 pages; £30.-

No. 17 DIRECTORY OF ON-GOING RESEARCH IN CANCER EPI-DEMIOLOGY 1977 (1977)
Edited by C.S. Muir & G. Wagner, 599 pages; out of print

Nρ. 18 ENVIRONMENTAL CARCINO-GENS - SELECTED METHODS OF ANALYSIS
Editor-in-Chief H. Egan
Vol. 1 - ANALYSIS OF VOLATILE NITROSAMINES IN FOOD (1978)
Edited by R. Preussmann, M. Castegnaro, E.A. Walker & A.E. Wassermann, 212 pages; £30.-

No. 19 ENVIRONMENTAL ASPECTS OF N-NITROSO COMPOUNDS (1978)
Edited by E.A. Walker, M. Castegnaro, L. Griciute & R.E. Lyle, 566 pages; £35.-

No. 20 NASOPHARYNGEAL CARCINOMA: ETIOLOGY AND CONTROL (1978)
Edited by G. de-Thé & Y. Ito, 610 pages; £35.-

No. 21 CANCER REGISTRATION AND ITS TECHNIQUES (1978)
Edited by R. MacLennan, C.S. Muir, R. Steinitz & A. Winkler, 235 pages; £11.95

No. 22 ENVIRONMENTAL CARCINO-
GENS - SELECTED METHODS OF
ANALYSIS
Editor-in-Chief H. Egan
Vol. 2 - METHODS FOR THE MEASURE-
MENT OF VINYL CHLORIDE IN
POLY(VINYL CHLORIDE), AIR, WATER
AND FOODSTUFFS (1978)
Edited by D.C.M. Squirrell & W. Thain,
142 pages; £35.-

No. 23 PATHOLOGY OF TUMOURS IN
LABORATORY ANIMALS. VOLUME II.
TUMOURS OF THE MOUSE (1979)
Editor-in-Chief V.S. Turusov, 669 pages;
£35.-

No. 24 ONCOGENESIS AND HERPES-
VIRUSES III (1978)
Edited by G. de-Thé, W. Henle & F. Rapp
Part 1, 580 pages; £20.-
Part 2, 522 pages; £20.-

No. 25 CARCINOGENIC RISKS -
STRATEGIES FOR INTERVENTION
(1979)
Edited by W. Davis & C. Rosenfeld,
283 pages; £20.-

No. 26 DIRECTORY OF ON-GOING
RESEARCH IN CANCER EPI-
DEMIOLOGY 1978 (1978)
Edited by C.S. Muir & G. Wagner,
550 pages; out of print

No. 27 MOLECULAR AND CELLULAR
ASPECTS OF CARCINOGEN
SCREENING TESTS (1980)
Edited by R. Montesano, H. Bartsch &
L. Tomatis, 371 pages; £20.-

No. 28 DIRECTORY OF ON-GOING
RESEARCH IN CANCER EPIDEMIOLOGY
1979 (1979)
Edited by C.S. Muir & G. Wagner,
672 pages; out of print

No. 29 ENVIRONMENTAL CARCINO-
GENS - SELECTED METHODS OF
ANALYSIS
Editor-in-Chief H. Egan
Vol. 3 - ANALYSIS OF POLYCYCLIC
AROMATIC HYDROCARBONS IN
ENVIRONMENTAL SAMPLES (1979)
Edited by M. Castegnaro, P. Bogovski,
H. Kunte & E.A. Walker, 240 pages; £17.50

No. 30 BIOLOGICAL EFFECTS OF
MINERAL FIBRES (1980)
Editor-in-Chief J.C. Wagner
Volume 1, 494 pages; £25.-
Volume 2, 513 pages; £25.-

No. 31 N-NITROSO COMPOUNDS:
ANALYSIS, FORMATION AND
OCCURRENCE (1980)
Edited by E.A. Walker, M. Castegnaro,
L. Griciute & M. Börzsönyi, 841 pages;
£30.-

No. 32 STATISTICAL METHODS IN
CANCER RESEARCH
Vol. 1. THE ANALYSIS OF CASE-
CONTROL STUDIES (1980)
By N.E. Breslow & N.E. Day, 338 pages;
£17.50

No. 33 HANDLING CHEMICAL
CARCINOGENS IN THE LABORATORY
- PROBLEMS OF SAFETY (1979)
Edited by R. Montesano, H. Bartsch,
E. Boyland, G. Della Porta, L. Fishbein,
R.A. Griesemer, A.B. Swan & L. Tomatis,
32 pages £3.95

No. 34 PATHOLOGY OF TUMOURS
IN LABORATORY ANIMALS. VOLUME
III. TUMOURS OF THE HAMSTER
(1982)
Editor-in-Chief V.S. Turusov,
461 pages; £30.-

No. 35 DIRECTORY OF ON-GOING
RESEARCH IN CANCER EPI-
DEMIOLOGY 1980 (1980)
Edited by C.S. Muir & G. Wagner,
660 pages; out of print

No. 36 CANCER MORTALITY BY
OCCUPATION AND SOCIAL CLASS
1851-1971 (1982)
By W.P.D. Logan, 253 pages £20.-

No. 37 LABORATORY DECONTAMI-
NATION AND DESTRUCTION OF
AFLATOXINS B_1, B_2, G_1, G_2 IN
LABORATORY WASTES (1980)
Edited by M. Castegnaro, D.C. Hunt,
E.B. Sansone, P.L. Schuller,
M.G. Siriwardana, G.M. Telling,
H.P. Van Egmond & E.A. Walker,
59 pages; £5.95

No. 38 DIRECTORY OF ON-GOING
RESEARCH IN CANCER EPI-
DEMIOLOGY 1981 (1981)
Edited by C.S. Muir & G. Wagner,
696 pages; out of print

No. 39 HOST FACTORS IN HUMAN
CARCINOGENESIS (1982)
Edited by H. Bartsch & B. Armstrong
583 pages; £35.-

No. 40 ENVIRONMENTAL CAR-
CINOGENS. SELECTED METHODS
OF ANALYSIS
Editor-in-Chief H. Egan
Vol. 4. SOME AROMATIC AMINES AND
AZO DYES IN THE GENERAL AND
INDUSTRIAL ENVIRONMENT (1981)
Edited by L. Fishbein, M. Castegnaro,
I.K. O'Neill & H. Bartsch, 347 pages;
£20.-

No. 41 N-NITROSO COMPOUNDS:
OCCURRENCE AND BIOLOGICAL
EFFECTS (1982)
Edited by H. Bartsch, I.K. O'Neill,
M. Castegnaro & M. Okada,
755 pages; £35.-

No. 42 CANCER INCIDENCE IN FIVE
CONTINENTS. VOLUME IV (1982)
Edited by J. Waterhouse, C. Muir,
K. Shanmugaratnam & J. Powell,
811 pages; £35.-

No. 43 LABORATORY DECONTAMI-
NATION AND DESTRUCTION OF
CARCINOGENS IN LABORATORY
WASTES: SOME N-NITROSAMINES
(1982) Edited by M. Castegnaro,
G. Eisenbrand, G. Ellen, L. Keefer,
D. Klein, E.B. Sansone, D. Spincer,
G. Telling & K. Webb, 73 pages £6.50

No. 44 ENVIRONMENTAL CAR-
CINOGENS. SELECTED METHODS
OF ANALYSIS
Editor-in-Chief H. Egan
Vol. 5. SOME MYCOTOXINS (1983)
Edited by L. Stoloff, M. Castegnaro,
P. Scott, I.K. O'Neill & H. Bartsch,
455 pages; £20.-

No. 45 ENVIRONMENTAL CAR-
CINOGENS. SELECTED METHODS
OF ANALYSIS
Editor-in-Chief H. Egan
Vol. 6: N-NITROSO COMPOUNDS
(1983)
Edited by R. Preussmann, I.K. O'Neill,
G. Eisenbrand, B. Spiegelhalder &
H. Bartsch, 508 pages; £20.-

No. 46 DIRECTORY OF ON-GOING
RESEARCH IN CANCER EPI-
DEMIOLOGY 1982 (1982)
Edited by C.S. Muir & G. Wagner,
722 pages; out of print

No. 47 CANCER INCIDENCE IN
SINGAPORE (1982)
Edited by K. Shanmugaratnam, H.P. Lee
& N.E. Day, 174 pages; £10.-

No. 48 CANCER INCIDENCE IN
THE USSR (1983) Second Revised
Edition
Edited by N.P. Napalkov,
G.F. Tserkovny, V.M. Merabishvili,
D.M. Parkin, M. Smans & C.S. Muir,
75 pages; £10.-

No. 49 LABORATORY DECONTAMI
NATION AND DESTRUCTION OF
CARCINOGENS IN LABORATORY
WASTES: SOME POLYCYCLIC
AROMATIC HYDROCARBONS (1983)
Edited by M. Castegnaro, G. Grimmer,
O. Hutzinger, W. Karcher, H. Kunte,
M. Lafontaine, E.B. Sansone, G. Telling
& S.P. Tucker, 81 pages; £7.95

No. 50 DIRECTORY OF ON-GOING
RESEARCH IN CANCER
EPIDEMIOLOGY 1983 (1983)
Edited by C.S. Muir & G. Wagner,
740 pages; out of print

No. 51 MODULATORS IN
EXPERIMENTAL CARCINO
GENESIS (1983)
Edited by R. Montesano &
V.S. Turusov, 307 pages; £25.-

No. 52 SECOND CANCER IN
RELATION TO RADIATION
TREATMENT FOR CERVICAL
CANCER: RESULTS OF A CANCER
REGISTRY COLLABORATION (1983)
Edited by N.E. Day & J.C. Boice, Jr,
207 pages; £17.50

No. 53 NICKEL IN THE HUMAN
ENVIRONMENT (1984)
Editor-in-Chief, F.W. Sunderman, Jr,
529 pages; £30.-

No. 54 LABORATORY
DECONTAMINATION AND
DESTRUCTION OF CARCINO-
GENS IN LABORATORY WASTES:
SOME HYDRAZINES (1983)
Edited by M. Castegnaro, G. Ellen,
M. Lafontaine, H.C. van der Plas,
E.B. Sansone & S.P. Tucker,
87 pages; £6.95

No. 55 LABORATORY
DECONTAMINATION AND
DESTRUCTION OF CARCINOGENS
IN LABORATORY WASTES: SOME
N-NITROSAMIDES (1983)
Edited by M. Castegnaro,
M. Benard, L.W. van Broekhoven,
D. Fine, R. Massey, E.B. Sansone,
P.L.R. Smith, B. Spiegelhalder,
A. Stacchini, G. Telling & J.J. Vallon,
65 pages; £6.95

No. 56 MODELS, MECHANISMS AND
ETIOLOGY OF TUMOUR PROMOTION
(1984)
Edited by M. Börszönyi, N.E. Day,
K. Lapis & H. Yamasaki, 532 pages,
£30.-

No. 57 *N*-NITROSO COMPOUNDS:
OCCURRENCE, BIOLOGICAL EFFECTS
AND RELEVANCE TO HUMAN
CANCER (1984)
Edited by I.K. O'Neill, R.C. von Borstel,
C.T. Miller, J. Long & H. Bartsch,
1013 pages, £75.-

No. 58 AGE-RELATED FACTORS
IN CARCINOGENESIS (1985)
Edited by A. Likhachev, V. Anisimov
& R. Montesano
288 pages; £20.-

No. 59 MONITORING HUMAN
EXPOSURE TO CARCINOGENIC AND
MUTAGENIC AGENTS (1985)
Edited by A. Berlin, M. Draper,
K. Hemminki & H. Vainio
457 pages, £25,-

No. 60 BURKITT'S LYMPHOMA: A
HUMAN CANCER MODEL (1985)
Edited by G. Lenoir, G. O'Conor
& C.L.M. Olweny
484 pages, £25,-

No. 61 LABORATORY DECONTAMI-
NATION AND DESTRUCTION OF
CARCINOGENS IN LABORATORY
WASTES: SOME HALOETHERS (1985)
Edited by M. Castegnaro,
M. Alvarez, M. Iovu, E.B. Sansone,
G.M. Telling & D.T. Williams
55 pages, £5.95

No. 62 DIRECTORY OF ON-GOING
RESEARCH IN CANCER EPI-
DEMIOLOGY 1984 (1984)
Edited by C.S. Muir & G.Wagner;
728 pages; £18.-

No. 63 VIRUS-ASSOCIATED CANCERS
IN AFRICA (1984)
Edited by A.O. Williams, G.T. O'Conor,
G.B. de-Thé & C.A. Johnson,
773 pages, £20.-

No. 64 LABORATORY DECONTAMI-
NATION AND DESTRUCTION OF
CARCINOGENS IN LABORATORY
WASTES: SOME AROMATIC AMINES
AND 4-NITROBIPHENYL (1985)
Edited by M. Castegnaro, J. Barek,
J. Dennis, G. Ellen, M. Klibanov,
M. Lafontaine, R. Mitchum,
P. Van Roosmalen, E.B. Sansone,
L.A. Sternson & M. Vahl
85 pages, £5.95

No. 65 INTERPRETATION OF
NEGATIVE EPIDEMIOLOGICAL
EVIDENCE FOR CARCINOGENICITY
Edited by N.J. Wald & R. Doll
(in press)

No. 66. THE ROLE OF THE REGISTRY
IN CANCER CONTROL
Edited by D.M. Parkin, G. Wagner
& C.S. Muir (in press)

No. 67. TRANSFORMATION ASSAY OF
ESTABLISHED CELL LINES:
MECHANISMS AND APPLICATIONS
Edited by T. Kakunaga & H. Yamasaki
(in press)

No. 68. ENVIRONMENTAL
CARCINOGENS — SELECTED
METHODS OF ANALYSIS
VOLUME 7 — SOME VOLATILE
HALOGENATED ALKANES AND
ALKENES
Edited by L. Fishbein & I.K. O'Neill
(in press)

No. 69. DIRECTORY OF ON-GOING
RESEARCH IN CANCER
EPIDEMIOLOGY 1985
Edited by C.S. Muir & G. Wagner
756 pages; £22.-

No. 70. THE ROLE OF CYCLIC NUCLEIC
ACID ADDUCTS IN CARCINOGENESIS
AND MUTAGENESIS
Edited by B. Singer & H. Bartsch
(in press)

No. 71 ENVIRONMENTAL CARCINOGENS.
SELECTED METHODS OF ANALYSIS
VOLUME 8. SOME METALS: As, Be, Cd,
Cr, Ni, Pb, Se, Zn
Edited by I.K. O'Neill, P. Schuller
& L. Fishbein (1985) (in preparation)

SCIENTIFIC PUBLICATIONS SERIES

No. 72. ATLAS OF CANCER IN
SCOTLAND 1975-1980: INCIDENCE AND
EPIDEMIOLOGICAL PERSPECTIVE
Edited by I. Kemp, P. Boyle, M. Smans
& C. Muir (1985)
282 pages; £30.-

No. 73. LABORATORY DECONTAMI-
NATION AND DESTRUCTION OF
CARCINOGENS IN LABORATORY
WASTES: SOME ANTINEOPLASTIC
AGENTS
Edited by M. Castegnaro, J. Adams,
M. Armour, J. Barek, J. Benvenuto,
C. Confalonieri, U. Goff, S. Ludeman,
D. Reed, E.B. Sansone & G. Telling
(1985) (in press)

NON-SERIAL PUBLICATIONS

(Available from IARC)

ALCOOL ET CANCER (1978)
by A.J. Tuyns (in French only)
42 pages; Fr.fr. 35.-; Sw.fr. 14.-

CANCER MORBIDITY AND CAUSES OF
DEATH AMONG DANISH BREWERY
WORKERS (1980) By O.M. Jensen
145 pages; US$ 25.00; Sw.fr. 45.-

IARC MONOGRAPHS ON THE EVALUATION OF THE CARCINOGENIC RISK OF CHEMICALS TO HUMANS

(English editions only)

(Available from WHO Sales Agents)

Volume 1
Some inorganic substances, chlorinated hydrocarbons, aromatic amines, N-nitroso compounds, and natural products (1972)
184 pp.; out of print

Volume 2
Some inorganic and organometallic compounds (1973)
181 pp.; out of print

Volume 3
Certain polycyclic aromatic hydrocarbons and heterocyclic compounds (1973)
271 pp.; out of print

Volume 4
Some aromatic amines, hydrazine and related substances, N-nitroso compounds and miscellaneous alkylating agents (1974)
286 pp.; US$7.20; Sw.fr. 18.-

Volume 5
Some organochlorine pesticides (1974)
241 pp.; out of print

Volume 6
Sex hormones (1974)
243 pp.; US$7.20; Sw.fr. 18.-

Volume 7
Some anti-thyroid and related substances, nitrofurans and industrial chemicals (1974)
326 pp.; US$12.80; Sw.fr. 32.-

Volume 8
Some aromatic azo compounds (1975)
357 pp.; US$14.40; Sw.fr. 36.-

Volume 9
Some aziridines, N-, S- and O-mustards and selenium (1975)
268 pp.; US$10.80; Sw.fr. 27.-

Volume 10
Some naturally occurring substances (1976)
353 pp.; US$15.00; Sw.fr. 38.-

Volume 11
Cadmium, nickel, some epoxides, miscellaneous industrial chemicals and general considerations on volatile anaesthetics (1976)
306 pp.; US$14.00; Sw.fr. 34.-

Volume 12
Some carbamates, thiocarbamates and carbazides (1976)
282 pp.; US$14.00; Sw.fr. 34.-

Volume 13
Some miscellaneous pharmaceutical substances (1977)
255 pp.; US$12.00; Sw.fr. 30.-

Volume 14
Asbestos (1977)
106 pp.; US$6.00; Sw.fr. 14.-

Volume 15
Some fumigants, the herbicides 2,4-D chlorinated dibenzodioxins and miscellaneous industrial chemicals (1977)
354 pp.; US$20.00; Sw.fr. 50.-

Volume 16
Some aromatic amines and related nitro compounds - hair dyes, colouring agents and miscellaneous industrial chemicals (1978)
400 pp.; US$20.00; Sw.fr. 50.-

Volume 17
Some N-nitroso compounds (1978)
365 pp.; US$25.00; Sw.fr. 50.-

Volume 18
Polychlorinated biphenyls and polybrominated biphenyls (1978)
140 pp.; US$13.00; Sw.fr. 20.-

Volume 19
Some monomers, plastics and synthetic elastomers, and acrolein (1979)
513 pp.; US$35.00; Sw.fr. 60.-

Volume 20
Some halogenated hydrocarbons (1979)
609 pp.; US$35.00; Sw.fr. 60.-

Volume 21
Sex hormones (II) (1979)
583 pp.; US$35.00; Sw.fr. 60.-

Volume 22
Some non-nutritive sweetening agents (1980)
208 pp.; US$15.00; Sw.fr. 25.-

IARC MONOGRAPHS SERIES

Volume 23
Some metals and metallic compounds (1980)
438 pp.; US$30.00; Sw.fr. 50.-

Volume 24
Some pharmaceutical drugs (1980)
337 pp.; US$25.00; Sw.fr. 40.-

Volume 25
Wood, leather and some associated industries (1981)
412 pp.; US$30.00; Sw.fr. 60.-

Volume 26
Some antineoplastic and immuno-suppressive agents (1981)
411 pp.; US$30.00; Sw.fr. 62.-

Volume 27
Some aromatic amines, anthraquinones and nitroso compounds, and inorganic fluorides used in drinking-water and dental preparations (1982)
341 pp.; US$25.00; Sw.fr. 40.-

Volume 28
The rubber industry (1982)
486 pp.; US$35.00; Sw.fr. 70.-

Volume 29
Some industrial chemicals and dyestuffs (1982)
416 pp.; US$30.00; Sw.fr. 60.-

Volume 30
Miscellaneous pesticides (1983)
424 pp; US$30.00; Sw.fr. 60.-

Volume 31
Some food additives, feed additives and naturally occurring substances (1983)
314 pp.; US$30.00; Sw.fr. 60.-

Volume 32
Polynuclear aromatic compounds, Part 1, Environmental and experimental data (1984)
477 pp.; US$30.00; Sw.fr. 60.-

Volume 33
Polynuclear aromatic compounds, Part 2, Carbon blacks, mineral oils and some nitroarene compounds (1984)
245 pp.; US$25.00; Sw.fr. 50.-

Volume 34
Polynuclear aromatic compounds, Part 3, Some complex industrial exposures in aluminium production, coal gasification, coke production, and iron and steel founding (1984)
219 pages; US$20.00; Sw.fr. 48.-

Volume 35
Polynuclear aromatic compounds, Part 4, Bitumens, coal-tars and derived products, shale-oils and soots (1985)
271 pages; US$25.00; Sw.fr. 70.-

Volume 36
Allyl Compounds, Aldehydes, Epoxides and Peroxides (1985)
369 pages; US$25.00; Sw.fr. 70.-

Volume 37
Tobacco habits other than smoking (1985)
(in preparation)

Volume 38
Tobacco smoking (1985)
(in preparation)

Supplement No. 1
Chemicals and industrial processes associated with cancer in humans (IARC Monographs, Volumes 1 to 20) (1979)
71 pp.; out of print

Supplement No. 2
Long-term and short-term screening assays for carcinogens: a critical appraisal (1980)
426 pp.; US$30.00; Sw.fr. 60.-

Supplement No. 3
Cross index of synonyms and trade names in Volumes 1 to 26 (1982)
199 pp.; US$30.00; Sw.fr. 60.-

Supplement No. 4
Chemicals, industrial processes and industries associated with cancer in humans (IARC Monographs, Volumes 1 to 29) (1982)
292 pp.; US$30.00; Sw.fr. 60.-

INFORMATION BULLETIN ON THE SURVEY OF CHEMICALS BEING TESTED FOR CARCINOGENICITY
No. 8 (1979)
Edited by M.-J. Ghess, H. Bartsch & L. Tomatis
604 pp.; US$20.00; Sw.fr. 40.-

INFORMATION BULLETIN ON THE SURVEY OF CHEMICALS BEING TESTED FOR CARCINOGENICITY
No. 9 (1981)
Edited by M.-J. Ghess, J.D. Wilbourn, H. Bartsch & L. Tomatis
294 pp.; US$20.00; Sw.fr. 41.-

INFORMATION BULLETIN ON THE
SURVEY OF CHEMICALS BEING
TESTED FOR CARCINOGENICITY
No. 10 (1982)
Edited by M.-J. Ghess, J.D. Wilbourn,
H. Bartsch
326 pp.; US$20.00; Sw.fr. 42.-

INFORMATION BULLETIN ON THE
SURVEY OF CHEMICALS BEING
TESTED FOR CARCINOGENICITY
No. 11 (1984)
Edited by M.-J. Ghess, J.D. Wilbourn,
H. Vainio & H. Bartsch
336 pp.; US$20.00; Sw.fr. 48.-